Civil Engineering Materials

Edited by
NEIL JACKSON
and
RAVINDRA K. DHIR

Fifth Edition

First edition 1976
Second edition 1980
Third edition 1983
Fourth edition 1988
Fifth edition 1996

Published by
MACMILLAN PRESS LTD
Houndmills, Basingstoke, Hampshire RG21 6XS
and London
Companies and representatives
throughout the world

ISBN 0–333–63683–X

A catalogue record for this book is available
from the British Library

Typeset by TTP International, Sutton, Surrey

Printed in Hong Kong

10   9   8   7   6   5   4
06   05   04   03   02   01   00

Contents

Preface

The importance of an understanding of the materials used in civil engineering and building projects is widely recognised, and this is reflected by the increasing emphasis being placed on the teaching of material properties at undergraduate level. This introductory textbook on materials satisfies a need for a single book covering the principal materials used in civil engineering and building works.

The aim has been to provide students with an authoritative text which will also serve as a valuable source of reference in their subsequent careers. In this fifth edition, with three new contributors, the Parts on Metals, Timber and Bituminous Materials have been rewritten, and all other Parts have been extensively revised to maintain an up-to-date coverage of all materials. The fundamental properties of soils continue to be covered in greater depth than usual, in recognition of the importance of soils as construction materials, and an additional chapter has been included to assist the reader in the transition from the study of soils as a material to the related topic of soil mechanics. Extensive references to all relevant British and other Standards are made throughout the book.

The treatment of material properties here is suitable for students studying for a degree or equivalent qualification in civil engineering, building technology, architecture and other related disciplines. The particular point in a course at which the study of civil engineering and building materials is introduced will depend on the course structure of the individual educational institution but would generally be during the first two years of a three or four year course. Similarly, the extent of further formal study of materials depends on the emphasis and structure of the course within a particular educational institution. However, it is not envisaged that further formal study of the basic material properties of metals, timber, concrete, polymer materials and bricks and blocks will be required, although the application of these materials within the general context of analysis and design might well continue throughout the remainder of a course. Further formal study of soils might normally be expected to continue, for civil engineering students, within the context of soil mechanics, with further formal study of bituminous materials only where highway materials are studied in later years. In this context, readers should recognise the need for continued study, whether this be of an informal or formal nature, throughout their subsequent careers if they are to maintain the highest professional standards, to which it is hoped they will aspire.

Throughout the book, the underlying theme is an emphasis on the factors affecting engineering decisions. It is hoped that this will promote an awareness in the reader of the importance of material behaviour in both design and construction, in order to ensure that the final product, whether structural, functional or purely decorative, will adequately fulfil the purpose for which it is provided.

N. Jackson
R.K. Dhir

Acknowledgements

The courtesy extended by the undernamed, and others named in the text, in permitting the reproduction of the material indicated, is gratefully acknowledged.

Material from BS 11: 1985 Specification for railway rails; BS 64: 1992 Specification for normal and high strength bolts and nuts for railway rail fishplates; BS 1377: 1990 Methods of test for soils for civil engineering purposes; BS 1471 – 1475 Specifications for wrought aluminium and aluminium alloys for general engineering purposes; BS 1490: 1988 Specification for aluminium and aluminium alloy ingots and castings for general engineering purposes; BS 1881: 1970 – 1972 Testing of concrete; BS 3921: 1985 Specification for clay bricks; BS 4190: 1967 Specification for ISO metric black hexagon bolts, screws and nuts; BS 4395 Specification for high strength friction grip bolts and associated nuts and washers for structural engineering; BS 5075 Concrete admixtures: 1982/1985; BS 5268 Structural use of timber: (1989 –1991); BS 5930: 1981 Code of practice for site investigations; BS 6031: 1981 Code of practice for earthworks; BS 7613: 1994 Specification for hot-rolled quenched and tempered weldable structural steels; BS 7668: 1994 Specification for weldable structural steels. Hot finished structural hollow sections in weather resistant steels; BS 8004: 1986 Code of practice for foundations; BS 8110: 1985 Structural use of concrete; BS 8118 Structural use of aluminium; BS EN 485-2: 1995 Aluminium and aluminium alloys – sheet, strip and plate; BS EN 10025: 1993 Hot-rolled products of non-alloy structural steels. Technical delivery conditions; BS EN 10113: 1993 Hot-rolled products in weldable fine grain structural steels; DD ENV 1992-1-1: 1992. EuroCode 2: Design of Concrete Structures, Part 1. General rules and rules for building; are reproduced with the permission of the British Standards Institution. Complete editions of the standards can be obtained by post from BSI Customer Services, 389 Chiswick High Road, London, W4 4AL (telephone: 0181 966 7000).

Table 27.1 is reproduced by permission of the American Society of Civil Engineers.

Figure 15.9 is reproduced by permission of Thomas Telford Services Limited and the Comité Euro-International du Beton.

European Standards

These will be issued initially as ENVs, for example ENV 1991. Use of an ENV will be voluntary for a few years, after which feedback will be reviewed and then appropriate revisions made to the ENV document prior to it being issued as a full European Standard, for example EN 196. As the life of the ENV documents is comparatively short, these are not included in the Reference sections throughout this book.

The reader is referred to the current *BSI Standards Catalogue* for details of the latest available European (and British) Standards.

List of Contributors

R.K. Dhir, BSc, PhD, CEng, MIMM, Hon. FICT, FGS (joint editor)
Professor of Concrete Technology, Department of Civil Engineering, University of Dundee

J.M. Edwards, BSc (Eng), CEng, MICE
Consultant, formerly with Shell International

C. Hall, MA, DPhil, CEng, FRSC, FIM
Scientific Advisor, Schlumberger Cambridge Research Ltd, Cambridge; Visiting Professor, Department of Building Engineering, UMIST

N. Jackson, BSc, PhD, CEng, MICE, MIStructE (joint editor)
Head, Department of Civil Engineering, Sultan Qaboos University, Sultanate of Oman

V.B. John, BSc, MSc, CEng, MIM
formerly Senior Lecturer in Materials, University of Westminster, London

D.G. McKinlay, BSc, PhD, CEng, FICE, FASCE, FGS, FRSE
Emeritus Professor, University of Strathclyde

J.G.L. Munday, BSc, PhD, MICE
Associate Head, School of Construction and Environment, University of Abertay, Dundee

1

Introduction

An understanding of the properties of materials is essential in both the *design and construction* phases of any civil engineering or building project if this is to prove satisfactory for its intended purpose. For the student reader it is believed that a few introductory remarks, in this context, might make the study of materials more meaningful in itself, rather than merely being a part of a required course of study.

Civil engineering and building projects include roads, railways, bridges, tunnels, dams, culverts, water and waste-water treatment plants, water distribution and drainage systems, coastal protection works, harbours, power stations, airports, industrial complexes and a wide range of building structures for residential, commercial, sports and leisure purposes. During the initial planning and design stages of a project the principal factors upon which any subsequent decision to proceed will be based include its economic viability and sociological and environmental impact.

During this initial or conceptual design stage, consideration is given to possible alternative locations and/or layouts of the associated works and to a preliminary assessment of suitable construction materials.

For building structures, alternative layouts and structural forms are studied together with the suitability of different materials for use in the structural elements for each of these. The decision as to which structural form and choice of materials are the most appropriate depends on a number of factors including, but not limited to, the cost, physical properties, durability and availability of materials and the ease and speed of construction. All of these affect the first cost (of construction) and/or the subsequent costs (of maintenance) during the design life of the structure, both of these being important considerations when assessing the economic viability of any project. It should be noted that the term *choice of materials* is used here to mean the choice of not only the generic names of materials (steel, concrete, aluminium, polymer, timber, brick, etc.) but also their specific type, composition and/or performance acceptance criteria.

To make sound decisions, in the above context, a designer must have the ability to assess all the factors likely to affect the performance of a material in its proposed location. For this a knowledge of structural analysis and design is necessary, together with an ability to identify all the environmental factors which might affect the material properties and behaviour in practice. Whilst this knowledge and

1

ability are necessary, they are not sufficient to ensure that an appropriate choice of materials is made. This requires also a knowledge and understanding of the properties of the materials themselves if the design process is to properly take into account the effects on the materials of the forces and the environmental conditions to which they will be subjected. The reader is referred to an interesting paper by Bloomer (1993) in which factors influencing the choice of materials for the construction of a ventilation stack are described.

Timber, metals, concrete, bituminous materials, polymers and bricks and blocks may all be found in buildings. Some of these may be used only in non-structural elements, whilst others may be used on their own or acting compositely with others in structural elements. The actual materials used in structural elements will depend both on the structural form and on other factors mentioned previously.

Consider the situation where, for architectural/aesthetic reasons, a large covered area with natural lighting and a tent-like structure is required. In this case the choice of materials is somewhat limited as some form of fabric structure is indicated. Suitable fabrics based on the use of polymers are available. The choice of an appropriate polymer-coated fabric to provide the required waterproofing and durability (resistance to weathering) will depend on the proposed design life of the structure. The shorter the design life, the less important the durability and so a cheaper fabric can be used. Fabrics are structurally useful only when loaded in tension and must therefore be supported, by other structural elements, so that their self-weight and all wind loads can be safely transmitted to the ground. A very small tent (structure), such as used by campers, might use canvas or other flexible material supported by timber poles or metal tubes and guy ropes. For the much greater loads associated with large fabric structures, reinforced concrete or steel masts with possibly steel cables would normally be used to provide the support structure for the fabric. To meet all design requirements satisfactorily a knowledge of the properties of all materials is essential.

Building structures generally lend themselves to a greater choice of structural form and of construction materials than in the above example, with metals, concrete, timber and masonry all finding a place as structural elements in appropriate situations. A sound decision, as to whether or not a situation is appropriate for a particular material, will depend, among other things, on the engineer having a thorough understanding of material behaviour and properties. Some of these properties may be quantified for general comparative purposes (see table 1.1), but all require an in-depth knowledge of the underlying factors which might influence them in a given situation.

In some circumstances the ready availability and relative cheapness of a particular material may dictate the most appropriate structural form. One example of this is the use of stone or brick masonry for the construction of short span bridges. Masonry has a relatively high compressive strength and low tensile strength. Recognition of this material property has led in the past to many short span bridge structures taking the form of a masonry arch (figure 1.1).

For long single span bridges, structural design considerations lead to the conclusion that some form of suspension structure (a suspension or cable-stayed bridge) is appropriate. In this form of structure the bridge deck is suspended from cables supported by towers (figures 1.2 and 1.3). Masonry has been used in the construction of supporting towers for relatively short span suspension bridges but practical considerations (material properties, size, weight, cost, etc.) make it

TABLE 1.1

Properties of some engineering materials

	Density (kg m⁻³)	Tensile modulus of elasticity, E (kN mm⁻²)	Tensile strength, f_t (N mm⁻²)	Coeff. of thermal expansion α (×10⁻⁶ K⁻¹)	Thermal conductivity, k (W m⁻¹ K⁻¹)
Cast irons	7 200	170	300–1000	12–13.5	33–45
Structural steels	7 850	210	350–700	12	52
Aluminium alloys	2 750	70	70–600	21–23	95–165
Lead	11 350	13	—	29	35
Scots pine (parallel to grain)	510	10	90–120	4	0.38
Concrete	2 400–2 500	20–50*	20–100* (3–15)**	9–12	1.8–2.5
Autoclaved aerated concrete (AAC – medium strength)	300–600	0.9–2.5*	1.8–4.0* (0.6–1.1)**	5–8	0.06–0.14
PVC	1 400	2.4–3.0	40–60	70	0.16
GRP: polyester	1 400–2 000	6–50	60–1000	10–30	0.2–0.4
Fired clay bricks	1 600–2 400	5–30*	5–100*	5–7	0.4–2.8
Concrete blocks	500–2 100	—	3–35*	8–12	0.1–2.3

* Compressive values.
** Modulus of rupture.
Note: All properties in table 1.1 are indicative of the probable range of values for each class of material. The mechanical properties (strength and E values) given are for short-term loading conditions. The properties of specific materials will be dependent on their composition, that is on the properties and proportions of their constituent materials, including moisture content. Their mechanical properties may additionally be dependent on their temperature and age and the rate of loading.

3

Figure 1.1 *Masonry arch bridge (Courtesy of A.W. Sawko)*

unsuitable for the extremely high supporting towers required for much longer spans. For these the use of steel and/or reinforced concrete as a construction material is more appropriate.

Naturally occurring materials are used very successfully in many parts of the world in the construction of buildings and unpaved/graded roads. In tropical climates, clays and soils containing clays (e.g. red laterite) have been used extensively for building both in the form of bricks and blocks and, when mixed with fibrous materials (e.g. straw) and water, as a direct building material. Temporary and semi-permanent graded roads have been constructed using natural gravels, red laterite and occasionally silty salt-laden sands, the two latter materials requiring watering and compaction during construction. More frequent maintenance than for paved roads is normally required but this is not usually a costly item in the areas concerned. The successful way in which these materials have been used is a direct result of those using them having become aware of the properties of the materials, and the ways in which these might be improved for a given purpose, through a trial-and-error (learning) process.

Figure 1.2 *Clifton suspension bridge, Bristol (Courtesy of F. Sawko)*

Figure 1.3 *Humber suspension bridge (Courtesy of F. Sawko)*

The use of composite materials, such as the clay mixed with fibrous materials referred to above, recognises the inadequacy of certain materials on their own to resist all the forces to which they will be subjected. Examples of these in use today include reinforced concrete (in which steel reinforcement is positioned to resist tensile forces), and glass fibre-reinforced plastic (GRP). Another notable example is the use of earth reinforcing systems in soil structures, in which steel (galvanized, stainless or coated steel), aluminium alloys, polymers and other materials are used, in various forms, as earth reinforcement (Jones, 1988).

Modern technology has resulted in a wide range of man-made materials being available for use in the construction industry today. How then is one to take advantage of this? Much information relating to the properties of all materials is already available but what if engineering judgement and intuition suggest possible advantages in using particular materials in a previously untried situation or manner? In such cases, a formal testing (research and development) programme to ascertain the appropriateness of innovative ideas prior to their incorporation in a construction project might, in general, be expected to take the place of the less formal trial-and-error (learning) process referred to earlier.

In practice, whilst the opportunity to learn from mistakes (preferably those of others) should never be missed, such opportunities will be minimised if all concerned keep abreast of the state of the art pertaining to all aspects of the work of the profession with which they are associated. For those directly involved in construction this clearly includes the properties, applications and behaviour of construction materials.

The following chapters include a comprehensive coverage of the structure, properties and applications of the principal construction materials in use today. These include metals, timber, concrete, bituminous materials, soils, polymers, bricks and blocks. The effects of the actual composition of a material on its properties and behaviour in service are explained as also is their relevance to decisions regarding the choice of suitable materials for use under given conditions. The emphasis throughout is on providing readers with a sound basic knowledge and understanding of materials, upon which their further studies can be based.

References

Bloomer, D.A. (1993) Stainless steel ventilation stack: WEP, Sellafield, *The Structural Engineer*, Vol. 71, No. 16, August, pp. 289–295.

Jones, C.J.F.P. (1988) *Earth Reinforcement and Soil Structures*, Butterworth, London.

I METALS

Introduction

The applications of metals in civil engineering are many and varied, ranging from their use as main structural materials to their use for fastenings and as bearing materials. As main structural materials, cast iron and wrought iron have been superseded by rolled-steel sections and some use has been made of wrought aluminium alloys. Steel is also of major importance for its use in reinforced and pre-stressed concrete. On a smaller scale, metals are used extensively for fastenings, such as nails, screws, bolts and rivets, for bearing surfaces in the expansion joints of bridges, and for decorative facings.

The properties of metals which make them unique among constructional materials are high stiffness and tensile strength, the ability to be formed into plate, sections and wire, and the weldability or ease of welding of those metals commonly used for constructional purposes. Other properties typical of metals are electrical conductivity, high thermal conductivity and metallic lustre, which are of importance in some circumstances. Perhaps the greatest disadvantage of the common metals and particularly steels, is the need to protect them from corrosion by moist conditions and the atmosphere, although weathering steels have been developed which offer much greater resistance to atmospheric corrosion.

When in service, metals frequently have to resist not only high tensile or compressive forces and corrosion, but also conditions of shock loading, low temperatures, constantly varying forces, or a combination of these effects. Most pure metals are relatively soft and weak, and do not meet the rigorous service requirements, except for those applications where the properties of high electrical conductivity or corrosion resistance are required. Normally one or more alloying elements are added to increase strength or to modify the properties in some other way. Metals and alloys are crystalline and the bulk material has a structure composed of myriads of small crystals, or grains. The grain structure, or microstructure, of any component is influenced by many factors, including the type of metal or alloy, methods of manufacture and heat treatment, and the final properties are highly dependent on structure. Micro- and macro-sized defects may be introduced into the structure during manufacture. In order to appreciate the properties and behaviour in service of metallic materials it is necessary to have some knowledge of the nature of the crystalline state and of the microstructure, and the coarser macrostructure, of these materials. Also, it is necessary to understand how these structures may be modified by alloying, deformation and heat treatments.

2

Structure of Metals

2.1 Atomic structure and bonding

Atoms of chemical elements consist of a very dense and compact nucleus surrounded by electrons. The nucleus is composed of two types of sub-atomic particles, protons and neutrons, each proton carrying a positive electrical charge. An electron possesses a negative electrical charge, equal in magnitude to the charge on a proton, but has a very small mass (the mass of an electron is 1/1836 of the mass of a proton) and the electrons are arranged around the nucleus in a series of well-defined shells or orbitals. The number of protons in the nucleus, known as the *atomic number*, Z, is characteristic of the element; for example, a carbon nucleus ($Z = 6$) contains 6 protons, an aluminium nucleus ($Z = 13$) has 13 protons and iron atoms ($Z = 26$) have 26 protons in their nuclei. In a neutral atom the number of orbital electrons will be equal to the number of nuclear protons. Atoms are extremely small and, typically, the diameter of many metal atoms is around 0.25 nm. This means that 1 mm^3 of a metal will contain some $3-9 \times 10^{19}$ atoms.

The orbiting electrons occupy discrete concentric shells or energy levels around a nucleus, with the electrons in the inner shells being more tightly bound to the nucleus than those in an outer shell. The number of electrons in the outermost electron shell has an important bearing on the properties of an element. If the outermost electron shell contains eight electrons it is completely filled and this confers a very high degree of stability to an atom. The *inert*, or *noble* gases have completely filled outer electron shells and these are extremely stable and inert chemically. All other elements, which have at least one incompletely filled shell, will take part in chemical reactions. It is the outermost shell electrons, the *valence* electrons, that are responsible for the chemical reactivity of elements, and atoms in combination attempt to achieve an outer shell configuration of eight electrons. Interatomic bonds, involving interaction between the valence electrons, may be formed in several ways, the three main types being the covalent bond, the ionic bond and the metallic bond. These are termed *primary* bonds and are relatively strong. There are, in addition, weaker secondary bonds, which may exist between molecules and are caused by electrostatic forces of attraction between polarised

9

molecules. All the substances that we are familiar with, both solid and fluid, are composed of very large aggregates of atoms and the properties of these materials derive in part from the type of bonding which occurs and the strengths of the bonds.

Covalent bond

The *covalent* bond is based on electron sharing and one covalent bond is created by a pair of valence electrons sharing a joint orbital around two atomic nuclei. The individual atoms of non-metallic elements bond to one another in this way, as for example in chlorine and oxygen. Chlorine atoms combine, in pairs, to form diatomic molecules. A chlorine atom has seven valence electrons and one electron from each atom enters into a joint orbital about two nuclei, thus giving each nucleus an effective complement of eight outer shell electrons. The pair of covalently bonded chlorine atoms is termed a *molecule*, symbolised as Cl_2. Similarly, oxygen atoms, each with six valence electrons, combine to form O_2 molecules by two electrons from each atom entering joint orbitals about two nuclei. The covalent bond between the atoms within the diatomic molecule is strong, but there is no primary bonding between the small discrete molecules and so the substance is gaseous at ordinary temperatures and pressures. Hydrogen and nitrogen are other examples of elements which form covalently bonded diatomic molecules. Carbon, with four outer shell electrons, is capable of forming four covalent bonds, and the shared orbitals point towards the corners of a regular tetrahedron, as shown in figure 2.1(a). The diamond form of carbon shows this structure, with each carbon atom being covalently bonded to four other carbon atoms. This tetrahedral arrangement extends in all directions to build up a symmetrical crystalline pattern (see figure 2.1(b)). Diamond is the hardest known mineral and possesses great

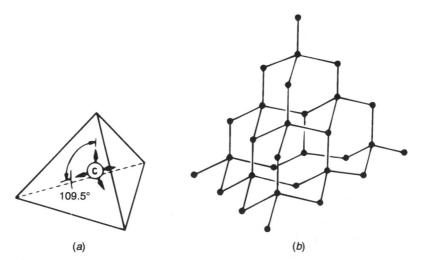

(a)   (b)

Figure 2.1 *(a) Covalent bonds of carbon directed towards the corners of a tetrahedron. (b) Structure of diamond*

strength. This is because a crystal of diamond is one gigantic molecule with strong covalent bonding throughout.

Covalent bonds can occur between atoms of different species. This may give rise to compounds composed of small discrete molecules such as methane, CH_4, carbon dioxide, CO_2, and water, H_2O, long chain molecules, as in polymers, or giant molecules (crystals), for example silica, SiO_2. Many oxides, carbides and nitrides of metals are covalently bonded crystalline substances. Owing to the high strength of the covalent bonding, these substances tend to be very hard and possess high melting points. In covalent bonds between dissimilar atoms there is usually an unequal sharing of bonding electrons between the two atoms, and this gives rise to some electrical polarisation which allows for the formation of weak secondary bonds between molecules. The covalent bond is the form of bonding which occurs in organic molecules (compounds composed principally of carbon and hydrogen) including natural and synthetic polymers. Polymer molecules are very large and an individual molecule may contain thousands of atoms, but although there is strong covalent bonding within the molecules only weaker secondary bonds exist between molecules. This means that polymeric materials are less strong, generally, than other solid materials. Wood is a naturally occurring polymer material and is discussed in Part II while plastics materials, synthetic polymers, are covered in Part VI.

Ionic bond

The ionic bond is also a strong bond but is based on electron transfer. Consider the elements sodium, Na, and chlorine, Cl. Sodium atoms possess one valence electron and the loss of this electron would expose a completely filled electron shell, while chlorine has seven valence electrons and by capturing one electron will achieve a filled outer shell containing eight electrons. By losing an electron, the sodium atom becomes out of balance electrically, and is said to be in an ionised state. Similarly, by gaining an additional electron, the chlorine atom has become ionised. An ion is a charged particle and the net charge may be positive or negative. Ions are symbolised by using the symbol for the chemical element with a superscript indicating both the number and sign of the electrical charge, for example Na^+, Cl^-, Ca^{2+}, O^{2-}. There will be a strong electrostatic attractive force between

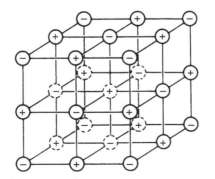

Figure 2.2 *Symmetrical arrangement of ions in NaCl structure*

ions of opposing charge and this allows the ions to pack closely together into symmetrical crystalline arrays. In sodium chloride, for example, the ions form into a cubic array with each Na^+ ion surrounded by six Cl^- ions, and vice versa, as shown in figure 2.2.

In an ionic crystal the charged ions occupy fixed sites and little, if any, movement of ions is possible even under the action of a high stress or in a strong electrical field. Ionic crystalline solids are hard, brittle and electrically insulating. Many ionically bonded solids are soluble in a polar solvent such as water and, in solution, the charge-carrying ions have mobility. Thus, in an electrical field they will move preferentially, constituting an electrical current. Conductive solutions of ionic salts are termed *electrolytes*.

Metallic bond

In the covalent and ionic bonds described above, the distribution of electrons is rearranged to provide each atomic nucleus with an external shell containing eight electrons. In the case of metals, where the number of valency electrons to each atom is small (usually one, two or three), it is not possible to satisfy fully the requirements of eight outer shell electrons per atom, and there is only a partial satisfying of this condition. In an assembly of metal atoms the principal interatomic forces acting will be a gravitational attraction between adjacent atomic nuclei and, when in close proximity, an electrostatic repulsion between the negatively charged electron shells. There is a stable separation distance at which these two opposing forces are equal in magnitude. (This stable separation distance can be regarded as an atomic diameter, and for the consideration of crystal structure it is feasible to consider metal atoms as hard spheres in contact.) At this equilibrium separation distance between atomic nuclei, the outer, or valence, electron shells of the atoms are in contact or overlap one another slightly. The outer shell electrons, at certain points in their orbits, are attracted as much by one nucleus as by another, and this means that the valence electrons follow complex paths around many nuclei. Thus, all valence electrons are shared by all the atoms in the assembly. This arrangement is termed the *metallic bond*. There are similarities to the covalent bond, in that valence electrons are shared, but the strength of the metallic bond is weaker than that of a true covalent bond. One can liken the metallic state to an arrangement of positive ions permeated by an electron cloud or gas. The extreme mobility of the valence electrons accounts for the high electrical and thermal conductivities of metals. The metallic bond is non-directional and this fact allows metal atoms to pack closely together in a regular crystalline array. Being composed of like atoms, as distinct from oppositely charged ions, the imposition of a high stress can cause a relative movement of atoms within a crystal. This means that most metals will plastically deform rather than rupture when a certain level of applied stress is exceeded. This is discussed in detail in Chapter 3. The majority of metals form closely packed crystalline arrangements and it is the nature of the metallic bond and the type of crystal structure that largely determine the strength, ductility and other properties of metals.

The main types of bonding have been described above but mixtures of bond types occur in many substances. In silicates, aluminosilicates and many other ceramics and glasses the bonding is part ionic and part covalent. For example, in dicalcium silicate, a constituent of Portland cement, ionic bonds exist between Ca^{2+} and SiO_4^{4-} ions but within the silicate ion (SiO_4^{4-}) the bonding is covalent. Clay minerals, complex aluminosilicates, tend to form small plate-like crystals. Unequal sharing of bonding electrons within the covalent bonds in the mineral causes some polarity. This permits the creation of weak secondary bonds between the small plate crystals. Intermetallic compounds are formed in some metallic alloy systems and this is an indication that some alloys contain bonds which are not entirely metallic.

2.2 The crystalline structure of metals

Solid metals and alloys are crystalline and the constituent atoms are arranged in a regular, repetitive pattern, referred to as a crystal space lattice. As stated in the previous section, metal atoms can be considered as hard solid spheres and these tend to pack closely into crystal forms of high symmetry, and the majority of metals crystallise in the cubic or hexagonal forms. The concept of a unit cell can be used to indicate the relative positions of atoms within a space lattice. The relationship between a unit cell and the space lattice is shown in figure 2.3. Three crystal types commonly found in metals are the body centred cubic (BCC), face centred cubic (FCC) and hexagonal close-packed (HCP), and the unit cells of these are illustrated in figure 2.4. Although the concept of the unit cell is necessary to describe the different types of space lattice, it is more realistic to consider the

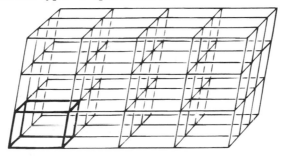

Figure 2.3 *Representation of part of a space lattice with a unit cell outlined*

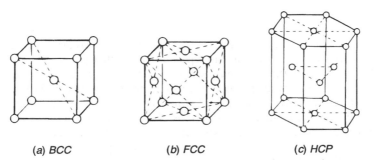

(a) BCC   (b) FCC   (c) HCP

Figure 2.4 *Unit cells (circles represent atom centres): (a) body centred cubic (BCC); (b) face centred cubic (FCC);. (c) hexagonal close-packed (HCP)*

packing of atoms in planes and the stacking of similar planes to give three-dimensional crystals. Plastic deformation in metals occurs by a process of slip. This slip, which is brought about by a shear stress, occurs across the most closely packed planes of atoms in the crystal and in the directions with the greatest line density of atoms. The process of slip is discussed in Chapter 3.

Body centred cubic system

In the body centred cubic system the most densely packed planes are the cube diagonal planes (figure 2.5(a)). Considering the atoms as uniform spheres in contact, the packing arrangement for atoms in a plane of this type is shown in figure 2.5(b). It will be seen that within this plane there are two directions, slip directions, in which atoms are in line contact. The three-dimensional crystal is built up by stacking a series of identical planes on one another in the manner shown in figure 2.5(c). The atoms in alternate planes, or layers, are all in line, giving an ABABAB stacking pattern. This is a high density of packing and the spheres occupy 68 per cent of the available space. A number of metals, including iron and chromium, crystallise in this form.

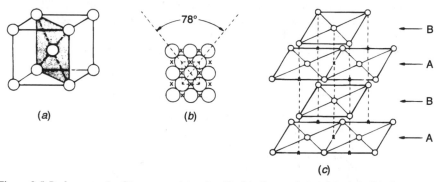

Figure 2.5 *Body centred cubic system: (a) unit cell with diagonal plane shaded; (b) plan view of a diagonal plane, the most densely packed in the BCC system – the positions of atomic centres in the second layer are marked x; (c) ABABAB type stacking sequence of diagonal planes*

Hexagonal close-packed system

The closest possible packing of spheres in a plane is hexagonal packing (figure 2.6), and the stacking of planes of this type on one another gives rise to the closest possible packing in three dimensions. It is, however, possible to build up a symmetrical three-dimensional structure from close-packed planes in two different ways. Figure 2.7 shows the centres of atoms arranged in a hexagonal plane pattern (plane A). When a second plane of atoms is stacked adjacent to the first, or A, plane, the centres of atoms in the second layer may be in either the position marked B or that marked C. If three relative spatial positions exist for close-packed layers, then symmetry in three dimensions can be obtained if the layers are

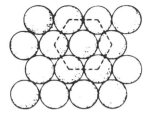

Figure 2.6 *Hexagonal packing of spheres in a plane*

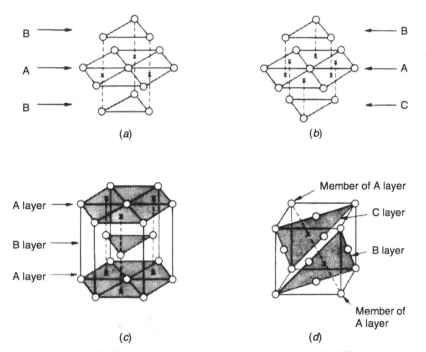

Figure 2.7 *Close-packed lattices: (a) ABAB type stacking sequence; (b) ABC type stacking sequence; (c) hexagonal close-packed crystal structure – ABAB stacking (basal plane shaded); (d) face centred cubic crystal system – ABCABC stacking (octahedral planes shaded)*

stacked in either an ABABAB sequence (figure 2.7(a)), or else in an ABCABC sequence (figure 2.7(b)).

In both cases the density of packing is the same, with 74 per cent of the total space filled with spheres. The ABABAB type stacking produces the crystal form known as hexagonal close-packed, and the ABCABC sequence gives rise to the face centred cubic type of crystal (figure 2.7(c), (d)). Although both of these types of crystal are made up of planes with a close hexagonal packing, the different stacking sequences give significant property differences. In the hexagonal close-packed system the basal planes, shown shaded in figure 2.7(c), are the only set of parallel close-packed planes.

Face centred cubic system

In the face centred cubic system the close-packed planes, shown shaded in figure 2.7(d), are the octahedral planes of the cube. Within a cube, because of the high symmetry, octahedral planes occur in several directions and there are four sets of parallel octahedral planes, all of close packing. This means that with four different sets of slip planes, plastic slip can occur more readily in the face centred cubic system than in the hexagonal close-packed system. Many metals crystallise in these close-packed systems, including magnesium and zinc (hexagonal close-packed) and aluminium, copper and nickel (face centred cubic).

2.3 Interstitial sites in crystals

The three types of crystal structure we have considered are densely packed structures. Nevertheless, even with a close packing of uniform spheres, there are small spaces, or interstices, between the spheres into which may be fitted atoms of smaller diameter. There are several types of interstice in crystals, the two most common being the tetrahedral and the octahedral. The former type is the central space between four atoms forming a tetrahedron, while in the latter the space is at the centre of six atoms which occupy the corners of an octahedral figure. The location of these interstitial sites in the body centred cubic and face centred cubic crystal systems is shown in figure 2.8. Interstitial sites are of importance in some

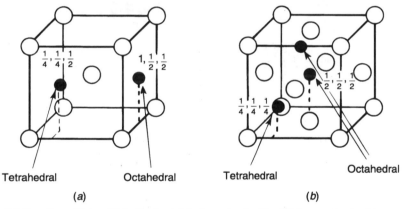

Figure 2.8 *Location of interstitial sites in: (a) body centred cubic; (b) face centred cubic*

alloys. In steels, for example, carbon atoms, with a much smaller diameter than iron atoms, can fit into interstitial spaces within the iron crystal lattice. The octahedral sites are the largest interstitial sites in both BCC and FCC structures. It can be calculated that the approximate diameters of these spaces are 0.036 nm in BCC or α-iron, and 0.103 nm in FCC or γ-iron, while the diameter of a carbon atom is about 0.15 nm. Although the size of carbon atoms is larger than the interstices, more than 1 per cent of carbon can be taken into interstitial solution in γ-iron, causing some strain in the lattice, but the solubility limit for carbon in α-iron is virtually zero. However, when a steel containing carbon in interstitial solution in the γ-phase is rapidly quenched from high temperature, the crystal structure changes from γ to α, trapping the carbon in enforced solution and causing the α lattice to be very highly strained. This puts the steel into an extremely hard, but highly brittle state (see also Chapter 4, section 4.6).

3

Deformation of Metals

3.1 Elastic deformation

The importance of metals as constructional materials is almost invariably related to their load-bearing capacity in tension, compression or shear and their ability to withstand limited deformation without fracture. Let us consider what happens when a load is applied to a metal. The imposition of an external load causes a balancing force to be set up within the material, and this internally acting force is termed a *stress*. The stress acting upon a material is defined as the force exerted per unit area. For example, for a force of 150 newtons (N) acting on a surface 10 mm square the stress is

$$\frac{\text{force}}{\text{area}} = \frac{150}{10 \times 10} = 1.5 \text{ N mm}^{-2} = 1.5 \text{ MN m}^{-2} = 1.5 \text{ MPa}$$

The types of stress normally considered are tensile, compressive and shear stresses, the two former types being referred to as direct stresses. When a material is in a state of stress its dimensions will be changed. A tensile stress will cause an extension of the length of the material while a compressive stress will shorten the length. A shear stress imparts a twist to the material. The dimensional change caused by a stress is termed *strain*. In direct tension or compression the strain is the ratio of the change in length to the original length. As a ratio, strain has no units and is simply a numerical value. In elastic behaviour the strain developed in a material, when the material is subject to a stress, is fully recovered immediately the stress is removed. In metallic materials, when the applied stress exceeds a critical level, the *elastic limit* or *yield stress*, the deformation behaviour of the material changes from elasticity to plasticity. Plastic strain is not recoverable and, when the deforming stress is removed, only the elastic portion of the strain will be recovered, leaving the material with some permanent change in dimensions, the plastic strain.

Robert Hooke, in 1678, enunciated his law stating that the strain developed in a material is directly proportional to the stress producing it. This law holds, at least within certain limits, for the stress range up to the elastic limit. For an elastic material, namely one that obeys Hooke's law, the ratio of stress to strain will be a constant for the material. For direct stresses acting in tension or compression the ratio direct stress, σ, to direct strain, ϵ, is equal to E. E is the material constant known as the *modulus of elasticity*, or *Young's modulus*. This constant has the dimensions of stress. There is a similar relationship between shear stress, τ, and shear strain, γ, and, within the elastic limit the ratio $\tau/\gamma = G$, a second constant for the material termed the *modulus of rigidity* or *shear modulus*.

Numerical values of the modulus of elasticity, E, for some common metals are given in table 3.1. E is a measure of the stiffness of a material, and the flexural stiffness (rigidity) of any component is directly proportional to the product EI (I being the *second moment of area* for the sectional shape). The value of E for aluminium is about one-third of the value of E for iron and steels. This simple comparison could indicate that aluminium would be unsuitable for creating stiff structures. However, the density of aluminium is also approximately one-third that of iron and consideration of specific moduli, namely E divided by density, shows that, as a structural material, aluminium can be compared favourably with steels.

TABLE 3.1
Modulus and density values for some metals

Metal	E ($kN\,mm^{-2}$)	Density ($kg\,m^{-3} \times 10^{-3}$)	Specific modulus ($Nm\,kg^{-1} \times 10^{-6}$)
Aluminium	70	2.7	25.9
Chromium	238	7.1	33.5
Cobalt	203	8.7	23.3
Copper	122	8.9	13.7
Iron	215	7.87	27.3
Magnesium	44	1.74	25.3
Nickel	208	8.9	23.4
Titanium	106	4.51	23.5
Zinc	92	7.14	12.9

3.2 Plastic deformation

As mentioned above, when stressed to a level beyond the elastic limit, metals will strain in a plastic manner. Some metals will begin to deform plastically at low values of stress and will yield to a very considerable extent before fracture occurs. These are termed *ductile*. Other metals and alloys show little plastic yielding before fracture and are termed *brittle* materials. Generally, plastic deformation in metals is by the process of slip across certain planes of atoms – slip planes – caused by the action of a shear stress. An early theory, the 'block slip' theory, postulated that slip took place by the relative movement of large blocks of atoms across slip planes within the crystal. The theoretical shear yield strength of a metal, τ_c, calculated on the basis of this theory is given by $\tau_c = G/2\pi$, where G is the shear modulus. The calculated value of τ_c for many metals is several hundred times greater that experimentally observed values. Present theories of plastic flow in metals are based on the existence of small structural defects, termed dislocations, within crystals. Plastic deformation is due to the movement of dislocations across the slip planes of a crystal under the action of an applied stress.

The slip planes within crystals are the planes with the highest degree of atomic packing, and the direction of slip within a slip plane is the direction of greatest atomic line density. The combination of a slip plane and a slip direction is termed a slip system. In the hexagonal close-packed system the planes with the greatest density of atomic packing are the basal planes, and within planes of this type there are three possible slip directions, as shown in figure 3.1(a). Within the hexagonal crystal space lattice there is only one series of parallel planes of this type, thus giving three slip systems for the hexagonal close-packed system. The most densely packed planes of atoms in the face centred cubic system are the octahedral planes and within each plane there are three possible directions of slip (figure 3.1(b)). There are four sets of octahedral planes in a cube, each set occurring at different inclinations. This gives a total of twelve slip systems for the face centred cubic crystal. In the body centred cubic system the most densely packed planes are cube diagonal planes, and within this type of plane there are two possible slip directions (figure 3.1(c)). With six different diagonal planes within a cube, and two directions per plane, the body centred cubic system also has a total of twelve slip systems.

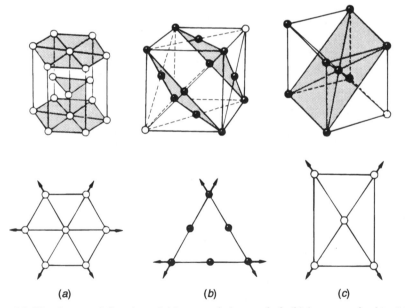

(a)   (b)   (c)

Figure 3.1 *Slip planes and directions: (a) hexagonal close packed; (b) face centred cubic; (c) body centred cubic*

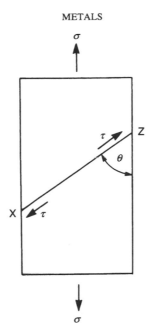

Figure 3.2 *Resolution of shear stress*

Plastic slip is initiated by the action of a shear stress on a slip plane and begins when the stress resolved on the slip plane in the slip direction reaches a certain value, termed the *critical resolved shear stress*. The magnitude of a resolved shear stress, τ, across a plane inclined at some angle θ to the axis of direct stress, σ (figure 3.2), is given by: $\tau = \frac{1}{2}\sigma \sin 2\theta$. The maximum value of this function occurs when $\theta = 45°$ ($\sin 2\theta = 1$) and there is no resolved shear stress when θ is either $0°$ or $90°$. It follows from this that when a crystal is subjected to a direct stress, plastic slip will occur if the slip planes are inclined to the axis of direct stress at some angle which favours the generation of a high resolved shear stress. Metals which crystallise in the hexagonal system possess only one set of parallel slip planes and tend to be brittle. Cubic crystalline metals, which contain twelve slip systems, are comparatively soft and ductile because, no matter from what direction a direct stress is applied, there is bound to be at least one slip system inclined in a highly favourable direction for slip. Generally, body centred cubic metals tend to be stronger and less ductile than face centred cubic metals, owing to the body centred cubic lattice being less densely packed than the face centred cubic.

3.3 Dislocations and point defects

Crystal space lattices are not perfectly crystalline but contain certain structural defects. These may be classified as either line or point defects.

Line defects

The line defects are known as *dislocations* and these may be classified as either edge-type or screw-type. An important property of a dislocation is its *Burger's vector, b,* which indicates the extent of lattice displacement caused by the dislocation. The Burger's vector also indicates the direction in which slip will occur. Figure 3.3 is a representation of an edge-type dislocation. An edge-type dislocation can be considered as an additional half-row of atoms within the lattice. If a regular square atom-to-atom circuit is described within a section of undislocated lattice, as shown at the bottom left of figure 3.3(a), it will be a complete closed circuit. If, however, one attempts to describe a similar circuit around the dislocated portion of the lattice, the start and finish points, S and F respectively, will not be coincident. The distances SF in figure 3.3(a) will be the Burger's vector, *b.* In an edge dislocation, the Burger's vector is normal to the dislocation line.

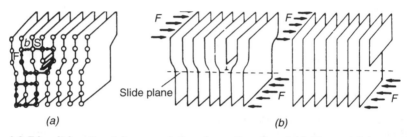

Figure 3.3 *Edge dislocation: (a) representation of a portion of crystal lattice containing an edge dislocation, with Burger's vector b; (b) application of shear force F causes the dislocation shown as ⊥ to move across the slip plane until it leaves this section of lattice, causing an increment of plastic deformation*

When a shearing stress is applied to a section of crystal lattice, and this stress is beyond the elastic limit, minor atomic movements will occur, causing the dislocation line to move through the lattice (see figure 3.3(b)). The magnitude of the stress necessary to initiate the movement of a dislocation across a slip plane is considerably less than that which would be required to bring about block slip. It will be seen that plastic flow occurs in the direction of the Burger's vector.

A diagrammatic representation of a screw dislocation is shown in figure 3.4. In this case a Burger's circuit describes a helical, or screw, path and the Burger's vector, *b,* is in the same direction as the line of the dislocation. It will be seen from figure 3.4(b), however, that the plastic flow under the action of a shearing stress still occurs in the direction of the Burger's vector.

In practice, dislocations are rarely of the pure edge- or pure screw-type but are mixed dislocations, that is, lines of dislocation containing both an edge and a screw component. However, both the magnitude and direction of the Burger's vector are constant for all points on any one dislocation line and slip is always in the direction of the Burger's vector. How many dislocations are there within metal crystals? The dislocation density in metal crystals is usually expressed as the number of dislocations that intersect a unit area of 1 mm^2. In the case of soft annealed metals the dislocation density, generally, is between 10^4 and 10^6/mm^2. This may seem a very large number but, to place it in proper context, the number of atoms

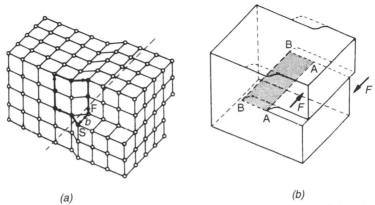

(a)   *(b)*

Figure 3.4 *Screw dislocation: (a) representation of a lattice containing a screw dislocation; (b) application of a shear force F will cause the dislocation line to move from AA to BB (slipped area is shaded)*

per unit area in a typical metal crystal is of the order of $10^{13}/mm^2$ and so it can be seen that dislocations in an annealed metal are widely separated.

Point defects

In addition to line defects, metal crystals invariably contain numerous point defects. The principal types of point defects which occur are the vacancy, the substitutional defect and the interstitial defect (figure 3.5). The latter two types are due to the presence of 'stranger' atoms of other elements which are present either as impurities or are deliberately added as alloying elements. The interstitial type occurs when the stranger atoms are very small in comparison with the atoms of the parent metal and fit into interstitial sites (see Chapter 2, section 2.3). These interstitial atoms, for example carbon in steels, are invariably larger than the interstice and, therefore, cause a positive strain to be developed in the crystal lattice. The substitutional type occurs when parent and stranger atoms are comparable in size and the stranger atoms occupy normal sites within the lattice. The stranger atoms may be either larger or smaller than the parent metal atoms and will create either positive or negative lattice strain. It will be noted that *all point defects generate strain within a crystal lattice*. Lattice strain will create irregularities and unevenness in the slip planes and it will require a higher stress to bring about dislocation movement. In other words, point defects strengthen metals.

3.4 Polycrystalline metals

The preceding sections have referred only to individual crystals but metallic structural materials are not single crystals. They are, instead, composed of many small crystals, or grains, and in many cases these small grains have a random orientation. If one considers the behaviour of a polycrystalline sample of a hexagonal close-packed metal, such as magnesium, in which each crystal grain will only possess one set of parallel slip planes, it will be apparent that while the slip planes in

CIVIL ENGINEERING MATERIALS

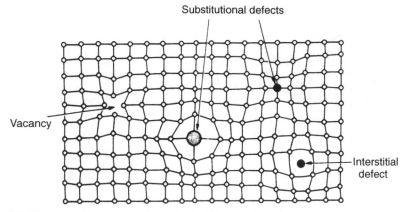

Figure 3.5 *Diagrammatic representation of point defects within a lattice*

some crystals might be favourably inclined for slip to occur readily, other crystals may not be aligned in a suitable direction. Plastic deformation, even of those favourably positioned grains, will be hindered, possibly even completely prevented, by the presence of adjacent crystals in unfavourable alignment with the axis of the imposed stress (figure 3.6(a)).

In cubic crystalline metals, possessing twelve slip systems, every individual crystal grain will have at least one slip system favourably inclined to the stress axis to permit slip and so the presence of adjacent grains does not greatly hinder plastic flow in any one grain. However, the crystal boundaries (grain boundaries) will hinder plastic deformation. These boundaries are not simple planes but are transition zones between adjacent crystals of differing orientation (figure 3.6(b)). These transition zones, which may be of several atoms in thickness, do not possess regular crystal planes and, in consequence, act as barriers to the movement of dis-

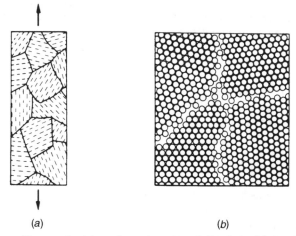

(a) (b)

Figure 3.6 *Polycrystalline metals: (a) random orientation of slip planes (shown dashed) in an HCP metal – some crystals are more favourably placed than others for plastic deformation; (b) representation of grains and boundaries*

locations. It follows from this that the grain boundaries in a pure metal tend to be stronger than the crystal grains. Consequently, a metal composed of a large number of very small grains will be stronger than a sample of the same metal containing a smaller number of larger crystal grains. The effect of crystal grain size on tensile yield strength has been quantified. The relationship between tensile yield strength in N mm^{-2}, σ_y, and average grain diameter in mm, d, is given by: $\sigma_y = \sigma_0 + Kd^{-1/2}$ with σ_0 and K being constants for the metal. This is the Petch relationship.

Although, generally, grain boundaries are strong, this is not always the case. Impurities or some alloying elements in a metal may segregate to the grain boundaries and in some cases this effect could produce weak or brittle boundaries.

4

Strengthening Mechanisms

The metallic materials used for structural work are not pure metals but are alloys containing two or more elements. There will be various interactive effects which affect both the number of phases present and the temperature at which phase changes occur. These can be shown by means of a *phase diagram*, which is simply a map of a system and shows the phases that should exist under equilibrium conditions for any particular combination of composition and temperature. Phase diagrams are important for the understanding of alloy systems because alloy structures, and hence alloy properties, will depend on the nature of the solid phase or phases formed. A brief introduction to phase diagrams and their use is given in section 4.1.

Structural engineering materials require to be both stiff and strong and also possess a measure of ductility if brittle failure is to be avoided. Stiffness (rigidity) is directly related to the modulus of elasticity, E, and cross-sectional shape of the member. The modulus of elasticity is an inherent property of a metal and is only marginally altered by processing treatments or alloying. The tensile, compressive and shear strengths of metals, however, are capable of being changed considerably by mechanical and thermal treatments and by the addition of alloying elements. Anything which disturbs the regularity of the slip planes will make the movement of dislocations more difficult and, thus, will increase the strength of the material. In general the increase in strength will be accompanied by a decrease in ductility but this is not always the case. The principal methods for strengthening metals are *strain hardening, solution strengthening* and *dispersion hardening*, and these are discussed in this chapter. These methods may be used, either singly or in conjunction, to produce strong materials.

4.1 Alloys and phase diagrams

An alloy is a mixture of two or more metals, although certain metallic alloys may be a mixture of a metal and a non-metal, a good example of this being alloys of iron and carbon, or steels. Most commercial alloys are based on systems in which the constituents form a completely homogeneous solution in the liquid state, but this does not mean that the liquid solution will solidify on cooling to give a homogeneous solid phase, or solid solution. There are several possibilities and on solidification the alloy components may remain completely soluble in one another, be partially soluble in one another, be totally insoluble in one another, or combine with each other to form one or more compounds. The structures of alloys and, hence, their properties will be dependent on the nature of the solid phase or phases which are formed. The various phases which occur within an alloy system can be shown by means of a phase diagram, which for a two constituent, or binary, alloy system is drawn with composition represented on a base line and temperature on the ordinate (see figure 4.2).

Solid solutions. It is possible for an alloy to consist of a single phase, that is all the atoms take up positions in a common crystal lattice with those of the alloying element occupying substitutional or interstitial sites in the lattice of the parent metal. This is termed a *solid solution.* A schematic representation of solid solutions is shown in figure 4.1. Generally, the arrangement of solute atoms in a substitutional solid solution is random but, in some instances, solutions of an ordered type may be formed. An ordered solid solution can exist only at either one fixed composition or over a narrow range of composition.

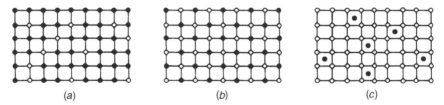

(a)   (b)   (c)

Figure 4.1 *Schematic representation of solid solutions: (a) substitutional (random); (b) substitutional (ordered); (c) interstitial*

Copper and nickel are two metals which form a continuous range of solid solutions and the phase diagram for the copper–nickel alloy system is shown in figure 4.2. The microstructure of any copper–nickel alloy is single phased.

Not all alloys are single-phase solid solutions. Certain conditions must be met for a solid solution to be formed, a major one being the relative size difference between the atoms of the two elements. If the sizes of the two types of atoms differ by more than 14 per cent then solid solution formation, if it occurs at all, will be extremely restricted. If the atoms of the solute element are very small in comparison with those of the solvent metal, a solution of the interstitial type may be formed. If both metals are of the same crystal type, say both FCC, and there is only a small size difference between the two atom types, then it is possible for complete solid solubility to occur over the whole range of compositions. Such is the case for alloys of copper and nickel. In many other commercial alloy systems there is only partial solid solubility. One example of this is the aluminium–magnesium system and a portion of the Al–Mg phase diagram is shown in figure 4.3. The maximum solubility of magnesium in aluminium under equilibrium conditions is 2 per cent at 20°C, rising to 15 per cent at 450°C. Solid solutions of magnesium in aluminium are referred to as α-phase. Alloy compositions beyond the limit of solubility

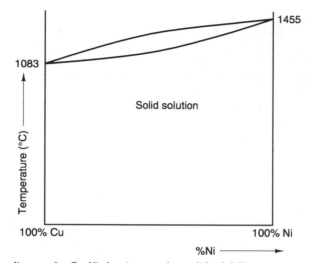

Figure 4.2 *Phase diagram for Cu–Ni showing complete solid solubility*

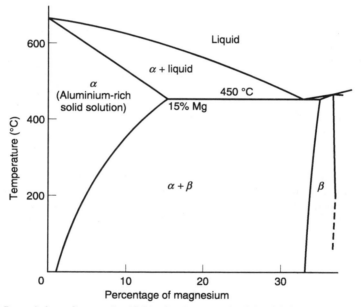

Figure 4.3 *Part of phase diagram for Al–Mg showing partial solid solubility*

will possess duplex structures with crystals of both α-phase and β-phase, β being a compound of formula Al_2Mg.

4.2 Strain hardening and recrystallisation

It was stated in the previous chapter that when a metal is stressed beyond its elastic limit, dislocations within the metal begin to move across slip planes and plastic deformation occurs. The movement of one dislocation through a lattice will give one increment of plastic deformation, namely a displacement of the order of one atomic spacing. Clearly, the movement of extremely large numbers of dislocations would be necessary to give a visible amount of plastic deformation. New dislocations are being generated continually during metal-working processes such as rolling and forging and the dislocation density in heavily cold worked metals is, typically, between 10^9 and 10^{10}/mm^2, compared with some 10^4 to 10^6/mm^2 in annealed material.

As plastic deformation occurs, the level of applied stress must be continually increased if the process is to continue. As more plastic deformation is given to the metal, the greater will be the force needed to continue deforming the metal. In other words, the metal is being strengthened. Dislocations cannot move freely throughout the whole of a crystalline material. Their progress will be impeded by barriers such as grain boundaries. A continued movement of the dislocations, that is, further plastic deformation can only occur if the stress is raised so that the dislocations can move across the potential barriers. Also, in cubic crystals, where slip may be occurring simultaneously on more than one intersecting slip plane, several dislocations may become entangled and thus interfere with one another's movement.

Cold working

After severe cold working, the array of dislocations within the metal will be in a highly tangled state, but there will be some areas that are almost dislocation-free while other areas will possess a very high dislocation density. The areas with a high dislocation density have a high degree of structural disorder and, eventually, these areas become the crystal grain boundaries of a pattern of small crystals created by the deformation process from the larger original crystals. Cold plastic deformation working processes cause the uniform crystal structure of an annealed metal to become distorted and fragmented. This phenomenon, known as strain, or work, hardening is utilised in practice for the strengthening of metals. The tensile strengths of some metals and alloys may be raised very considerably by this method but strain hardening is always accompanied by a decrease in ductility (see table 4.1). There are many types of cold deformation processes available for the forming of metals including cold rolling, for sheet, strip and plate, and drawing, for wire, rod and tubing.

TABLE 4.1
Effect of cold working on strength and ductility

Material	Condition	Typical properties		
		Hardness H_D	Tensile strength $(N\ mm^{-2})$	Per cent elongation on 50 mm
Low carbon steel (BS 970: 040A10)	Annealed	90–100	300–350	30–40
	Cold worked	245–255	850–880	5
Pure aluminium (BS EN 485-2: 1200)	Annealed	30–35	80–90	35–40
	Cold worked	54–62	140–160	3–5

If attempts are made to continue cold working beyond the point where maximum hardness and strength are achieved, cracks will develop in the material and failure will occur. The start of failure may be due to a number of dislocations on the same slip plane being forced together at a major barrier, such as a grain boundary, by a large applied stress, thus creating a void or internal crack.

A metal which has been cold worked will contain residual internal stresses. During the plastic deformation process all the crystal grains of the original material will have been strained both elastically and plastically. When the deforming force is removed not all the elastic strain will be recovered. Complete recovery of the elastic strain in an individual crystal grain will be prevented by the rigidity of surrounding crystals and there will be some locked-in elastic strain in the material. When a cold worked, or strain hardened, material is heated, changes will begin to occur. The changes are in three stages: stress relief, recrystallisation and grain growth.

Stress relief. As the temperature of the material is raised so the vibrational energies of individual atoms are increased, permitting some movement of atoms. Comparatively minor atomic movements occur and these result in the removal of residual stresses associated with the locked-in elastic strains. This change, which occurs at comparatively low temperatures, has a negligible effect on the strength, hardness and ductility of the material, and the microstructure of the metal is unchanged in its appearance.

Recrystallisation. When the temperature is raised above a certain threshold value, the recrystallisation temperature, the process of recrystallisation begins. New unstressed crystals begin to form and grow from nuclei until the whole of the material has a structure of unstressed polygonal crystals (figure 4.4). This change in structure is accompanied by a reduction in hardness, strength and brittleness to the original values prior to plastic deformation. The driving force for the process is the release of the strain energy stored in the zones of high dislocation density. The temperature at which recrystallisation occurs is, for a pure metal, within the range from one-third to one-half of the melting temperature (K). The recrystallisation temperature is not, however, constant for any material as its value is affected by the amount of plastic deformation, and is lower for very heavily cold worked metals than for samples of the same material which have received small amounts of plastic deformation. The recrystallisation temperature is also affected by composition. The presence of impurities or alloying elements will increase the recrystallisation temperature of the material.

Grain growth. If the temperature is raised to some value above the recrystallisation temperature, a process of grain growth may occur after the completion of recrystallisation. Some crystal grains will grow in size at the expense of others by a process of grain boundary migration. Again, the driving force for grain growth is the release of boundary surface energy as the amount of total grain boundary surface is reduced.

The industrial process of *annealing* is a heat treatment for work hardened materials which allows recrystallisation to take place with a consequent softening. Usually, a fine, uniform crystal grain structure is required for optimum properties,

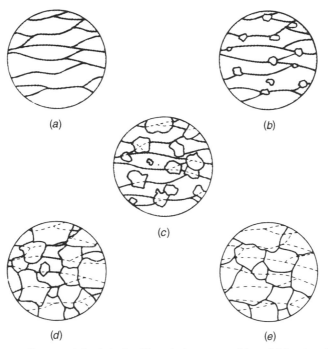

Figure 4.4 *Recrystallisation: (a) original cold worked structure; (b) nuclei forming; (c) and (d) new grains forming from nuclei; (e) recrystallisation complete*

and the time and temperature of the annealing treatment are carefully controlled to give complete recrystallisation without excessive grain growth. It should be noted that there are several different types of heat treatments for the softening, or annealing, of steels and a number of these are based on a different principle from the straightforward recrystallisation of a cold worked structure (refer to section 4.7).

Cold working and annealing treatments are used mainly for pure metals and single-phase alloys. When a two-phase alloy is cold worked, both phases are broken up and elongated in the direction of working. The two cold worked phases may have different recrystallisation temperatures and in some cases the annealing treatment may only cause full recrystallisation of one of the phases. This can be seen in cold worked low carbon steels which have been given a *process anneal* treatment (see section 4.7) where one phase, pearlite, appears in the structure as elongated particles or bands.

Hot working

Plastic deformation processes carried out at temperatures above the recrystallisation temperature are referred to as hot working. There is no strain hardening effect with hot working as there is an almost immediate recrystallisation following the plastic deformation. Hot working processes such as hot rolling, forging and extrusion are used for the initial plastic deformation of large ingots in industrial metal processing and for the production of such products as structural sections, rails, rolled plate and bar. Metals are much more plastic at hot working temperatures and the deformation forces needed are lower than those for cold deformation. Close control of the hot work finishing temperature is necessary so that, after the final deformation, the temperature is sufficient for recrystallisation but not high enough to give grain growth. In the case of steels, the hot work finishing temperature is usually just above the critical temperature range of the steel and cooling in air from this temperature produces a fine grained *normalised* structure (for an explanation of critical temperatures and normalising refer to sections 4.6 and 4.7).

4.3 Solution strengthening

An alloy is a mixture of two or more metals or a mixture of a metal and a non-metallic element, a good example of the latter being alloys of iron and carbon, namely steels. As mentioned in chapter 3, section 3.3, the presence of interstitial and substitutional atoms within a crystal lattice causes lattice strain which hinders the movement of dislocations, thereby producing a strengthening effect. This is termed *solution strengthening* and is a major factor in the development of high strength in many alloy systems.

For an alloy system showing complete solid solubility, such as the copper–nickel system (figure 4.2), the addition of nickel to copper will create lattice strain, thus increasing the strength of the copper. Similarly, the addition of copper to nickel will increase the strength of nickel. The strongest alloy of the two metals possesses a strength considerably higher than that of either pure copper or pure nickel. Some property values for alloys in this system are given in table 4.2. Values for both the annealed and work hardened conditions are given to illustrate the point that two different strengthening mechanisms can be used in conjunction.

The aluminium–magnesium alloy system exhibits partial solid solubility (refer to figure 4.3) and there are a number of commercially useful aluminium alloys containing magnesium. In practice, commercial alloys containing up to 5 per cent of magnesium possess a single-phase structure of α solid solution at room temperature with the β phase only separating out if the alloys are cooled extremely slowly from high temperature. Typical mechanical properties of some alloys in this system are given in table 4.2 and it will be seen that the presence of 5 per cent of magnesium increases the strength of aluminium by more than 200 per cent. Again, strain hardening can be used to increase the strength further, but notice that in the case of the alloy containing 5 per cent of magnesium only a small percentage strength increase can be obtained by strain or work hardening. This is because the presence of 5 per cent of magnesium creates such a high strain in the aluminium lattice that only a small amount of additional strain induced by plastic deformation can be accommodated without the lattice breaking up.

4.4 Eutectics

The term *eutectic* (from Greek) means lowest melting point, and many alloy systems contain eutectics. A good example is the lead–tin system (figure 4.5).

It will be seen from figure 4.5 that there is only limited solid solubility and that alloy structures will consist of a mixture of α (lead-rich) and β (tin-rich) solid

TABLE 4.2

Solution strengthening: typical properties of some copper–nickel and aluminium–magnesium solid solutions in both the annealed and work hardened conditions.

Alloy	Condition	Hardness H_D	Tensile strength $(N\ mm^{-2})$	Percentage elongation on 50 mm
A. *Copper–nickel alloys*				
Pure copper	Annealed	45	210	55
	Work hardened	115	390	5
95% Cu; 5% Ni	Annealed	65	260	50
	Work hardened	130	465	5
80% Cu; 20% Ni	Annealed	75	340	45
	Work hardened	165	540	5
70% Cu; 30% Ni	Annealed	80	355	45
	Work hardened	175	650	5
60% Cu; 40% Ni	Annealed	90	390	45
	Work hardened	190	660	5
30% Cu; 70% Ni	Annealed	120	540	45
	Work hardened	220	720	8
Pure nickel	Annealed	95	450	50
	Work hardened	200	650	10
B. *Aluminium–magnesium alloys*				
1200 (Al 99% pure)	Annealed	32	85	40
	Work hardened	60	155	5
5251 (Al + 2.25% Mg)	Annealed	70	185	25
	Work hardened	100	265	4
5154A (Al + 3.5% Mg)	Annealed	85	220	18
	Work hardened	115	300	8
5056A (Al + 5.0% Mg)	Annealed	105	280	18
	Work hardened	120	310	8

solutions. The eutectic point occurs at a composition of 62 per cent tin and a temperature of 183°C. A liquid alloy of this composition will solidify at 183°C with both α and β phases forming simultaneously and giving a structure composed of a mixture of very small crystals of α and β, the eutectic mixture. The structures of other alloy compositions will consist of comparatively large primary crystals of either α or β solid solutions, depending on whether the alloy contains less than, or more than, 62 per cent tin, respectively, together with amounts of eutectic mixture of α and β. Many of the lead–tin alloys are used as soft solders. Both lead and tin are low strength metals with tensile strengths of less than 16 N mm^{-2} but the eutectic alloy has a tensile strength of more than 50 N mm^{-2} and this increase in strength is largely due to the very fine structure of the eutectic mixture. Several two-phased high-strength alloys, where the high strength is due to the formation of the fine grained eutectic structures, have been developed for applications in mechanical engineering.

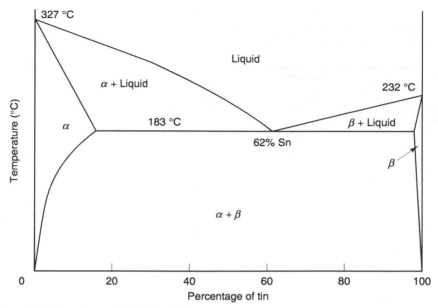

Figure 4.5 *Phase diagram for lead and tin*

4.5 Dispersion strengthening

The properties of a two-phase alloy depend to a large extent on the size and distribution of the second phase. If the second phase is present as comparatively large crystals randomly distributed throughout the structure, there will be little, if any, strengthening, but if the second phase consists of extremely small particles widely dispersed throughout the lattice of the parent metal then a considerable strengthening effect can occur. This is termed dispersion strengthening. The effect can be achieved in two ways. A finely divided insoluble phase may be dispersed throughout the material during manufacture and this method has been used to produce nickel alloys strengthened by thorium oxide for use in high-temperature applications. The other method, known as precipitation hardening or age hardening, is to develop a high-strength structure by heat treatment. The types of alloys that may respond to this type of treatment are those in which partial solid solubility occurs and where there is a wide difference between the high-temperature and low-temperature solid solubility limits. The aluminium-rich end of the aluminium–copper phase diagram is of this type (figure 4.6).

Aluminium can hold up to 5.7 per cent, by weight, of copper in solution at 548°C but only 0.2 per cent, by weight, of copper in solid solution at room temperature. If an alloy containing, say, 5 per cent, by weight, of copper is slowly cooled from high temperature, the microstructure will be two-phase with crystals of the compound $CuAl_2$ precipitated at aluminium grain boundaries. If, alternatively, the alloy is heated to a high temperature and rapidly cooled, the structure will consist of a supersaturated solid solution of copper in aluminium. Such a solution is unstable, or metastable, and there will be a tendency for copper atoms to diffuse through the aluminium lattice prior to the precipitation of $CuAl_2$ particles. The first result of the diffusion of copper atoms is to increase the concentration of

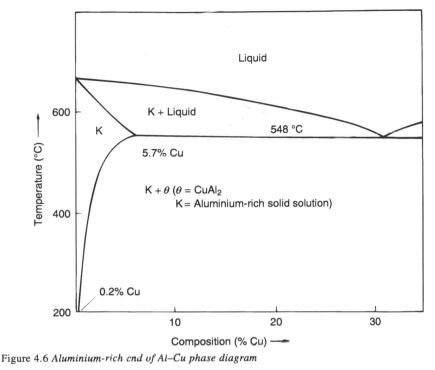

Figure 4.6 *Aluminium-rich end of Al–Cu phase diagram*

copper in many localised zones within the aluminium lattice. There is a very high degree of strain in the parent lattice in these zones of high copper concentration. As the solid solution lattice is still continuous and fully coherent, these zones are termed areas of coherent precipitate. These widely dispersed regions of high strain offer considerable resistance to the passage of dislocations and, hence, produce a major strengthening effect. If the diffusion of copper proceeds fully then true precipitate particles of $CuAl_2$, which are non-coherent with the aluminium lattice, will begin forming, and the strain in the lattice starts to be released. Consequently, during this stage, the strength of the material begins to reduce. When this phenomenon occurs it is termed *overageing*.

With some aluminium–copper alloys the diffusion of copper, and consequent hardening, will occur slowly at room temperature following a solution heat treatment but in others, and in some other alloy systems, hardening does not take place spontaneously at room temperature. In these latter cases, the alloy has to be heated to some comparatively low temperature before the diffusion process will occur. When an alloy slowly hardens with time on being kept at ordinary temperatures after receiving a solution heat treatment, it is termed an age-hardening alloy. If, after solution heat treatment, it is necessary to heat the alloy to bring about hardening, the second heat treatment is termed precipitation heat treatment, even though it is stopped before true precipitation and overageing can occur. Figure 4.7 shows the relationship for hardness with ageing temperature and time for an alloy of aluminium and copper. Typically, for such an alloy the tensile strength immediately after solution heat treatment would be about 280 N mm^{-2}, increasing after

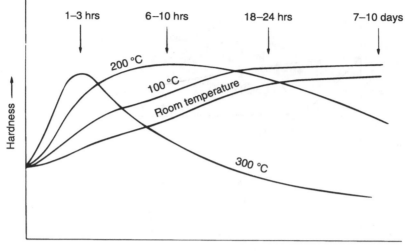

Figure 4.7 *Hardness–time relationships for an aluminium–copper age hardening alloy*

precipitation hardening to almost 500 N mm^{-2}. Other alloy systems which respond to precipitation hardening treatments include copper–beryllium, magnesium alloys containing thorium, zirconium or zinc, some titanium alloys, some nickel alloys and a group of alloy steels termed maraging steels.

4.6 Eutectoids and the Fe–C phase diagram

As stated in section 4.4, a eutectic structure is formed in some alloy systems when a liquid solution, on freezing, solidifies simultaneously into two separate solid phases. An analogous situation is the formation of a *eutectoid*, but in this case it is structural change which occurs entirely within the solid state and a solid solution transforms during cooling into a mixture of two new solid phases. The phase diagram for the iron–carbon system (figure 4.8) contains both a eutectic and a eutectoid. The diagram may appear at first sight to be extremely complex but, in practice, it can be viewed in two sections. Steels contain up to 1.5 per cent carbon and cast irons contain between 2 and 4.5 per cent of carbon. For the consideration of steels, therefore, it is only necessary to use the left-hand portion of the diagram, which contains the eutectoid, while only the high carbon end of the diagram, containing the eutectic, need be considered for cast irons.

Iron is an element which exhibits allotropy, that is it can exist in two crystal forms, body centred cubic and face centred cubic. The crystal structure of pure iron at ordinary temperatures is body centred cubic. This form is known as α-iron, or ferrite, and it is stable at all temperatures up to 908°C. On heating through 908°C the crystal structure of iron changes to face centred cubic. This form is known as γ-iron, or austenite. γ-iron changes into the body centred cubic form again on heating through 1388°C and this high temperature form of ferrite is referred to as δ-iron. δ-iron is the stable form up to the melting temperature of pure iron at 1535°C. When carbon is added to iron, it can enter into intersitial solid solution in the iron lattice to a limited extent. Carbon, in excess of the solubility limit, will combine with iron to form an extremely hard and brittle compound, cementite, with the formula Fe₃C.

It will be seen from figure 4.8 that ferrite cannot hold carbon in solution to any great extent, the limits being 0.04 per cent of carbon at 723°C and 0.006 per cent of carbon at 200°C. Austenite, however, can hold a considerable amount of carbon in solid solution, ranging from 0.87 per cent at 723°C to 1.7 per cent at 1130°C. The eutectoid occurs at a carbon content of 0.87 per cent and at a temperature of 723°C.

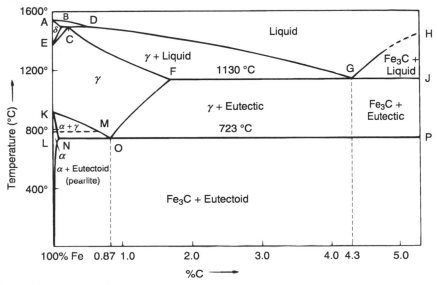

Figure 4.8 *The iron–carbon phase diagram*

Factors influencing the structure of steels

At a high temperature, say 1000°C, a steel containing less than 0.87 per cent of carbon will have a wholly austenite structure with all carbon held in solid solution. When the steel is cooled, the austenite will begin to transform into ferrite when the phase transformation boundary (line KO) is crossed. Ferrite can hold very little carbon in solid solution and so the remaining austenite becomes enriched in dissolved carbon. As the temperature falls further, more ferrite is formed, and the composition of the remaining austenite increases in carbon content, following the line KO towards the eutectoid point O. At, or just below, the eutectoid temperature, all remaining austenite transforms into the eutectoid mixture. This mixture, known as *pearlite*, consists of alternate layers of ferrite and cementite. Similarly, if a steel containing more than 0.87 per cent carbon is cooled from a high temperature, structural changes occur. In this case it is cementite which separates first from austenite, when the phase line OF is crossed, reducing the amount of carbon dissolved in austenite, with all remaining austenite transforming into pearlite when the eutectoid temperature is reached. Ferrite is soft and ductile, cementite is very hard and brittle, and the eutectoid mixture, pearlite, is hard and strong, but of low ductility.

The *rate of cooling* through the transformation temperatures has a major effect on the nature of the pearlite formed. Slow cooling will give a coarse pearlite structure while rapid cooling will produce a fine pearlite, with extremely thin alternate layers, or lamellae, of ferrite and cementite, and this affects the properties of the steel. The tensile strength of ferrite is about 280–300 N mm^{-2}, that of coarse pearlite is about 700 N mm^{-2}, while that of fine pearlite formed by rapid cooling may be as high as 1300 N mm^{-2}. If a steel is cooled extremely rapidly there will be insufficient time for austenite to transform into pearlite and, instead, it changes into a body centred lattice with all the carbon trapped in interstitial solid solution. Theoretically, ferrite can hold virtually no carbon in solid solution and this structure is highly strained and distorted by the large amount of carbon in enforced solution into a body centred tetragonal form, referred to as α' or *martensite*. Martensite is extremely hard and brittle but these properties can be usefully modified by subsequent heat treatment.

Because the structures and, hence, properties of steels are highly dependent on the rate of cooling through the γ to α transformation temperatures, these temperatures are referred to as *critical temperatures*. The lower critical temperature, or A_1 corresponds to the line NOP on the phase diagram, while the upper critical temperature, or A_3, corresponds to the line KO. Commercial steels, which always contain amounts of manganese, silicon and possibly other elements such as chromium, nickel, etc., will have critical temperatures which may be higher or lower than those shown on the phase diagram for pure iron and carbon.

Cast irons

Cast irons contain between 2 and 4.5 per cent of carbon and, according to the phase diagram (figure 4.8), should solidify giving structures with either primary austenite and eutectic mixture or primary cementite plus eutectic, the eutectic mixture being of austenite and cementite. Then, according to the phase diagram, austenite will transform into the eutectoid pearlite on cooling through the critical temperatures. This is the case only for a small number of iron compositions with a low silicon content. Irons with these structures are termed *white irons*, and they are very hard but also very brittle. The majority of cast irons contain some or all of the carbon content present in the structure in the form of graphite and these are known as *grey irons*. The structure of a cast iron is determined by a number of factors, including the composition and the rate of solidification. The presence of some elements, particularly silicon, and a slow rate of solidification favour the formation of graphite. Most cast iron compositions contain between 1 and 3 per cent of silicon and solidification rates in sand moulds are relatively slow. During solidification either primary austenite or primary graphite flakes, depending on the carbon content, separate out with, at the eutectic temperature, formation of a graphite/austenite eutectic mixture. In many, but not all, cases the austenite transforms into pearlite on cooling through the critical temperatures. The final structure of the cast iron will consist of graphite + pearlite, graphite + pearlite + ferrite, or graphite + ferrite. The higher the carbon and silicon contents, the greater will be the tendency for development of a purely graphite and ferrite structure.

In general, grey cast irons have low tensile strengths and are brittle, owing to the presence of flake graphite. They possess, however, comparable compressive strengths to steels. The tensile strengths of common grey cast irons lie in the range 100 to 350 N mm^{-2} but irons of higher strength and in some cases possessing reasonable ductility, can be produced by a variety of methods. Close control of composition and special treatments to the melt immediately prior to casting can produce structures with very small graphite flakes which will give a strength increase. Innoculation of the melt with a small amount of cerium or magnesium will cause graphite to solidify in the form of spheroidal nodules, rather than as flakes. The presence of spheroidal graphite improves both strength and fracture toughness, and tensile strengths of up to 540 N mm^{-2} can be achieved. The properties may be improved further by making alloying additions. Malleable cast irons can be made by a prolonged annealing of certain white irons. Graphitisation occurs during the heat treatment and small compact rosettes of graphite are formed in a matrix of ferrite and pearlite. Malleable cast irons possess tensile strengths in the range 300–700 N mm^{-2} combined with some ductility and toughness.

4.7 Heat treatment of steels

The micro-structures and, hence, the properties of steels are determined by both composition and the heat treatment given, in particular the rate of cooling through the critical temperature range. Unlike most non-ferrous materials, where an annealing treatment involves heating to just above the recrystallisation temperature of the alloy, there are several different types of annealing treatments available for steels, as well as hardening and tempering treatments. These various heat treatments, and their applicability, are described below.

Full annealing.   This is the treatment which will put a steel with less than the eutectoid composition of carbon into the softest possible condition. It involves heating the steel at 30–50°C above the upper critical temperature followed by slow cooling within a furnace. This produces a coarse pearlite structure. Full annealing is a lengthy and expensive process and when a reasonably soft and ductile condition is required and it is not essential for the steel to be in its softest state, then the treatment known as normalising is used.

Normalising.   This involves heating the steel at 30–50°C above the upper critical temperature followed by cooling in still air. The faster rate of cooling than with full annealing gives a finer pearlite structure and the steel, while relatively soft and ductile, is somewhat harder than full annealed material. Hot worked products, such as structural steel sections, plain reinforcement bar and forgings, are allowed to cool in air from the hot work finishing temperature, which is just above the upper critical temperature, and are, therefore, in the normalised condition.

Spheroidise annealing.   Steels containing more than the eutectoid content of carbon are not softened by either full annealing or normalising treatments since the cementite which first separates from austenite during cooling forms as a continuous network around the austenite grain boundaries. This will have a major embrittling effect. These steels can be softened by heating them at a temperature just below the critical temperature. During this treatment the cementite reforms into small spheroidal shaped particles, and this greatly toughens the material.

Process annealing or sub-critical annealing.   This is a process often used for softening cold worked low carbon steels. The structures of these steels contain about 90 per cent ferrite and 10 per cent pearlite. The recrystallisation temperature of cold worked ferrite is about 500°C and annealing at about 550°C will completely recrystallise ferrite, even though the pearlite will be largely unaffected. Full annealing of such a steel would involve temperatures just above 900°C with subsequent high cost.

Hardening.   A steel is hardened by rapidly quenching the steel, from a high temperature, into water or oil. The very rapid cooling through the critical temperatures causes the formation of martensite. The heat treatment temperature, prior to quenching, is 30–50°C above the upper critical temperature for steels of less than the eutectoid carbon content and 30–50°C above the lower critical temperature for steels with a greater percentage carbon than the eutectoid composition. (It is not practicable to harden, by quenching, plain carbon steels containing less than 0.3 per cent carbon, although some low carbon alloy steels may be hardened.)

Tempering.   Hardened steels may be tempered by heating within the range 200–700°C. This treatment will remove internal stresses created by quenching, reduce the hardness, and increase the toughness of the steel. The higher the tempering temperature, the greater will be the reduction in hardness and the increase in toughness.

The various heat treatments described above apply to both plain carbon and low alloy steels. *Low carbon steels*, which contain up to 0.3 per cent carbon, are generally used in the normalised, cold worked, or cold worked and process annealed conditions. *Medium carbon steels* are those which contain between 0.3 and 0.6 per cent carbon, and these may be hardened and tempered. *High carbon steels* (or *tool steels*) contain more than 0.6 per cent carbon, and are always used in the hardened and tempered condition.

5

Behaviour in Service

A material, component or structure is deemed to have failed when its ability to fully satisfy the original design function ceases. This may be the result of a variety of causes. Failure may be due to fracture, either partial or complete, plastic buckling, dimensional change with time, loss of material by corrosion, erosion or abrasive wear, or an alteration of properties and characteristics with time as a result of environmental or other effects. These various modes of failure may be dependent on stress, time, temperature, environment, or a combination of any of these. Most civil engineering structures are required to have a long life, and it is important that the designer is aware of the possible modes of failure of materials and of the effects of the various stress and environmental factors on the behaviour of constructional materials, so that the materials best suited to the design requirement can be selected.

5.1 Failure by yielding

Metals show both elastic and plastic behaviour. As the level of stress is increased so the amount of elastic strain increases in direct proportion to the stress applied up to a certain limit, the *elastic limit*. When the elastic limit is exceeded, permanent plastic yielding occurs. Many steels possess a definite yield point, shown by a sharp discontinuity on a tensile force–extension curve, but in most metals there is a gradual transition from elastic to plastic deformation modes. Some metals have low elastic limits, and are highly ductile with considerable plastic deformation before fracture. Most high-strength materials have high yield strengths and show little plastic deformation, with final fracture occurring at a stress level which is little higher than the yield strength. To minimise the possibility of excessive deformation or failure by yielding in service, it is usual in design to limit the maximum stress under service load to the yield or proof strength divided by a safety factor, the latter typically being about two.

The tensile properties of metals can be determined in the laboratory with comparative ease, and the results of standard tensile tests are of considerable use to the designer.

5.2 The tensile test

There are standardised procedures for the tensile testing of metals and these, together with the recommended dimensions for tensile test pieces, are contained in BS EN 10002-1 (the testing of cast iron is covered in a separate publication, BS 1452).

The general shapes of force–extension curves are given on figure 5.1. The initial strain is elastic, but beyond point Y (figure 5.1(a)) or point E (figure 5.1(b)) strain is plastic. Point U represents the maximum force applied during the test, and this value of load is used to determine the tensile strength of the material. Point F marks the point of fracture. Localised necking down of the test piece cross-section occurs beyond point U and, although the applied force on the test piece decreases, the true stress acting on the test piece, taking into account the reducing cross-sectional area, continues to increase until fracture occurs. The following information is determined in a routine tensile test.

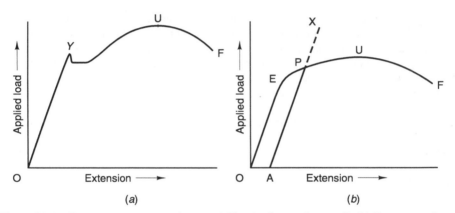

Figure 5.1 *(a) Force-extension curve showing yield point (low carbon steel). (b) Force–extension curve with no yield point, and showing construction for determination of proof force*

Tensile strength (formerly termed ultimate tensile strength). This is based on the maximum load sustained by the test piece when the latter is tested to destruction, and corresponds to point U in figure 5.1(a) and (b). The numerical value of tensile strength is given by

$$\text{Tensile strength} = \frac{\text{maximum load applied}}{\text{original cross-sectional area}}$$

The units in which tensile strength is normally quoted are newtons per millimetre^2 (N mm^{-2}) or megapascals (MPa). Numerically these values are equal.

Yield point or yield strength. There is a sharp discontinuity in the force–extension curve for wrought iron and many steels, and the material will yield suddenly with little or no increase in the applied force necessary. The extent of this sudden yielding may be up to 7 per cent. The yield strength of the material is given by

$$\text{Yield strength} = \frac{\text{applied load at yield point}}{\text{original cross-sectional area}}$$

Proof strength or offset yield strength. Most metals, unlike steels, show a smooth transition from elastic to plastic deformation. It can be difficult to determine the value of the elastic limit with exactitude and, instead, the parameter proof strength is measured. This is the level of stress required to produce some specified small amount of plastic deformation. Generally, the amount of plastic strain specified is 0.2 per cent (nominal strain = 0.002) but a 0.1 per cent value is specified in some cases. Frequently, proof strength is referred to by the alternative term *offset yield strength*, or simply *yield strength*. In figure 5.1(b) the force–extension curve for some metal is shown as curve OEUF. Point A, corresponding to an extension of 0.2 per cent, is marked, and the line AX is drawn parallel to OE. Point P denotes the 0.2 per cent proof force. The value for the 0.2 per cent proof strength is given by

$$\text{0.2 per cent proof strength} = \frac{\text{0.2 per cent proof force}}{\text{original cross-sectional area}}$$

Modulus of elasticity (Young's modulus). E may be calculated from the slope of the elastic portion of the force–extension curve. E is given by

$$\text{slope} \times (\text{gauge length/cross-sectional area})$$

The units in which E is quoted may be kilonewtons/millimetre^2 (kN mm^{-2}) or gigapascals (GPa).

Percentage elongation. The percentage elongation value for a material is a measure of its ductility. A specified length, the gauge length, is marked on the test piece before testing. After fracture, the two portions of the test piece are placed together and the distance between the gauge marks remeasured. The amount of extension, expressed as a percentage of the original gauge length, is quoted as the elongation value. For an elongation value to have validity, the fracture must occur in the central section of the gauge length and the gauge length must be specified, for example, *the percentage elongation on 50 mm = 20 per cent.* As the amount of plastic deformation is greatest close to the point of fracture, the elongation value for any particular material will be higher if measured over a short gauge length than if measured over a long gauge length.

Percentage reduction of area. Percentage reduction of area is also a measure of ductility and, often, is quoted instead of a percentage elongation value for test pieces with a circular cross-section. There is merit in this as the reduction of area value is largely independent of test piece dimensions. It is given by

$$\text{Percentage reduction of area} = \left(\frac{A_0 - A}{A_0} \right) \times 100$$

where A_0 is the original cross-sectional area and A is the cross-sectional area at the point of fracture.

5.3 Failure by fracture

The terms tough, ductile, brittle, or fatigue are frequently used to describe the fracture behaviour of a material. In a tough, or ductile fracture, failure is preceded by a considerable amount of plastic deformation and considerable energy is required to bring about fracture. On the other hand, in a brittle, or non-ductile fracture, there is little or no plastic deformation prior to failure and the energy input to cause fracture is relatively low. The type of fracture which occurs is dependent largely on the nature and condition of the material, but it is also affected by many other factors including the type of stress applied, the rate of application of stress, temperature and environmental conditions, component geometry and surface condition, and the nature and size of internal flaws. In some circumstances, components made from materials which, generally, would be expected to fail by yielding if overstressed may suddenly fracture in a brittle manner, with fracture growing at sonic speed from some existing crack or flaw, even though the average level of stress within the material is well below the design stress.

Most commercial materials contain internal cracks and other defects, and these can act as points of local stress concentration. When a material is subject to a stress, any cracks within it can propagate by one of two methods, either by ductile tearing or by cleavage. The former process applies to ductile metals and considerable energy is absorbed in plastically deforming the metal just ahead of a growing crack. Plastic deformation tends to blunt the tip of a crack, consequently reducing its stress-concentrating effect. However, severe plastic deformation tends to create small cavities in the deformed area ahead of the crack tip and these eventually link up, thus extending the length of the crack. Little, if any, plastic deformation can occur ahead of a crack tip to blunt the crack in a brittle metal, and if the local stress at the crack tip exceeds the cleavage fracture stress of the metal, interatomic bonds are ruptured and the crack develops catastrophically as a fast brittle fracture.

Fracture mechanics

Fracture mechanics is the study of the relationships between crack geometry, material strength and toughness, and stress systems, as they affect the fracture characteristics of a material. The aim of fracture mechanics is to determine the critical size of a crack or other defect necessary for fast fracture to occur, that is, catastrophic crack propagation and failure under service loading conditions. An analysis of the critical stress to cause crack propagation in brittle materials was made by Griffith who proposed the following relationship for plane stress conditions

$$\sigma_c = \left(\frac{2\gamma E}{\pi a}\right)^{1/2}$$

where σ_c is the critical stress for fracture, γ is the fracture surface energy per unit area, E is the modulus of elasticity and a is one-half the crack length.

The Griffith relationship applies only to brittle materials and was modified by Orowan and Irwin to apply to metals and take account of the plastic flow which occurs at a crack tip at the beginning of crack development. With the assumption that the size of the plastic zone at the crack tip is very small, the expression becomes, for plain stress conditions

$$\sigma_c = \left(\frac{2(\gamma + \gamma_p)\, E}{\pi a}\right)^{1/2}$$

where γ_p is the work done in plastic deformation per unit area of crack extension. The equation is generally written in the form

$$\sigma_c = \left(\frac{G_c E}{\pi a}\right)^{1/2}$$

The term G_c where $G_c = 2(\gamma + \gamma_p)$ is a fundamental property of a material and is termed the toughness, or critical strain energy release rate, and has the units Jm^{-2}.

There are three possible ways in which a stress may be applied to a crack so as to cause an extension of a crack, with mode I (as shown in figure 5.2) being the

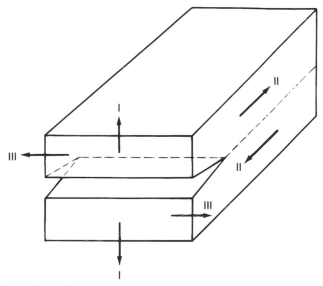

Figure 5.2 *Crack opening modes*

easiest. The toughness values for the three modes, termed G_{Ic}, G_{IIc} and G_{IIIc} respectively, will differ, with G_{Ic} having the lowest value.

Irwin noted that stresses at a crack tip were proportional to $(\pi a)^{-1/2}$ and proposed a parameter, K, the *stress intensity factor*. For an elliptical shaped flaw under plane stress conditions in a plate of infinite size, $K = \sigma(\pi a)^{1/2}$ and the general form of the equation can be written as $K = \sigma M(\pi a)^{1/2}$ where M is a geometry factor to take account of varying flaw shapes and finite boundaries. Sudden fast fracture will occur when K reaches some critical value K_c. This critical stress intensity factor, $K_c = (G_c E)^{1/2}$ and, hence, is a constant of the material. It is termed the *fracture toughness* of the material and has the units MPa $m^{1/2}$. The value generally quoted for a material is K_{Ic}, the value for mode I crack opening. Values of both toughness, G_{Ic}, and fracture toughness, K_{Ic}, for some metals are given in table 5.1. There are standardised laboratory tests for the determination of fracture toughness, K_{Ic}, and these are detailed in BS 7448.

TABLE 5.1

Toughness and fracture toughness values for some metals

Material	Toughness G_{Ic} $(kJ\ m^{-2})$	Fracture toughness K_{Ic} $(MPa\ m^{1/2})$
Steels	30–135	80–170
Cast irons	0.2–3	6–20
Aluminium alloys	0.4–70	5–70
Copper alloys	10–100	30–120
Nickel alloys	50–110	100–150
Titanium alloys	20–100	50–100

No material is completely free from defects, but it is essential that any crack-like defects that are present are relatively harmless. By using values of fracture toughness, K_{Ic}, it is possible to calculate the size of defect necessary to initiate fast fracture at a given value of applied stress or, alternatively, the level of stress required to cause failure in a component containing a defect of known size. A high value of fracture toughness is an indication that the resistance to fracture will be high. In general, as the hardness and tensile yield stress of a metal increase, so its fracture toughness decreases. This is because metals with high yield strengths tend to have low values of ductility, and their ability to plastically deform ahead of a crack, with a consequent blunting of a crack tip, is small.

The general equation $\sigma_c = K_{Ic}/M(\pi a)^{1/2}$ shows that σ_c, the critical stress for fast fracture, decreases as the size of a flaw, a, increases. This relationship is shown graphically in figure 5.3. Most metals and alloys deform plastically when the yield stress, σ_y, is exceeded. The value of yield stress, σ_y, is also marked in figure 5.3 and it can be seen that there is a critical flaw size, below which plastic yielding will occur. If the material contains a flaw larger than this critical size, it will fail by fast fracture before the yield stress is reached.

Quantitative non-destructive testing methods can be used to determine the location and sizes of flaws in many materials. Thus it is possible to decide whether the dimensions of a discovered defect are below the critical value and 'safe', or above the critical size which would lead to catastrophic failure in service. Flaws in materials subject to fluctuating stresses in service can grow in size as a result of *fatigue*. Similarly, corrosion effects can produce crack growth with time, and a flaw initially considered 'safe' may become unstable and reach critical dimensions after a period of time in service. Many engineering fabrications, such as aircraft structures, bridges and pressure vessels, are examined at regular intervals using quantitative non-destructive testing techniques, so that components may be taken out of service, or the damage rectified, before a crack has reached critical dimensions. The failure of a structure by fast fracture can have disastrous consequences.

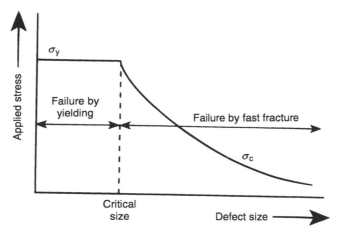

Figure 5.3 *Plot of defect size against* σ_c *and* σ_y *for a material*

Fracture mechanics is a very useful tool for assessing the risk potential of cracks in structures and for determining a 'safe maximum' or permissible crack size in a component. This is invaluable when designing structures to be 'fracture safe'. The techniques can be used with confidence for very high-strength materials where fracture occurs under conditions that are almost totally elastic, but greater care must be taken where plastic deformation occurs.

5.4 Effects of temperature and strain rate

A change in temperature may have a significant effect on the fracture toughness of a material. An increase in temperature will render dislocations in metals more mobile and, hence, will cause a reduction in yield strength. Conversely, a reduction in temperature will give an increased yield strength with an accompanying decrease in ductility. Many metals, those possessing body centred cubic or hexagonal crystal structures, show a sharp transition from ductile to brittle behaviour as the temperature is lowered. This does not apply to face centred cubic metals, which retain ductile behaviour at all temperatures. The cleavage fracture stress of a metal also varies with temperature, and increases slightly with reducing temperature. Figure 5.4(a) shows the variation of both yield stress, σ_y, and cleavage fracture stress, σ_f, for a typical body centred cubic metal. Curve (ii) shows the yield stress at low strain rates. At temperatures below T_1, σ_f is lower than σ_y and failure by brittle cleavage will occur, while above T_1 ductile behaviour occurs. The value of T_1 is about $-170°C$ for low carbon steels and, at low rates of strain, they remain ductile down to this temperature.

At high rates of strain, as under impact loading conditions, the yield stress of a metal is effectively increased and curve (iii) in figure 5.4 applies. Under these conditions the transition between ductile and brittle behaviour occurs at a higher temperature, T_2, than for low strain rates. Between T_1 and T_2 the metal will be brittle under impact conditions but ductile for slow strain rates. The transition temperature, T_2, for low carbon steels, for impact loading, is about $0°C$.

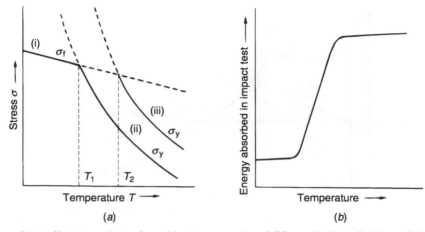

Figure 5.4 *(a) Variation of σ_f and σ_y with temperature for a BCC metal; (i) σ_f –T; (ii) σ_y –T for slow strain rate; (iii) σ_y –T for fast strain rate. (b) Notch-impact energy against temperature showing ductile–brittle transition*

Figure 5.4(b) shows the relationship between notch-impact toughness, as measured in a Charpy notch-impact test, and temperature for a body centred cubic metal. The transition from tough to brittle behaviour occurs over a narrow range of temperature. In the case of low carbon steels, the energy absorbed in breaking standard-size Charpy test pieces is about 80 J (1 MJ m^{-2}) at +10°C but only about 10 J (125 kJ m^{-2}) at −10°C. This dramatic change in behaviour around 0°C means that some structural steels cannot be used for structures where the ambient temperatures are likely to fall to values well below 0°C. The ductile–brittle transition temperature is not the same for all grades of steel and is affected by composition. It is reduced by manganese and nickel, but increased by carbon, nitrogen and phosphorus.

Notch-impact testing was mentioned in the previous paragraph and, in this type of test, test pieces containing milled notches are broken by a heavy pendulum swinging through an arc and the energy required to break a standard-size test piece is measured. The main tests of this type are the Charpy and the Izod tests, and the standard procedures for these are given in BS EN 10045-1 and BS 131: Part 1 respectively. In this type of testing a triaxial stress system is developed in the test piece. With stresses acting in three directions, the shear stresses which normally lead to plastic deformation tend to cancel each other, leading to an increase in the yield stress and a decrease in ductility. This effect, coupled with the very high strain rate under impact loading conditions, makes the impact test results indicative of the behaviour of the material under some of the most severe stress conditions likely to be encountered. Notch-impact tests are used to reveal a tendency to brittleness in materials, but it should be noted that the parameter of toughness measured in a notch-impact test is not the same as the toughness G_{Ic}, or the fracture toughness, K_{Ic}, of a material.

5.5 Fatigue of metals

When a metal component or structure is subjected to repeated or cyclic stresses it may eventually fail, even though the maximum stress in any one stress cycle is considerably less than the fracture stress of the material. This type of failure is termed *fatigue failure*. Failure by fatigue is a fairly common occurrence, as many components are subject to alternating or fluctuating loads during their service life. Fatigue failure is the result of processes of crack nucleation and growth or, for components which may contain cracks introduced during manufacture, growth only. Generally, the appearance of a fatigue fracture surface is unmistakable, consisting of two distinct zones, a smooth portion, often showing conchoidal or 'mussel shell' markings, and the final fast fracture zone (figure 5.5). The marks which are visible in the smooth portion indicate stages in the growth of the fatigue crack and the final fracture is usually of the cleavage type.

Figure 5.5 *Fatigue failure of a large shaft showing the two zones*

Failure by fatigue takes two forms, *low-cycle fatigue* and *high-cycle fatigue*. In the former type, the maximum stress in any cycle is greater than the yield stress, although less than the static tensile strength of the metal, and the number of stress cycles necessary to cause failure is low, generally less than 1000. High-cycle fatigue is the result of cyclic stressing when the maximum stress in a cycle is less than the yield stress, and 10^5 or 10^6 cycles may be required for failure.

Stress cycles are described as *alternating* when the mean stress is zero, as *repeating* when the minimum stress is zero, and as *fluctuating* when neither the mean stress nor the minimum stress is zero (figure 5.6). Much fatigue testing is carried out using alternating stress cycles and the results of many tests are expressed in the form of an S–N plot, where S is the maximum stress in a cycle and N is the number of cycles to failure (figure 5.7). Most steels show an S–N curve of type (i), with a very definite *fatigue limit* or *endurance strength*, usually with a value of about one-half of the tensile strength, and below which fatigue failure should not occur. Most other metals and alloys show fatigue curves of type (ii) with no definite fatigue limit. With this type it is only possible to design for a limited life, and a life of 10^6 or 10^7 cycles is often used. In many practical situations,

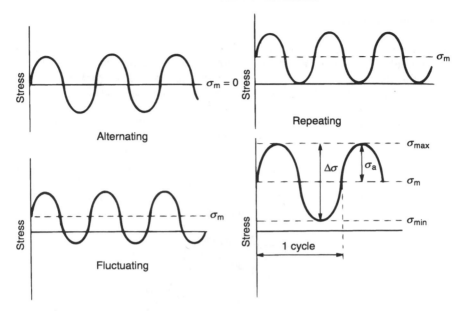

Figure 5.6 *Types of stress cycle*

structures may be subject to more than one type of cyclical loading system, with different ranges of stress and where the mean stress is not zero. Several empirical 'laws' have been proposed, relating stress ranges, mean stresses and fatigue life. These relationships are not very precise but are sometimes used in preliminary design calculations.

Although, in high-cycle fatigue, the maximum stresses are below the yield strength of the material, it has been established that some plastic slip takes place as a result of cyclic stressing. The formation of slip bands causes a roughening of

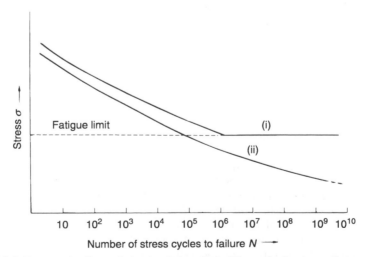

Figure 5.7 *S–N curves for (i) metal showing fatigue limit, (ii) metals showing no fatigue limit*

the surface. The ridges and crevices formed are extremely small, being of the order of 1 μm in size, but once a crevice has formed on a highly polished surface it can act as a point of stress concentration and the point of initiation for a fatigue crack, which can then propagate through the material. The propagation of cracks occurs by one of two methods, as described in section 5.3, namely by ductile tearing in the case of ductile metals, or by cleavage for brittle metals.

Factors affecting fatigue strength

Many factors affect the fatigue strength of a material, including surface condition, component design and the nature of the environment. Fatigue test pieces are generally prepared with a highly polished surface, and this condition gives the best fatigue performance. Highly polished steel specimens will have a fatigue limit which is approximately one-half of the tensile strength. The presence of scratches or notches, or a ground or turned surface, will provide stress-raising features for the initiation of fatigue cracks and, thus, reduce the fatigue strength. Similarly, geometrical features in components, such as sectional changes with small fillet radii, keyways and oil holes in shafts, are all stress-concentration features and are often points at which fatigue commences. The effect of a notch, scratch, or other stress-raising feature is not the same for all materials. As in the case of brittle failure, a ductile metal is much less sensitive to the presence of surface flaws than a brittle material. As crack initiation and growth are largely a result of tensile stress, the fatigue resistance of a component can be increased if the surface layers contain residual compressive stresses. These can be induced by peening or by surface hardening treatments, such as flame or induction hardening, case carburising and nitriding.

If a structure is subject to cyclical stressing in conditions in which corrosion can occur, then not only is the fatigue life very greatly reduced but also the rate of corrosion is increased. This combined effect is termed *corrosion fatigue*. An effect of corrosion may be to form surface pits, or other irregularities, which act as stress raisers and initiate fatigue cracks. The cyclical stressing may rupture any passive films which might form and, thus, accelerate corrosion. For most materials, including many steels, there is no fatigue limit in a corrosive environment, and failure will eventually occur, even at very low levels of stress.

5.6 Creep and relaxation

Creep is the continued slow straining of a material under the influence of a constant force. The amount of strain developed in a material is a function of stress, time and temperature. A typical strain–time creep curve at a constant stress and constant temperature is shown in figure 5.8.

The rate of secondary or steady-state creep at a constant stress increases exponentially with temperature according to the relationship

$$d\epsilon/dt = K \exp(-B/T)$$

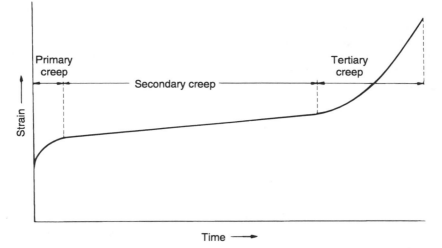

Figure 5.8 *Typical creep curve*

where $d\epsilon/dt$ is strain rate, T is temperature in Kelvin, and K and B are constants for the material. At a constant temperature, the creep rate is stress dependent and follows a power law, the relationship being

$$d\epsilon/dt = C\,\sigma^n$$

where σ is the stress, and C and n are constants. For many metals the value of the time exponent, n, lies between 3 and 8. The above two relationships can be combined to give a general creep equation

$$d\epsilon/dt = A\,\sigma^n \exp(-B/T)$$

where $A = K \times C$.

The creep of metals is mainly a high-temperature phenomenon and assumes significance at temperatures above $0.3T_m$, where T_m is the melting temperature of the metal in Kelvin. As such, it is not of great importance to civil and structural engineers.

Relaxation is a phenomenon related to creep, and is the slow reduction of stress with time in a material which is subject to a constant value of strain. The rate of relaxation is exponential and the *relaxation time* is the time required for the level of stress to fall to $1/e$ of the original value. The general relationship for relaxation at a constant temperature is

$$\sigma_t = \sigma_0 \exp(-At)$$

where σ_0 is the initial stress, σ_t is the stress after time t, and A is a constant for the material.

Examples of relaxation occurring within materials subject to constant strain over long periods of time in service are the levels of stress in tensioned bolts, and in the bars and wires in prestressed concrete. A bolt is strained to a particular level when the nut is tightened. During service the total strain in the bolt remains constant but, if the temperature is such as to permit some creep, then stress relaxation will occur within the bolt and the nut will require to be retightened at intervals. Prestressing steel must have a high proof strength and low stress relaxation characteristics as the prestressing force should not fall below 80 per cent of its initial value during the service life of the member.

5.7 Corrosion

The corrosion of metals may be broadly classified into dry corrosion and wet corrosion. Usually, the former is a direct reaction between the surface metal and atmospheric oxygen, while the latter involves a series of electrochemical reactions in the presence of an aqueous electrolyte. Direct oxidation is not a problem with most metals, except at greatly elevated temperatures when the oxidation rates are high. At ordinary temperatures, the oxidation rate for most metals is low and, frequently, when an oxide layer is formed it can be protective. Aluminium and chromium, for example, have high affinities for oxygen but the oxide layer which forms on the surface of the metal is compact and impervious, and offers protection by preventing further oxygen from reaching the metal. In the case of aluminium, the degree of protection can be improved by increasing the thickness of the aluminium oxide layer by anodising, an electrolytic oxidation process. When an alloy steel contains more than 12 per cent of chromium, a continuous film of chromic oxide forms, rapidly covering the whole of the steel surface. Steels containing more than 12 per cent of chromium, protected by a self-healing passive film of chromic oxide are, therefore, corrosion resistant or *stainless*.

Wet corrosion is the result of electrochemical, or galvanic, action. Any metal in contact with an electrolyte will ionise to some extent, releasing free electrons which remain in the metal, and creating ions which pass into solution in the electrolyte. A typical reaction is

$$Zn \rightarrow Zn^{2+} + 2e^-$$

Therefore, when equilibrium has been established, the metal will possess a certain electrical charge. Different metals ionise to differing extents and therefore will possess differing electrical charges when each is in equilibrium with its ions. The electrical potential of a charged metal cannot be measured directly, but the potential difference between the metal and some reference electrode can be measured. A standard hydrogen electrode is used as the reference point and the potential difference between this and a metal in equilibrium with its ions is termed the *Standard Electrode Potential*. A list of these for some metals is given in table 5.2.

Galvanic cells

A galvanic cell will be established if two dissimilar metals are connected through an electrolyte and also external to the electrolyte. Consider such a cell between zinc and copper, as shown in figure 5.9. The potential difference of this cell will be 1.10 V and, under this driving force, there will be a flow of electrons through

TABLE 5.2
Standard Electrode Potentials

Base metals	Metal	Ion	Electrode potential (V)	Anodic
	Sodium	Na^+	−2.71	
	Magnesium	Mg^{2+}	−2.38	
	Aluminium	Al^{3+}	−1.67	
	Zinc	Zn^{2+}	−0.76	
	Chromium	Cr^{2+}	−0.56	
	Iron	Fe^{2+}	−0.44	
	Cadmium	Cd^{2+}	−0.40	
	Cobalt	Co^{2+}	−0.28	
	Nickel	Ni^{2+}	−0.25	
	Tin	Sn^{2+}	−0.14	
	Lead	Pb^{2+}	−0.13	
	Hydrogen	H^+	0.000	
	Copper	Cu^{2+}	+0.34	
	Mercury	Hg^{2+}	+0.79	
	Silver	Ag^+	+0.80	
	Platinum	Pt^{2+}	+1.20	
Noble metals	Gold	Au^+	+1.80	Cathodic

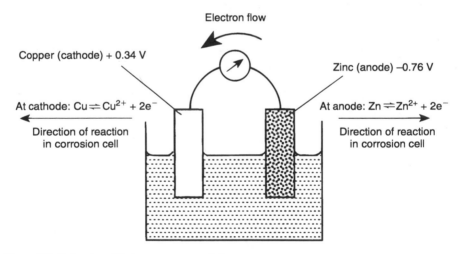

Figure 5.9 *Galvanic cell between copper and zinc*

the external circuit from zinc to copper. At the zinc anode the reaction $Zn \rightarrow Zn^{2+} + 2e^-$ will be thrown out of equilibrium as electrons flow from the zinc, and ionisation of zinc will be accelerated in an attempt to restore equilibrium, thus the zinc will be dissolved or corroded. Conversely, at the copper cathode, the ionisation of copper will be suppressed as excess electrons flow into it. Referring to table 5.2, if any two metals are linked to form a galvanic cell, then the one which is higher in the table will be anodic with respect to the other and will tend to corrode. The driving force will be the potential difference in the cell.

TABLE 5.3
Galvanic series in sea water

Anodic end:	Magnesium
	Magnesium alloys
	Zinc
	Galvanised steel
	Aluminium
	Cadmium
	Mild steel
	Wrought iron
	Cast iron
	Austenitic stainless steel (active)
	Lead
	Tin
	Muntz metal
	Nickel
	Alpha brass
	Copper
	70/30 Cupronickel
	Austenitic stainless steel (passive)
	Titanium
Cathodic end:	Gold

The order in which the metals are placed in table 5.2 is termed the *electromotive series* and, while the values given in the table are of interest, it should be noted that potential differences between metals in equilibrium with their ions may differ from the standard values if the metals are immersed in some electrolyte other than that used to determine standard values. Also, the order in which metals appear in a table may change when immersed in other electrolytes, for example sea water. The order of metals, when determined for some other electrolyte, is termed a *galvanic series*, and the galvanic series for some metals and alloys in sea water is given in table 5.3.

Generally, in a galvanic cell, it will be the anode which is corroded while the cathode is protected against corrosion. There are a few exceptions to this and cathodic corrosion, a type of corrosion associated with the formation of alkaline corrosion products, may occur in some instances. Galvanic, or corrosion, cells can be established in several ways, the main types being: (a) *composition cells*, (b) *stress cells* and (c) *concentration cells*.

Composition cells

An obvious example of a composition cell is two different metals in direct electrical contact and also joined by an electrolyte. For this reason, different metals should not be in contact in situations where corrosion conditions may occur. If two different metals have to be joined in such circumstances, then there should be a layer of electrically insulating material between them. Composition type galvanic cells can also occur within a single piece of metal. In any alloy with a two-phase structure, one phase may be anodic with respect to the other. An example of this is pearlite in steels, where ferrite is anodic with respect to cementite. A film of moisture on the steel surface is then all that is necessary to establish galvanic microcells. Generally, pure metals or single-phase alloys possess better corrosion resistance than impure metals or multi-phase alloys.

Stress cells

Atoms within a stressed metal will tend to ionise to a greater extent than atoms of the same metal in an unstressed condition and, so, the stressed metal will be anodic with respect to the unstressed material. Stress type galvanic cells can be established in a component or structure where the stress distribution is uneven. Randomly arranged grain boundary atoms tend to ionise more rapidly than atoms in the regular crystal lattice, so that a grain boundary tends to be anodic with respect to a crystal grain. The potential differences due to stress variations, or between grains and grain boundaries, are usually small but some alloys are particularly prone to stress corrosion failure, which often is a grain boundary (intercrystalline) type of corrosion, causing rapid growth of a crack through the material.

Concentration cells

A metal in contact with a concentrated electrolyte will not ionise to as great an extent as when it is in contact with a dilute solution. If one piece of metal is in contact with an electrolyte of varying concentration, those portions in contact with dilute electrolyte will be anodic with respect to portions in contact with more concentrated electrolyte. This type of situation may occur with electrolytes flowing in pipes or ducts. Another, and most important, type of concentration galvanic cell occurs when there are variations in dissolved oxygen content through an electrolyte. Metal in contact with electrolyte low in dissolved oxygen will be anodic with respect to metal in contact with electrolyte rich in dissolved oxygen. A major reason for this is that in areas rich in dissolved oxygen, a cathode-type reaction which absorbs electrons can occur. This reaction is

$$2H_2O + O_2 + 4e^- \rightarrow 4(OH)^-$$

The electrons needed for this reaction will be supplied from another part of the metal where there is less oxygen, and so such areas act as anodes. This is one of the main features of the *rusting* of irons and steels. Rust is the hydrated oxide of iron, $Fe(OH)_3$. Both oxygen and water are required for the formation of rust, and steels will not rust in either a dry atmosphere or when immersed in oxygen-free water.

Consider the situation with a drop of water on the surface of a piece of steel, figure 5.10. There will be a greater amount of oxygen available near the edge of the drop and this area will become cathodic. Hydroxyl ions will be formed according to the reaction

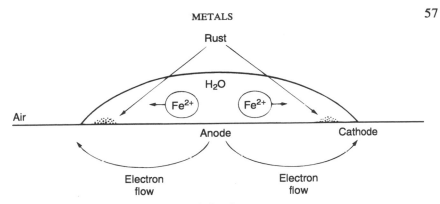

Figure 5.10 *Oxygen concentration cell on steel, forming rust*

$$2H_2O + O_2 + 4e^- \rightarrow 4(OH)^-$$

The anode area will be under the centre of the drop, and electrons will be generated here according to the reaction

$$Fe \rightarrow Fe^{2+} + 2e^-$$

The Fe^{2+} cations will move through the electrolyte towards the cathode and the $(OH)^-$ anions will move towards the anode. The cations are very much smaller than the $(OH)^-$ ions and possess a greater mobility. The two ion types meet close to the cathode areas to form insoluble ferrous hydroxide, $Fe(OH)_2$, which settles as a deposit on the steel surface. This subsequently oxidises to rust, $Fe(OH)_3$.

A rust deposit is porous and, once it has formed, the steel surface beneath the deposit will become deficient in oxygen and, hence, anodic with respect to bare metal. This means that corrosion of the steel can continue undisturbed below the covering of rust. Similarly, any covered or hidden area to which moisture can gain admittance, for example, a dirt covered surface, a crack or crevice, or the gap between overlapping rivetted or bolted plates, can become an anode (figure 5.11). In the last case mentioned, the incidence of hidden corrosion can be minimised by a water-tight caulking of all seams. Corrosion in oxygen concentration cells is more severe in salt water than in fresh water because of the higher conductivity of the former electrolyte. A good example of the severe corrosion effects that can occur, particularly in sea water, is *waterline corrosion*. The oxygen concentration at the water/air interface will be higher than at lower levels in the water, rendering the steel at the waterline cathodic and with anode areas existing below the water surface. The most rapid corrosion will take place just below the waterline, with solution of the anodic areas, while the greatest rust deposit will form at the waterline.

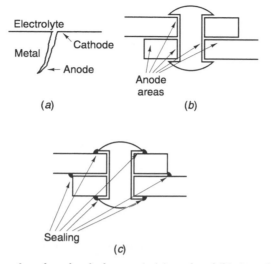

Figure 5.11 *Formation of anode and cathode areas in (a) crack and (b) rivetted joint; (c) areas to be sealed to minimise this type of corrosion*

Polarisation

In most instances, once a galvanic cell has been established, there is a tendency for *polarisation* to occur, namely, a reduction in the current flowing within the corrosion cell. If the diffusion rate of ions is relatively slow, there will be an increasing concentration of cations around the anode and a corresponding decrease in the cation concentration at the cathode, and these effects combine to reduce the current flow in the cell. However, because this type of polarisation is controlled by ion diffusion rates, it follows that anything that will increase ion diffusion, such as an increase in temperature or agitation of the electrolyte, will reduce the polarisation effect and allow high corrosion rates to be maintained. Another effect that can cause polarisation is termed *passivation*. Passivation occurs when products of corrosion reactions form an impervious layer over an anode surface, and this can often occur in oxidising conditions. Metals such as aluminium, titanium and stainless steels readily form passive films under these conditions, but such protective films may be destroyed in a chemically reducing environment and the metals will then become active for corrosion.

Corrosion failure

The corrosion of a metal may lead to total failure. When the effective thickness of a load-bearing structure is decreased, the level of stress within the material is raised and this increase could accelerate the corrosion in a localised area to such a point that the level of stress is high enough to cause failure. Some types of corrosion that could lead to early failure are described briefly below.

Pitting corrosion. This is a highly localised form of corrosion and, usually, is associated with a local breakdown of an otherwise protective film. Pits can form where there is a discontinuity in the passive oxide layer and they can also occur, owing to a differential aeration effect, where an external deposit has settled on a metal surface. Frequently the source for corrosion pits is a non-homogeneous metal and the presence of inclusions of an impurity may cause breaks in the passive coating. A corrosion product often forms over the surface of a pit, but corrosion continues beneath this. Pitting corrosion can occur on any metal, but it is much more common with metals and alloys that possess a high resistance passive film. The presence of the film tends to prevent the spread of corrosion to a wider area and the attack is concentrated in a series of small localities. Stainless steels are susceptible to this form of corrosion in the presence of chloride solutions.

Stress corrosion. This is a phenomenon in which there is the initiation of a crack and then propagation of the crack through the material as the result of the combined effects of a stress and a corrosive environment. Generally, stress corrosion is found only in alloys, rather than in pure metals, and it occurs only in specific environments, for example, the cracking of brass in ammonia or the cracking of aluminium alloys in chloride solutions. The stress within the material must have a tensile component, and both stress and specific corrosive environment must be present for cracks to form and propagate. The stress corrosion cracks are usually of an intercrystalline type, but transcrystalline cracking can occur in some instances.

Corrosion fatigue. This type of failure is due to the initiation and propagation of cracks in a material that is subject to alternating or fluctuating cycles of load and a corrosive environment. Even when conditions are only mildly corrosive, the combined effects will accelerate the formation and growth of fatigue cracks, thus greatly reducing the life of the structure.

5.8 Corrosion protection

Corrosion can never be eliminated entirely but it is important that all necessary steps be taken to minimise the incidence of corrosion. It is, therefore, vital that adequate preventative measures be taken and this means careful material selection, not having different metals in direct contact, paying close attention to design details to avoid the creation of crevices and the collection of water in parts of a structure, as well as the use of protective measures. Much of the rapid corrosion of car bodywork which was prevalent in the 1960s and 1970s was due as much to the presence of many areas in which water could be trapped as to, often, less than adequate protective measures. Many methods are used to minimise the risk of corrosion and these may be grouped into the following categories: (a) protective coatings, (b) cathodic protection and (c) the use of inhibitors.

Protective coatings

The object of a protective coating is to isolate the metal from its environment and there are many types of protective coating available, ranging from paints and bituminous coatings, through plastic and vitreous enamel coatings, to a variety of metallic coatings applied by hot dipping, electroplating, metal spraying, or other methods. One of the more common forms of protection is a paint layer. Paints are easily applied, either by brush or spray. On drying, paints form a semi-elastic resin coating, the pigment in the paint serving both to colour and strengthen the semi-elastic coat. Most paints age and weather, particularly in strong sunlight, and the layer either becomes porous or flakes off the surface. Also, paint surfaces are fairly soft and may be easily scratched and damaged. Once oxygen and moisture can reach the metal surface, corrosion can begin and it can even occur beneath portions of the paint-covered surface. Paints for use on ferrous metals may contain active ingredients such as red lead or zinc chromate, both of which tend to make the metal surface passive by the formation of a fine protective oxide layer, or metal powders such as aluminium or zinc. These two metals are anodic to iron and, when part of a paint coating, will confer a measure of electrochemical protection to the iron or steel.

Metal components can be given a continuous protective layer of a plastic by dipping the heated metal part into a bath containing a thermoplastic powder. The method is used mainly for the protection of steel wire trays for refrigerators and freezers. Vitreous enamelling produces a hard, wear-resistant coating, although this is brittle and damaged easily if the component is subject to mechanical shock. The method is used mainly for parts that will be subjected to moderate heating, for example, as components in electric and gas cookers. These two methods are not suitable for the protection of stress-bearing structures, but they have been used in the manufacture of decorative steel cladding panels for some buildings.

Metallic coatings are frequently used for the protection of metals, particularly steels. The metal coating could be of a metal that is anodic, with respect to the steel, such as aluminium or zinc, but is itself corrosion resistant because of the presence of a passive surface film. This type of coating will afford some protection to the steel, even if there are breaks in the coating, as the coating is anodic and will corrode in preference to the steel. The application of a zinc coating to a steel is termed *galvanising*, and galvanised steel parts offer good resistance to corrosion under many conditions. Other coatings, such as tin, are cathodic with respect to steel and, if the coating is broken, the exposed steel will become the anode and corrosion will take place rapidly. Cadmium, which has an electrode potential very close to that of iron, is often used as a protective electroplated layer on parts such as nuts, bolts, washers and screws. The protective layer could be broken by abrasion during the act of tightening a nut on a bolt but, if this happens and a galvanic cell is formed, the e.m.f. of the cell is small (0.04 V) and the rate of corrosion will be very low. Chromium, in the form of a thin electroplated layer on steel, is frequently used for decorative purposes for its bright and shiny appearance. Chromium should never be plated directly on to steel, since the layer is porous and, on its own, does not offer protection. A three-stage process is usually used. Firstly, an electrodeposited layer of copper is applied, followed by a thin layer of nickel before the final deposit of chromium, although in some cases the copper is omitted.

Cathodic protection

Cathodic protection is widely used to protect structures in which the effects of severe corrosion could be serious or where the renewal of a corroded structure could prove very expensive. It is particularly useful for the hulls of ships, steel piers and harbour installations, buried steelwork in building foundations and underground pipe-lines. A block of a highly anodic material, magnesium, aluminium or zinc, is placed adjacent to the structure to be protected and connected to it electrically, a galvanic cell is created deliberately and the sacrificial anode will corrode, protecting the structure (figure 5.12).

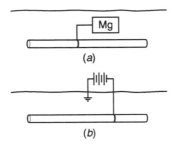

Figure 5.12 *Cathodic protection of buried pipe-line with (a) magnesium anode, (b) impressed voltage*

The sacrificial anodes are inspected periodically and renewed when necessary. A variant on the use of a sacrificial anode of another metal is to use an impressed voltage. The steelwork will be protected as long as it receives a steady supply of electrons. When using a direct current supply as a means of protection, it is common practice to bury scrap iron or steel adjacent to the structure, with connections that render the scrap metal anodic and the steel requiring protection cathodic. The full code of practice for the cathodic protection of structures is detailed in BS 7361 Part 1.

Underground pipework is usually given a thick bituminous coating before installation but this may be damaged during pipe laying, and cathodic protection is used to minimise corrosion of any exposed surface. Uncoated pipes are not used, as the very large surface area would result in too rapid a solution of sacrificial anode material. The risk of corrosion will vary considerably with soil type. In low-risk conditions, anodes may be widely separated, but in high-risk situations, anodes would be placed relatively close to one another. Corrosion would be minimal in dry, sandy soils but there could be a high risk in wet clays. Another variable is the presence of bacteria in soils or clays with a sulphate content. Some bacteria cause the breakdown of sulphates with liberation of oxygen, and this could lead to the formation of oxygen concentration cells. There is also a major corrosion risk at the junction of two different soil types.

Passive coatings

Passive coatings can be developed on steel surfaces by chemical methods. The treatment of a clean steel surface with a solution of phosphoric acid, zinc phosphate and manganese phosphate will produce a thin passive phosphate film. Frequently, the phosphate treatment is not used on its own but as the basis for painting. The paint layer will key in well to a phosphate coating. Another type of chemical treatment is the addition of passifying agents and inhibitors to corrosive fluids. This method is widely used in cooling systems where cooling water is recirculated. Chromate salts are often used as inhibitors in car radiator systems and, when corrosion starts, an insoluble iron chromate is formed which is deposited on the steel surface, restricting further corrosion.

BS 5493 details the whole range of corrosion-resistant protective coatings which can be applied to steel structures, with information on the effectiveness and durability of these coatings in various types of corrosion regime.

Pure aluminium owes its normally high corrosion resistance to the thin film of aluminium oxide which forms on any fresh metal surface, but it is subject to pitting corrosion in marine atmospheres. The aluminium–magnesium alloys, however, possess superior corrosion-resistant properties in marine conditions. Most aluminium alloys, except the high-strength heat treatable Al–Cu and Al–Zn alloys, have a very good resistance to corrosion and this can be further improved by anodising to thicken the aluminium oxide layer. The high-strength precipitation-hardened alloys are protected in one of two ways. Generally, sheet and plate products are clad with pure aluminium. The alloy ingot is sandwiched between two plates of pure aluminium and the assembly is hot rolled. The final rolled sheet, or plate, has a thin layer of pure aluminium on either side and the product, essentially, has the strength of the strong alloy combined with the corrosion resistance of pure aluminium. Castings, forgings and extruded sections cannot be protected in this way and are treated by immersion in chromate or dichromate solutions. A highly protective surface layer is formed but this layer is an excellent basis for paint and, often, the treated surfaces are painted to improve protection.

5.9 The role of non-destructive testing

Most materials contain defects which have been introduced at some stage in their manufacture and fabrication into structures. Castings may contain inclusions, porosity or cracks caused by stresses generated during solidification and cooling. Hot and cold worked products may possess cracks developed during the working process or subsequent heat treatment. Slag inclusions and porosity may be formed during welding, and residual stresses following welding may be sufficiently high to initiate cracking if the appropriate stress relief treatment is not given. The importance of cracks in relation to fast fracture was highlighted in section 5.3. In section 5.5 it was shown that cracks can be initiated and grow in size through the effects of cyclical or fluctuating stresses, or because of corrosion effects. It becomes necessary, therefore, to have reliable systems for detecting, characterising and sizing defects. A number of non-destructive testing methods have been developed to detect both surface and internal flaws in materials and structures. The

Figure 5.13 *Ultrasonic inspection of butt welds in a pipe (Courtesy of Wells–Krautkramer Ltd)*

basic principles and major features of the main systems are given in table 5.4. The range of equipment available is extensive and includes compact and portable units which can be used both inside a test laboratory and out on site. Figure 5.13 shows a pipe-line weld being checked using ultrasonics, and figure 5.14 is taken from a radiograph of a weld in stainless steel containing entrapped argon.

The role of non-destructive testing is to guarantee with a known level of confidence that cracks corresponding to a critical size for fracture at the design load are absent from a component or structure. It may be necessary to guarantee, with

TABLE 5.4
Principal non-destructive test methods

System	Features	Applicability
Visual inspection probes	Detection of defects which break the surface, surface corrosion, etc.	Interior of ducts, pipes and assemblies
Liquid penetrant	Detection of defects which break the surface	Can be used for any metal, many plastics, glass and glazed ceramics
Magnetic particle	Detection of defects which break the surface and sub-surface defects close to the surface	Can only be used for ferro-magnetic materials (most steels and irons)
Electrical methods (eddy currents)	Detection of surface defects and some sub-surface defects. Can also be used to measure the thickness of non-conductive coatings, e.g. paint on a metal	Can be used for any metal
Ultrasonic testing	Detection of internal defects but can also detect surface flaws	Can be used for most materials
Radiography	Detection of internal defects, surface defects and to check correctness of assemblies	Can be used for most materials but there are limitations on the maximum material thickness

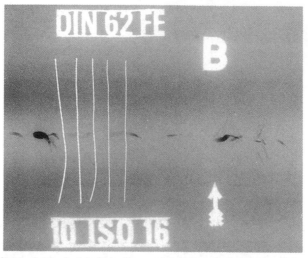

Figure 5.14 *Radiograph of a butt weld in 13 mm stainless steel, showing pockets of argon (Courtesy of Kodak Ltd)*

confidence, that cracks much smaller than the critical size are absent also in components subject to fatigue loading or corrosive environments, so that some minimum specified service life can be achieved before catastrophic failures occur and, in many situations, it will be necessary to inspect at periodic intervals, or continuously monitor, to ensure that flaws do not grow to critical size.

6

Metallic Materials in Construction

It is customary to refer to two main categories of metals and alloys, ferrous and non-ferrous. The term ferrous covers iron and its alloys, while all the 70 or so other metals are classified as non-ferrous. The element iron occupies a very special position because of its availability, comparatively low cost, and the wide range of very useful alloys obtained from alloying iron with carbon and other elements. In fact, well over 90 per cent of the total world consumption of metals is in the form of steels and cast irons. Only a small number of non-ferrous metals are produced in large quantity. Although very many metallic alloys have developed for engineering use, the number of alloys used in the construction industry is comparatively small. Only those materials of direct interest to civil engineers are discussed in this chapter, and alloys such as tool steels, magnesium, nickel and titanium alloys are not covered. Information on these and many other alloys can be obtained, if required, by consulting other texts and reference books.

6.1 Steels

Steels are essentially alloys of iron and carbon but they always contain other elements, either as impurities or alloying elements. The iron, as produced from the smelting of iron ore in a blast furnace, is impure and, in steel-making processes, the impurities in blast furnace iron are removed by oxidation. At the end of the refining process, the melt will contain only small amounts of carbon, silicon, manganese, sulphur and phosphorus, usually less than 0.05 per cent of each, but will be in a highly oxidised state. With the exception of low carbon *rimming* steels, which are not used as structural materials, the liquid steel must be deoxidised. Silicon and manganese are generally used as deoxidisers, and sufficient is added to give the steel a residual content of these elements as they help to give increased strength. Frequently aluminium, an element with a very high affinity for oxygen, is also used for deoxidation of the liquid steel and this element promotes a fine grain structure. After deoxidation is complete, anthracite is added to bring the carbon content up to the desired level. Other additions may be made, for example, small amounts of niobium and/or vanadium which promote very fine crystal grain

65

structures. The molten steel is then ready to be cast into ingots for hot working, or made into castings. Fully deoxidised steel is termed *killed steel*.

Plain carbon or *non-alloy steel* is defined as steel containing up to 1.6 per cent of carbon, together with not more than 0.6 per cent of silicon and not more than 1.6 per cent of manganese, and only traces of other elements. It follows from this that an *alloy steel* is one that contains either silicon or manganese in amounts greater than those stated above, or that contains any other element or elements as the result of deliberate alloying additions.

Alloying additions are made to steels to enhance the properties. Most alloying elements enter solid solution in the iron, giving a strengthening effect with little loss of ductility. Some elements improve the resistance to corrosion, and additions of copper and chromium are made to give the categories known as *weathering steels*. When more than 12 per cent of chromium is present, the steel becomes highly corrosion resistant, or stainless. A large addition of nickel or manganese will render a steel austenitic at all temperatures, and alloy steels containing more than 12 per cent of chromium and more than 7 per cent of nickel give the group of materials known as austenitic stainless steels. Small additions of the elements niobium or vanadium promote very fine crystal grain structures in a steel. All elements retard the structural changes which occur when a steel is cooled through the critical temperature range (refer to Chapter 4, section 4.6), making it easier to form fine pearlite and martensite; for example, in steels containing 1.5 per cent of manganese, martensite may be formed even when the steel is cooled in air. This effect permits the formation of brittle martensite and subsequent crack initiation in rail steels under certain circumstances.

In civil and structural engineering the vast majority of the steels used are plain carbon steels, or slightly modified plain carbon steels. Figure 6.1 shows how the hardness and strength of normalised plain carbon steels tend to increase with increasing carbon content, while ductility and notch-impact toughness decrease.

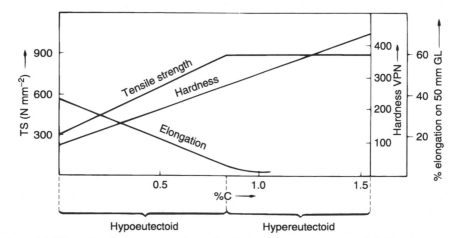

Figure 6.1 *Effect of carbon content on the mechanical properties of normalised plain carbon steels.*

6.2 Wrought steels

Wrought steels are those shaped by hot or cold working processes, and many engineering components are produced by these methods. In the case of some components, machining operations may also be needed. Wrought steel components may be put directly into service in the as-manufactured condition, or they may be heat treated or surface treated, depending on the grade of steel or the application. Many structural steel grades are used in the normalised condition, but quenching and tempering is specified for some. Generally, for hot rolled steels, the hot work finishing temperature is just above the upper critical temperature, permitting full recrystallisation of deformed austenite with little grain growth. However, the specifications for structural steels permit the use of a hot work finishing temperature below the upper critical temperature for some grades. This lower temperature allows very little recrystallisation of austenite to occur, resulting in a stronger material.

Structural steels

The design of steel structures is based primarily on the yield strength or proof strength of the steel, but other properties including ductility, toughness (both at normal and sub-zero temperatures) and weldability are also important. Weldability is important because welded structures, through ease of fabrication, can produce weight and cost savings compared with bolted or rivetted structures. The relative importance of the properties of ductility and toughness will vary, depending on the exact nature of the application. Figure 6.1 shows that increasing the carbon content of a steel will give an increased strength but a reduction in ductility and notch-impact strength. The weldability of a steel also reduces as the carbon content decreases, becoming poor when the carbon content, or the carbon equivalent value (CEV), exceeds 0.5 per cent. The CEV for a steel is given by

$$\text{CEV} = \text{C} + \frac{\text{Mn}}{6} + \frac{\text{Cr} + \text{Mo} + \text{V}}{5} + \frac{\text{Ni} + \text{Cu}}{15}$$

The carbon content of weldable structural steels is kept below 0.25 per cent as a compromise between the various conflicting requirements. The relevant standards specify the maximum CEV permitted for each steel and the maximum values lie between 0.35 and 0.51, depending on the strength grade, sectional thickness and application.

Non-alloy steels

These contain up to 1.6 per cent carbon together with manganese, silicon, sulphur and phosphorus, the two latter elements being impurities which can cause embrittlement and these are controlled to low levels, generally less than 0.05 per cent. Manganese gives an improvement in strength and a reduction in the notch-ductile transition temperature. Silicon also improves the strength but the silicon content should not exceed 0.6 per cent, otherwise there is a possibility that some graphitisation of the Fe_3C may occur. Formerly, many structural steels were plain carbon steels containing 0.2–0.3 per cent carbon, 0.3–0.5 per cent silicon and 0.5–1.0 per cent manganese. These steels had some deficiencies. These were:

1. low toughness and notch ductile–brittle transition temperatures at or just below room temperature;
2. low yield strength; and
3. poor atmospheric corrosion resistance.

A good toughness will ensure that catastrophic failure from points of stress concentration is less likely to occur. Materials, transport and erection costs will be reduced if the yield strength is high, since thinner sections can be used, provided that the steel possesses sufficient ductility to permit the production of thin sections. Maintenance costs will be reduced if the steel possesses a good corrosion resistance. Modified low carbon steels have been, and continue to be, developed in order to achieve these benefits. These are covered in the following paragraphs. There are several British Standards covering structural steels. The system of structural steel designation is common to the whole of the European Union. Each steel is identified by a 7 to 9 character code, the first character, S, signifying that the steel is for structural applications. This is followed by three digits indicating the minimum yield strength (in $N\ mm^{-2}$) for the smallest thickness. The final characters give coded information on impact properties, delivery condition and type of product.

Notch ductile steels

The ductile–brittle transition temperature of a steel can be lowered by a reduction in the carbon content or an increase in the manganese content, or both. The ductile–brittle transition temperature is also lowered by reducing the ferrite grain size. This may be achieved by treating the molten steel with aluminium. Aluminium nitride particles form at grain boundaries and inhibit grain growth during processing.

Micro-alloyed steels

The grain size of the ferrite formed by the transformation of austenite during cooling is a key factor in the control of strength and toughness of low carbon steels. The strongest and toughest structural steels have been developed by making micro-alloying additions of the elements aluminium, niobium and vanadium, either singly or in combination, to give grain refinement. Niobium and vanadium, elements which form carbo-nitrides, can also introduce a capacity for the steel to respond to a precipitation hardening treatment. It is also the practice, in many cases, to process the steel by controlled rolling with a low finishing temperature so as to achieve the maximum refinement of the ferrite grain structure. This type of processing is more commonly used for thin plates. It is less easy to apply to structural sections because of the difficulty of ensuring a uniform reduction throughout the section, with a consequent variation in grain size and properties

TABLE 6.1
Composition and properties of some structural steels (from BS 7613, BS EN 10025 and BS EN 10113)

Steel	Chemical composition							Tensile strength (N mm⁻²)	Yield strength (minimum) (N mm⁻²) Thickness of specimen (mm)				% elong. 80 mm GL	Charpy value (minimum)	
	C max.	Si	Mn max.	P max.	S max.	Nb	V		up to 16	16–40	46–63	63–100		Temp. (°C)	Energy (J)
S235J2	0.19	0.6 (max.)	1.5	0.05	0.05	—	—	340–370	235	225	215	215	15–24	−20	27
S275J2	0.21	0.6 (max.)	1.6	0.05	0.05	—	—	410–560	275	265	255	245	12–20	−20	27
S355J2	0.23	0.6 (max.)	1.7	0.05	0.05	—	—	490–630	355	345	335	325	12–20	−20	27
S390J6Q	0.16	0.1–0.5	1.5	0.025	0.025	0.003–0.08	0.003–0.10	490–640	390	390†	390*	—	17	−60	27
S450J6Q	0.16	0.1–0.5	1.5	0.025	0.025	0.003–0.08	0.003–0.10	550–700	450	430†	415*	—	17	−60	27

† For thickness 16–25 mm.
* For thickness 25–40 mm.
GL = gauge length.

TABLE 6.2
Composition and properties of weather resistant steels (from BS 7668)

Steel	Chemical composition								TS min. ($N\,mm^{-2}$)	Yield strength (minimum) ($N\,mm^{-2}$) Thickness of specimen (mm)			% elong. 80 mm GL	Charpy value (minimum)	
	C	Si	Mn	P	S	Cr	Ni	Cu		up to 12	13-25	26-40		Temp. (°C)	Energy (J)
	max.				max.		max.								
S345J0WPH	0.12	0.25–0.75	0.6 max.	0.07–0.15	0.05	0.3–1.25	0.65	0.25–0.55	480	345	325	325	21	0	27
S345J0WH	0.19	0.15–0.50	0.9–1.25	0.04 max.	0.05	0.4–0.7	—	0.25–0.4	480	345	345	345	21	0	27
S345J0GWH	0.23	0.15–0.50	0.9–1.45	0.04 max.	0.05	0.4–0.7	—	0.25–0.4	480	345	345	345	21	0	27

Note: Steels S345J0WH and S345J0GWH also contain 0.01–0.06% Al and 0.02–0.1% V.
TS = tensile strength; GL = gauge length.

within the section. It could lead to, for example, different properties in the flanges and web of a beam.

The compositions and properties of some structural steels are given in table 6.1. It will be seen from the table that the minimum yield strength values vary with sectional thickness, being higher for thinner sections. Most structural steels are used in the normalised condition of heat treatment. During normalising, cooling in still air, thin sections will cool through the critical temperature range more rapidly than thick sections. Hence, thin sections will possess a finer grain structure and, thus, higher strengths, than thick sections of the same material.

Corrosion protection and weather resistant steels

The atmospheric corrosion resistance of steel structures can be increased by several measures, including painting, metallic coatings and cathodic protection, as discussed in Chapter 5, section 5.8. It is also possible to manufacture steels with an enhanced resistance to corrosion by making alloying additions of copper and chromium. These steels, termed *weather resistant steels*, are detailed in BS 7668 and BS EN 10155. There are three steels listed, and one of these also contains some nickel and an increased phosphorus content. The composition and properties of the weather resistant steels are summarised in table 6.2.

After exposure to the weather these steels develop an adherent protective oxide film instead of normal flaky rust, and after about two years this possesses a pleasing purplish-copper colour. The surface of the material is treated by shot-blasting after manufacture to ensure a uniform weathering. The cost of weather resistant steels is about 20 per cent greater than normal structural steels, but this may be offset by savings in weight, protective treatment and maintenance.

Selection of structural steel

The selection of a steel for a specific application is determined by several factors. These are:

1. the tensile yield strength required;
2. the toughness, ductility and other properties required;
3. the availability and cost; and
4. arbitrary local conditions as may be imposed by specifications and codes of practice.

The significance of factor (3) is that a particular steel-making plant can only supply the grades of steel which are within the capability of that plant. The plant may lack the steel-making facilities to produce a particular grade, or not possess the required heat treatment capacity or a rolling mill capable of giving a controlled rolling treatment. However, the plant may be able to supply an alternative grade suitable for the application.

Typical applications for structural steels are bridges, gantries, electrical power-supply pylons, industrial and high-rise buildings, offshore structures and pipelines. The range of structural steels for offshore structures is detailed in a separate specification, BS 7191. In general, such steels are similar to those detailed in BS

7613, although some grades are modified substantially from those given in BS 7613.

Building Regulations require that load-bearing members in structures possess specified fire resistance times, as measured in tests according to BS 476 Part 21. The fire resistance times required vary from 0.5 to 4 hours, depending on the size of building and its proposed use. In the test, the structural member is loaded to its design stress and heated at a specified rate until its deflection under load reaches a limiting value, at which time it is considered to have failed. Mathematical models have been developed to predict the changing temperature distribution in sections under test and then, from a knowledge of the mechanical properties of the steel at elevated temperatures, to predict the deflection of a structural member at any instant during the test. The use of models provides an inexpensive method for estimating the fire resistance of sections too large to be tested by available equipment, and also for studying the effects of section size, design stress and new design concepts on the cost of any required additional fire protection.

Buried steel pipe-work may suffer from corrosion and the material should be protected. Commonly used forms of protection include bitumen, polymer resins, pitch impregnated and polymer wrapping tapes, cement and reinforced concrete. However, some damage to any coating may occur during pipe-laying, exposing areas of unprotected steel. Cathodic protection (see Chapter 5, section 5.8) is used to prevent corrosion at breaks in the coating. The nature of the ground has a major influence. There is good drainage in sand or limestone and this gives aerobic conditions with a good oxygen supply. Anaerobic conditions, with little oxygen, can occur in wet clays and, in these conditions, bacteria can cause the breakdown of sulphates, leading to a high corrosion risk. This may be prevented by surrounding the pipe-line with chalk or limestone to give aerobic conditions. BS 8010 is the specification covering pipe materials, coatings and cathodic protection. Part 2.8 of this publication is for overland pipe-lines and Part 3 refers to undersea pipe-lines.

Concrete reinforcement

Plain concrete, whilst strong in compression, is relatively weak in tension and is therefore generally unsuitable for structural use. For this reason, most concrete structures are of *reinforced concrete* in which steel bars or wires are positioned in the concrete to resist the tensile forces, with the concrete resisting the compressive forces. Thus, for example, the main steel reinforcement will be positioned near the bottom of a simply supported reinforced concrete beam.

BS 4449, which applies equally to hot rolled and cold worked steel bars, specifies the required properties for plain round steel bar in grade 250 and deformed high yield steel bars in grade 460 for use as reinforcement. The grade number is the *characteristic strength* of the steel in N mm^{-2} and is the yield or 0.2 per cent proof strength, below which not more than 5 per cent of the test results, for the material supplied, shall fall.

BS 4482 specifies the required properties for cold reduced plain and deformed steel wire in grade 460 for use as reinforcement and for manufacturing steel fabric.

The exact composition of the steel to meet the specified properties is determined by the manufacturer with the proviso that the carbon content should not exceed

0.25 per cent and the carbon equivalent value (CEV) should not exceed 0.42 for grade 250 and grade 460 (BS 4482) and 0.51 for grade 460 (BS 4449). Additionally, limits are placed on the maximum sulphur, phosphorus and nitrogen contents; typically, for grade 250 steel these are 0.060, 0.060 and 0.012 per cent respectively.

Concrete is a highly alkaline material and forms a passive oxide film on the surface of the steel reinforcement, which protects the steel from corrosion. Carbonation, that is the reaction between carbon dioxide, moisture and $Ca(OH)_2$, can however reduce the alkalinity of the concrete and, if it penetrates as far as the steel, destroy this protective passive oxide film. Any subsequent corrosion of the steel will inevitably have an adverse effect on the durability of the associated reinforced concrete structure (see Chapter 15).

In a highly corrosive environment, or in the absence of any practical alternative protective measures, the use of either carbon steel bars with a protective epoxy coating (BS 7295) or austenitic stainless steel bars (BS 6744) as reinforcement may well prove to be economical over the design life of a structure, despite the initial higher cost of these materials. For example, stainless steel reinforcement has been used in ground anchors and in bridge parapets subjected to splashing by salt water.

Prestressing steels

Prestressed concrete refers to concrete members whose sections are subjected to compressive force, by prestressing, before being put into service. The tensile stresses normally associated with the subsequent application of service load do not then result in the concrete being put into tension, but merely reduce the compressive stresses in the concrete induced by the compressive force resulting from the prestressing operations. Prestressing may be achieved by pre-tensioning or by post-tensioning.

In pre-tensioning, the prestressing (steel) tendons are strained in tension, concrete is cast around them and, when the concrete is capable of resisting the prestressing forces, the external straining force used to tension the steel is removed. As the steel attempts to return to its original (unstressed) length, the bond between the steel and concrete restrains this movement and results in the concrete being put into compression, the associated compressive force being equal to the residual tension in the steel.

In post-tensioning, a similar condition is achieved by threading the prestressing tendons through ducts cast into the concrete. When the concrete has hardened the tendons are strained, using external jacks, with the prestressing (compressive) force being transferred to the concrete through mechanical anchorages at the ends of the tendons.

Over a long period of time the initial stress in the pre-stressing steel will decrease, partly owing the loss of the steel strain resulting from concrete creep and shrinkage (see Chapter 14) and partly owing to stress relaxation within the steel itself (see Chapter 5, section 5.6). To ensure that the effective prestressing force does not fall below about 80 per cent of its initial value the prestressing steel must have a high proof strength, so that it is capable of being subjected to a high initial elastic strain, and also low stress-relaxation properties. A very much higher tensile strength than for low carbon steel is therefore required, and for this reason prestressing steels have a high carbon content, generally in the range 0.6–0.9 per cent carbon (see figure 6.1). Strengths can be further increased by cold working.

BS 4486 specifies the required properties of smooth and ribbed high tensile alloy steel bars for use in prestressed concrete. These include a nominal 0.1 per cent proof strength of 835 N mm^{-2} with a nominal tensile strength of 1030 N mm^{-2}.

BS 5896 specifies the required properties of cold-drawn steel wire, these including 0.1 per cent proof strengths in the range 1300–1470 N mm^{-2} and tensile strengths in the range 1570–1860 N mm^{-2}.

When exceptional corrosion resistance is essential, stainless steel may be specified. This was considered necessary for the post-tensioning bars used in the concrete foundations of York Minster during structural restoration work carried out in the early 1970s.

Bolt steels

Many structures are fabricated, at least in part, using bolts. Figure 6.2 shows a bolted structure in which some of its component parts have been fabricated using shop welds. Bolts must have adequate strength and toughness and, when they will be subjected to fluctuating loads, adequate fatigue strength. In this context, for coastal and estuarine sites, the possibility of corrosion fatigue (see Chapter 5, section 5.5) must be considered.

Figure 6.2 *Assembly of bolted structure (Courtesy of Billington Structures Ltd)*

Black bolts

Black bolts are manufactured from carbon steels containing not more than 0.5 per cent carbon by either hot forging or cold forging. The transfer of loads in the connected members is by shear in the bolts and bearing on the bolts and the connected members.

Three grades of black bolts are specified in BS 4190. Tensile strengths and yield strengths for each of these, based on BS 4190, are given in table 6.3. Nuts for black bolts are produced in two grades, namely grade 4 for use with grades 4.6 and 4.8 bolts and grade 6 for use with grade 6.9 bolts.

TABLE 6.3
Properties of bolt steels (adapted from BS 4190 and BS 4395)

Grade	Minimum tensile strength ($N\,mm^{-2}$)	Minimum yield strength ($N\,mm^{-2}$)	Minimum 0.2% proof strength ($N\,mm^{-2}$)	
A. Black bolts				
4.6	392	235	—	Hot forged
4.8	392	314	—	Cold forged
6.9	588	—	529	Cold forged
B. High-strength friction grip bolts				
General	725–825*	560–635*	—	Quenched and tempered
Higher	980	880	—	Quenched and tempered

*Values dependent on bolt size.

High-strength friction grip bolts

These are high tensile steel bolts, used in conjunction with high tensile steel nuts and quenched and tempered steel washers, which are tightened to a predetermined shank tension to provide a clamping force so that transfer of loads in the connected member takes place by *friction* between the parts, and not by shear and bearing.

High-strength bolts are usually manufactured from forged Cr–Mo or Ni–Cr–Mo low alloy steels, and the required properties are obtained by quenching and tempering. BS 4395 specifies two grades of high-strength friction grip bolts (see table 6.3). Nuts for these bolts have specified proof loads based for general grade nuts on a stress of $1000\ N\ mm^{-2}$, on the equivalent stressed area of the corresponding bolt, and for higher grade nuts on a stress of $1176\ N\ mm^{-2}$.

Bridge wire

The mechanical properties required for suspension bridge cables are high strength, toughness and fatigue resistance. This latter property is important because of the fluctuations of stress caused by variable traffic and wind loadings. Corrosion resistance is also important. Suitable cables are produced from cold drawn steel containing 0.75–0.85 per cent carbon and 0.5–0.7 per cent manganese, produced to have a minimum tensile strength of 1600 N mm^{-2}. The wires are galvanised to give a heavy coating of zinc for good resistance to corrosion.

Cladding steels

Steels are used frequently as architectural cladding panels for curtain walls and other external walls, and for covering internal and external structural columns. Such cladding is not a stressed part of the structure but it should be strong and rigid so that cost and weight can be reduced by the use of thin sheet. The steel selected must possess a measure of ductility, to permit forming of the thin sheet, good corrosion resistance and a pleasing appearance. These requirements can be satisfied by low carbon steel, provided that it is suitably protected against corrosion, and by stainless steels.

Low carbon steel sheet is protected against corrosion by either vitreous enamel or plastic coatings. A vitreous enamel finish is produced by fusing the coating on to the steel surface at 800°C. The coating is on both sides of the sheet, encasing it completely in a hard and inert glass. This finish is available in a wide range of attractive and durable colours, and it is also possible to produce multi-colour designs to form part of a mural. The most usual plastic coating is of PVC and, because this has a tendency to be porous and is soft and not scratch resistant, the steel sheet should be galvanised before application of the coating. A limited range of colours is available with PVC coated steels.

Both high chromium ferritic and chromium–nickel austenitic stainless steels are used in sheet form as cladding materials. The materials normally specified are steel 430S17, containing 17 per cent chromium, steel 304S12, containing 18 per cent chromium and 10 per cent nickel, and steel 316S12, with 18 per cent chromium, 10 per cent nickel and 2.75 per cent molybdenum. The last mentioned is preferred for external use as the presence of molybdenum gives an improved resistance to corrosion, particularly with respect to acids and chlorides. The cost of stainless steels is considerably higher than that of plain carbon steels but this is partially offset by the fact that no protective treatments or surface coatings are needed. Stainless steel sheet can be produced in a number of attractive finishes ranging from plain to textured and from matt to polished.

It is also possible to obtain plain carbon steel sheet with a thin coating of an austenitic stainless steel. This offers the corrosion resistance and durability of a stainless steel at a lower cost. Another type of cladding material available is *Terne*, which is austenitic stainless steel sheet with a thin coating of lead. This material, which is much stronger than lead sheet, is used as a roof cladding material and gives the traditional appearance of lead roofing.

Railway steels

Railway structures, such as bridges and signal gantries, are constructed from the normal range of structural steels discussed earlier. Rails need to be hard and strong to resist wear. Also, the passing of trains gives high dynamic loadings and the rails should be resistant to fatigue. Rail wear will be greatest on lines with a high traffic density, on sharp curves and at switches and cross-overs. High hardness and strength can be achieved by using medium carbon or high carbon steels with a relatively high manganese content. There are three grades of rail steel, classified as *normal, wear resistant A* and *wear resistant B*, listed in BS 11 and details of these are given in table 6.4. The rail sections are produced by hot rolling and used in the normalised condition. Fishplates are made by cutting from hot rolled sections of a medium carbon steel, and two types of bolt, normal and high strength, are specified in BS 64 for joining rail sections. These bolts may be made by hot or cold forging and heat treated, if necessary, to produce the required properties. Rail cross-overs, where wear is very heavy, are made in the form of castings in steel BW10 (BS 3100).

The most common cause of failure of rails is fatigue. Fatigue cracks are initiated on the running surface of long welded track lengths and at the bolt hole nearest to the rail end in bolted rails. Surface cracks develop where there is temporary loss of adhesion between a driving wheel and the rail. Friction causes intense local heating and the subsequent rapid cooling causes brittle martensite formation in the surface layers. The stresses caused by following wheels can create small cracks in this brittle layer and repeated loading cycles will cause these cracks to grow. Rails are checked for fatigue cracks, using non-destructive inspection methods, at regular intervals and replaced when faults are found.

TABLE 6.4

Properties of railway steels (from BS 11 and BS 64)

Type	Composition (%)			0.2% proof strength $(N\ mm^{-2})$	Tensile strength $(N\ mm^{-2})$	Elongation % on $5.650\ S_0^{1/2}$
	C	Si	Mn			
Rail steels						
Normal	0.45–0.6	0.05–0.35	0.95–1.25	—	710	9
Wear resistant A	0.65–0.8	0.1 –0.5	0.8 –1.3	—	880	8
Wear resistant B	0.55–0.75	0.1 –0.5	1.3 –1.7	—	880	8
Fishplate steel	0.3 –0.45	0.35 max.	0.9 max.	—	550–650	18
Bolt steels						
Normal	Composition not specified			—	550–700	15
High strength				835	1000–1100	9

Steel forgings

Rolling and drawing processes are only capable of producing wrought products of uniform cross-section. The bulk of steel usage in civil engineering is in the form of rolled or drawn products but some components of complex shape are also required. These shapes may be formed by either forging or casting. Forging involves the hot working of a material by a series of hammer blows. One form of heavy forging is smith's forging, using a plain hammer and anvil, in which the master smith controls both the position of and the force of each hammer blow. This technique is used for individual and large forgings. The other type is closed die forging, in which both hammer and anvil possess shaped recesses. The hot metal stock is forced to flow into the shape of the die. Owing to the high cost of forging dies, this manufacturing process is only economic when large quantities of a component are required. The design of a forging and the deformation sequence must be such that an adequate amount of plastic deformation is given to all parts of the component and the direction of metal flow is controlled to give optimum structure and properties.

Plain carbon steels are adequate for most forgings used in civil engineering and BS 29 covers recommended forging steels with tensile strengths in the range 485–695 N mm^{-2}. Alloy forging steels with tensile strengths in the range 600–1250 N mm^{-2} are covered in BS 4670.

6.3 Steel and iron castings

Casting may provide an alternative to forging for the production of complex shapes. A major process used is sand casting, in which the liquid iron or steel is poured into a prepared shaped cavity in a mould made from sand. Sand moulds are made by hard ramming moulding sand around a pattern. The pattern, in the shape of the finished component, is made from hard wood, but other materials may also be used. After the pattern has been removed, moulded sand cores may be placed in position within the mould if a hollow casting is to be made. Moulding sand contains a binder which holds the grains together so that the mould shape is retained when the pattern is removed and is strong enough to withstand the inflow of liquid metal. Naturally occurring greensand, a sand–clay mix, is used as foundry moulding sand but synthetic sand/binder mixes are also available. The rate of solidification of metal within a sand mould is relatively slow and sand castings tend to have a coarse grain size. In addition a sand casting is likely to possess some porosity and show some compositional segregation, so that the mechanical properties of sand castings are inferior to those of forgings. Casting is likely to be selected when the number of components required is so small that the high cost of forging dies is not warranted, the shape is too complex for forging, or components with adequate properties can be produced more cheaply. Typical applications for castings are as bridge bearing blocks, pipes and pipe fittings such as elbows, tees, etc. and manhole covers and frames.

The casting of iron pipes deserves a special mention because, generally, pipes are cast by the centrifugal or spin casting process. In this process a cylindrical mould shell, lined with moulding sand, is rotated rapidly about its longitudinal axis and liquid metal poured in. The liquid metal will spread evenly along the length of the mould under the action of centrifugal force. Rotational speeds are such as to give accelerations of the order of 60g (g refers to the acceleration due to gravity). The mechanical properties of a centrifugal casting are superior to those of a normal sand casting as the centrifuging action ensures that many impurities, which are often less dense than the iron or steel, segregate to the bore of the pipe, from where they may be removed easily by machining. Similarly, any gases will also escape to the bore, leaving a sound, dense casting.

Cast steels

BS 3100 covers many grades of steels for the manufacture of castings for general engineering purposes. Many of these grades are unlikely to be of interest to civil engineers as they are designed for service at elevated temperatures, to operate in severe corrosive environments, or required to possess special magnetic characteristics. The various steels are designated by a coding consisting of two letters and a number. The first letter in the code may be either A, for carbon steels, or B, for alloy steels. The second letter in the code gives the main characteristics, namely, L = low temperature toughness, W = wear resistant, T = high tensile strength and M = special magnetic characteristics.

One cast steel of particular interest is steel BW 10 (Hadfield's manganese steel). This steel possesses exceptional resistance to impact and abrasive wear. It contains 1.0–1.35 per cent carbon and 11–13 per cent manganese, is austenitic in structure and is non-magnetic. The heat treatment given is heating at 1000°C, to ensure that all carbon is in solution in austenite, followed by water quenching in order to prevent any precipitation of the carbide, Mn_3C, occurring at austenite grain boundaries, as such precipitation would have an embrittling effect. Any attempt to abrade or deform the surface of this material will result in hardening the surface by causing the breakdown of austenite into a very hard martensite. This only occurs to a depth of about 100 atoms ($\approx$ 30 nm), giving exceptional resistance to abrasion but leaving the main material comparatively soft and very tough. This steel finds application for road-breaker bits, rock-crusher jaws, blades and teeth for earth-moving equipment and diggers, caterpillar track links and railway points and cross-overs.

Cast irons

The effects of graphite flakes in ordinary grey cast irons, mentioned in Chapter 4, section 4.6, give them low ductility and toughness and limit their tensile strengths to 350 N mm^{-2}. Despite this, they are used where tensile strength and toughness requirements are not stringent and where advantage can be taken of their other properties. These are relatively low cost, and have good compressive strength, good castability and machinability, better corrosion resistance than mild steel and a reasonable fatigue strength. This last property is due to the high damping capacity of these materials, which effectively minimises the build-up of vibrational stresses. BS 1452 covers eight grades of flake cast irons with minimum tensile strength values ranging from 100 to 350 N mm^{-2}. Typical applications for flake cast irons include manhole covers and frames, gulley drains, rainwater pipes, gutters and fittings, and components of bridge bearings.

Where greater strength and toughness than are possible with flake irons are required, then spheroidal graphite or malleable irons can be used. BS 2789 covers a range of 11 spheroidal cast irons with minimum tensile strength values ranging from 350 to 900 N mm^{-2} and minimum values of 0.2 per cent proof strength from 220 to 600 N mm^{-2}. The various irons are designated by a number which indicates the minimum tensile strength and the minimum percentage elongation value for the material. For example, iron 450/10 has a minimum tensile strength of 450 N

mm^{-2} and a minimum value of percentage elongation of 10 per cent. Typical applications for spheroidal graphite iron castings include pipes and pipe fittings, bridge bearing blocks and comminutor drums for sewage. There is a British Standard, BS 4772, specifically for ductile iron castings for pipes and fittings. The iron castings must possess a minimum tensile strength of 420 N mm^{-2}, a minimum 0.2 per cent proof strength value of 300 N mm^{-2}, and an elongation value of between 5 and 10 per cent. This standard also gives details of the surface protection requirements, specifying that the pipes be galvanised with spray coated zinc to a minimum covering of 130 g m^{-2}.

Thirteen grades of malleable cast iron, some whiteheart, some blackheart and some pearlitic types, and with tensile strengths in the range from 300 to 700 N mm^{-2} are listed in BS 6681.

6.4 Aluminium and its alloys

Aluminium and its alloys possess a number of attractive properties, principally low density and a high resistance to corrosion. Aluminum has a face centred cubic structure, is highly ductile and can be shaped easily by a variety of methods. Pure aluminium has a low strength (about 70 N mm^{-2}) but, by alloying with other elements, the strength can be increased considerably to give alloys with very good strength to weight ratios; thus whilst the cost per tonne of aluminium is considerably higher than that of carbon steel, the actual cost differential in a structure may be quite small. Erection costs for aluminium structures may be lower than those for steel structures, because individual components are lighter, and maintenance costs will be considerably less than for steel, because of the high corrosion resistance. Taking these factors into account, there are numerous situations where the choice of aluminium alloys can be justified on both technical and economic grounds. Aluminium alloys are used for structural and architectural applications, including bridges, roof structures and curtain walling.

High purity aluminium (99.5 per cent, or greater) is too weak to be used for many purposes, but because it is soft and highly malleable it is used as a flashing material in building construction. The material commonly termed pure aluminium (alloy 1200) is really an aluminium–iron alloy containing about 0.5 per cent of iron. Its only use in construction is, in the form of sheet, as architectural panelling. The alloys of aluminium are subdivided into non heat-treatable and heat-treatable categories. Those in the first group can be strengthened by cold working, and the only type of heat treatment that can be given is an annealing to soften work hardened material. The heat treatable alloys, on the other hand, can be strengthened by solution heat treatment and age or precipitation hardening (refer to Chapter 4, section 4.5). Alloys of the Al–Mg, Al–Mn and Al–Si systems are not heat-treatable, while the principal heat-treatable alloy systems are Al–Cu, Al–Mg–Si and Al–Zn.

All wrought aluminium alloys are identified by a four-digit number, and, these are listed and detailed in BS EN 485-2, in five consecutive specifications, BS 1471 to 1475, and in BS 4300. The 1xxx series are the various grades of pure aluminium, the 2xxx series are heat-treatable Al–Cu alloys, and the 3xxx and 5xxx series are non-heat-treatable Al–Mn and Al–Mg alloys, respectively. When magnesium and silicon are present together, they combine to form the compound Mg_2Si. The phase diagram between aluminium and the compound Mg_2Si is similar to the Al–Cu diagram (figure 4.6) and the Al–Mg_2Si alloys are heat-treatable. This alloy series is designated 6xxx. The 7xxx series alloys are also heat-treatable and are in the Al–Zn series. Aluminium casting alloys are listed in BS 1490 and are designated by the letters LM, followed by a number.

Only some of the wide range of aluminium alloys made are used in the construction industry, and these are listed in BS 8118 and BS 1161. The main characteristics and applications of these alloys are given in tables 6.5 and 6.6.

The strong wrought aluminium alloys are viable alternatives to steels for bridges and the roof structures of buildings. Usually, the final choice of material and design will be made on economic grounds, and the lower long-term maintenance costs of a corrosion resistant aluminium structure may offset the higher initial material cost. There could also be reduced operating costs in the case of movable bridges, as the lighter structure would require less power to move it. Although the density of aluminium is only about one-third that of steel, the weight saving in an aluminium structure is not as great as might be expected at first sight, as the value of E is also about one-third that for steel and deeper sections are required in aluminium to give the required stiffness. In general, weight savings of the order of 50 per cent can be achieved with aluminium, as compared with steel. Considerable use has been made of aluminium alloys for the construction of portable military bridging units.

One of the main aluminium alloys used for structural purposes is the heat-treatable alloy 6082 in the T6 (fully heat treated) temper. Usually, structural members in this alloy are joined by bolting or rivetting as, when welded, there is a significant loss of strength in the weld and heat affected zones. This effect applies to all the heat-treatable aluminium alloys as, during welding, non-coherent precipitate particles are formed (see Chapter 4, section 4.5). Several alloys, including 6082, are specified for the manufacture of bolts and rivets. The aluminium–magnesium alloy 5083 has a good strength and is a useful structural alloy, particularly for structures in marine environments. As a non heat-treatable alloy, this material can be welded successfully, with little loss of strength in the weld zone.

Alloy 6082 is also the alloy specified in BS 1139 for the manufacture of scaffold tubing. The tubing is made by a hot extrusion process. Scaffolding couplers may be made by casting or forging from several alloys, including 6082, 2014 and LM25. The specification permits the use of both cold drawn and extruded tube in a range of alloys, including 6061, 6063 and 6082, for the construction of lightweight prefabricated access and working towers.

The heat-treatable alloy 6063, which is less strong than 6082, is used mainly for architectural purposes, for example, as extruded sections for mullions and glazing bars in curtain walling systems. The aluminium–manganese alloy 3103 is widely used as building sheet but the somewhat stronger aluminium–manganese–magnesium alloy 3015 is specified for the manufacture of profiled sheet for construction. Frequently, these products are anodised to improve the resistance to corrosion.

The resistance of aluminium alloys to dust, dirt and moisture makes them very suitable for many items of street furniture, including railings, posts and road signs.

TABLE 6.5

Properties of wrought aluminium alloys used in construction (adapted from BS EN 485-2, BS 1471–1475 and BS 8118)

Alloy	Composition	Temper*	0.2% proof strength $(N\ mm^{-2})$	Tensile strength $(N\ mm^{-2})$ min.	max.	% elong. on 50 mm	Forms†	
1200	99.0% Al (minimum)	0		70	105	20–30	S	Sheet material for architectural use only
		H14		105	140	3–6		
		H18		140		2–4		
3103	1.25% Mn	0		90	130	20–25	S	Sheet for building panels
		H18		175		2–4		
3105	0.6% Mn 0.5% Mg	0		110	155	16–20	S	Stronger than 3103. Used for profiled sheet
		H18	190	215		1–2		
5083	4.5% Mg	0		275	350	12–16	S, P, E, T, F	Good strength structural alloy. Little loss of strength on welding
		H24	270	345	405	4–6		
5154A	3.5% Mg	0	85	215	275	12–18	S, P, E, T, F	Good corrosion resistance in marine environments
		H24	225	275	325	4–6		
5251	2.0% Mg	0	60	160	200	18–20	S, P, T, F	Good in marine environment
		H28	215	255	285	2–4		
5454	2.75% Mg, 0.15% Cr	0	80	215	285	12–18	S, P, E	Good strength structural alloy. Good corrosion resistance in marine environment
		H24	200	270	325	3–6		
6082	1.0% Si, 0.9% Mg, 0.7% Mn	0			170	16	S, P, E, T, F	Good structural materials but care needed in design because of loss of strength in welded joints. Scaffolding tube
		T4	120	190		13		
		T6	255	295		8		
6061	0.6% Si, 1.0% Mg, 0.25% Cu	T4	115	190		18	E, T	As for 6082
		T6	240	280		7		
6063	0.4% Si, 0.7% Mg	0			140	16	E, T, F	Lower strength than 6082 and 6061. Main uses, architectural, curtain walling, glazing bars
		T4	70	130		14		
		T6	160	195		6		
7020	1.2% Mg, 4.5% Zn	T4	170	280		10	S, P, E	Good strength but sensitive to stress corrosion
		T6	270	320		8		
2014	4.5% Cu, 0.7% Si 0.5% Mg, 0.8% Mn	0	110		235	14–16	S, P, E, T	High strength. Consultation with manufacturers recommended before using this alloy
		T4	225	400		13–14		
		T6	380	440		6–8		

KEY: *Temper: 0 = annealed, H14 = partially work hardened, H18 = fully work hardened, H24 and H28 = work hardened and partially annealed, T4 = solution heat treated and naturally aged, T6 = solution heat treated and precipitation hardened.

†Form: S = sheet, P = plate, E = extruded sections and tube, T = cold drawn tube, F = forgings.

TABLE 6.6
Properties of cast aluminium alloys used in construction
(adapted from BS 1490)

Alloy	Composition	Condition	Tensile strength ($N\ mm^{-2}$)		% Elongation (5.65 $S_0^{1/2}$)	
			Sand or investment cast	Chill cast	Sand or investment cast	Chill cast
LM5	4.5% Mg, 0.5% Mn	As cast	140	170	3	5
LM6	11% Si	As cast	160	190	5	7
LM25	0.4% Cu, 7.0% Si	As cast	130	160	2	3
		T6	230	280	—	2

6.5 Copper and its alloys

Copper is one of the oldest metals known to man and both it and its alloy with tin, bronze, has been worked for more than 5000 years. Copper, with its high electrical and thermal conductivities, ease of working and good resistance to atmospheric corrosion, is a valuable engineering metal. It can also be alloyed with many other metals to produce a wide range of useful alloys although, because of the present high cost of the metal, copper and its alloys are being replaced by cheaper materials, such as aluminium and thermoplastics, for many applications. Copper and copper alloys are not strictly civil engineering materials, but are used widely in buildings (see BS 2870 to BS 2875).

Copper

There are several grades of pure copper, including tough pitch copper (C101 and C102) and phosphorus deoxidised copper (C106). Tough pitch copper, with a minimum purity of 99.90 per cent, contains about 0.03 per cent of oxygen in the form of the oxide Cu_2O. This material has a high conductivity but is not suitable for welding or brazing, because of the oxygen content. It is used in building for power cables, electrical wiring and lightning conductors. Copper is commonly deoxidised by adding phosphorus, in the form of a copper–phosphorus alloy. Phosphorus deoxidised copper (C106) has a minimum purity of 99.85 per cent copper and contains about 0.03 per cent residual phosphorus. Copper sheet is used as architectural cladding and for supported roofing, as flashing, as a damp course material, and for the manufacture of water tanks and cisterns, while copper tubing is used for gas and water supplies and in central heating installations. Copper used for roofing, flashing and building cladding develops a green patina, owing to reactions with the atmosphere causing the formation of copper salts. Care needs to be taken in plumbing installations to avoid galvanic coupling between copper and galvanised steel pipework and tanks. When joints have to be made between copper and steel, the two metals must be electrically insulated from each other.

Copper alloys

Copper can be alloyed with many metals, but only brasses, copper–zinc alloys, and bronzes, copper–tin alloys, are used to any extent in the building and construction industries. Brasses are, primarily, alloys of copper and zinc but may also contain small amounts of other elements for the enhancement of properties. For zinc contents up to 37 per cent the alloys possess a single phase α (face centred cubic) structure and are highly ductile. These are cold working alloys and are produced in the forms of sheet, strip and tubing. At zinc contents between 37 and 47 per cent the alloys possess a duplex structure of α and β phases. These (α + β) alloys are not cold worked and are used in the hot worked condition or for castings. Brass plumbing fittings are castings in brass containing 60 per cent copper and 40 per cent zinc.

Brass has been used for building cladding and curtain walling but its use for this type of application is not an unqualified success. Toner (1969) describes the application of brasses for the curtain walling of the engineering and science building of the former Polytechnic of Central London, now the University of Westminster. Hot extruded sections of a high tensile (α + β) brass containing 60 per cent copper and 40 per cent zinc, modified by small additions of tin, iron and aluminium, to improve corrosion resistance, were used for the window sections. This alloy has a tensile strength in the range 455–530 N mm^{-2} and a 0.1 per cent proof stress in the range 215–278 N mm^{-2}, so that the manufacture of thin but rigid window sections was possible. The underwindow spandrel panels were of composite construction and of glass fibre, asbestos and aluminium foil, faced with thin sheets of an α brass containing 85 per cent copper and 15 per cent zinc. It was considered that the development of an even natural patina would eliminate the need for protective treatment and costly maintenance. When originally constructed in 1970, the building was of pleasing appearance and an Architectural Association prize was awarded for it. However, as weathering occurred, not always evenly, the building, in the opinion of the author, soon took on a dull and rather shabby appearance with the brass acquiring an uneven matt brown patina with occasional streaks of green corrosion products. This situation remained until 1993 when it was decided to improve the appearance. The work, which took a year to complete, involved cleaning off all the surface grime and corrosion products followed by spray painting all the external metalwork with several layers of acrylic paint.

Bronzes, alloys of copper and tin, may possess either single-phase α structures or duplex (α + δ) structures. The cold working α bronzes, containing up to 7 per cent tin, are not used in building or civil engineering, but there is some limited use made of the 10 per cent tin two-phase alloy. A phosphor-bronze with 10 per cent tin and up to 1 per cent phosphorus to improve the strength and castability is an excellent bearing material capable of sustaining high loadings. Castings in this material are used for the centre bearings of swing bridges and also for components of the expansion bearings of static bridges. These bearings require good lubrication and in recent years rubbers have been in competition with bronze for static bridge bearings, requiring less maintenance.

6.6 Other non-ferrous metals

Zinc

Zinc is a metal with a low melting point, 419°C, and a hexagonal closed packed crystal structure. Normally, metals with hexagonal crystal structures tend to be brittle and difficult to cold work but zinc can be plastically deformed with ease at temperatures in the range 100–150°C. Sheet zinc can be produced by 'warm rolling' the metal at these temperatures. The rolled zinc sheet possesses directional properties, being somewhat stronger but having less ductility transverse to the rolling direction than parallel to the direction of rolling. Sheet zinc is used as a building flashing and damp course material, and BS 6561 gives two grades of zinc for these purposes. Alloy type A is zinc with small additions of titanium and copper, while alloy type B is zinc with about 0.7 per cent of lead.

The low melting temperature of zinc makes it an ideal metal for die casting purposes. The die casting alloys contain 4 per cent of aluminium and, in some cases, 1 or 2 per cent of copper. These alloying additions increase the tensile strength to about 250 N mm^{-2}. Zinc alloy die castings are used, among other things, for window, door and bathroom fittings.

A major use for zinc is for galvanising, that is the coating with zinc of steel sheet, structural sections, pipes and so on, as a protection against corrosion. The zinc coating may be applied to a clean surface by one of several methods, these being hot dipping into a bath of liquid zinc, electroplating, metal spraying and sherardising. This last process involves heating the steel parts, together with zinc dust, in a sealed container at a temperature in the range of 350–375°C. Zinc may also be used as a sacrificial anode for the cathodic protection of steel structures (see Chapter 5, section 5.8).

Lead

Lead is a soft and malleable metal with an excellent resistance to corrosion, and the applications of lead sheet and strip in building include flashing, damp-proof courses, sound proofing and sealing in puttyless glazing systems. Lead possesses a high density ($11\,340$ kg m^{-3}) and, because of this, is a very suitable material as a radiation shielding material. Formerly, lead was used widely for water and waste disposal pipe-work but, because if its toxicity, it has been replaced by other materials for these applications.

Lead is a major constituent in soft solders, which are essentially lead–tin alloys but often contain small amounts of antimony. The soft solders possess melting temperatures in the range 180–260°C and are used for making low strength joints. Many joints in copper pipe-work in building plumbing systems are made using Yorkshire fittings. These are copper pipe fittings which include rings of a soft solder and, after assembly of the pipe and fitting, it is only necessary to heat the fitting with a gas torch to melt the solder and effect a leak-proof joint.

Lead is also a constituent of the fusible alloys, with melting points below 100°C. A relevant application of one of these alloys is in the fusible plugs of automatic sprinkler systems for fire prevention. A suitable alloy for this purpose is Wood's alloy, containing 24 per cent lead, 50 per cent bismuth, 14 per cent tin and 12 per cent cadmium, which has a melting point of 71°C.

6.7 Principles of metal joining

The permanent assembly of individual manufactured components is an important aspect of fabrication and construction and there are many ways for accomplishing this including the use of fastenings such as bolts and rivets, the use of adhesives, and by soldering, brazing and welding. The making of bolted, riveted and adhesive bonded joints will not be discussed here. In soldering and brazing processes, the joints are effected by means of a filler material which is molten at temperatures below the melting temperature of the metals being joined; whereas in fusion welding (see section 6.8), some portion of the metals to be joined is melted during the process and, where a filler metal is used, this is generally of the same or similar composition to the metals being joined. Pressure and electric resistance welding processes do not require any filler material.

Soldering and brazing

In soldering and brazing the parts to be joined are prepared and fitted together and, at a comparatively low temperature, the liquid filler is drawn by capillary action into the small gap between them. A major advantage of soldering and brazing operations is that there is little heat distortion of parts and little change, if any, to the microstructure of the parent metal. Soft solders, mainly lead–tin alloys, have melting temperatures in the range 180–260°C, are comparatively weak and not used for making joints in structural materials. The brazing alloys, or hard solders, have melting temperatures in the range 690–890°C and possess tensile strengths in the range 400–500 N mm^{-2}, and can be used for making high strength joints in steels. Brasses containing between 40 and 50 per cent of zinc are widely used as hard solders and the alloys may contain small amounts of silicon or silver, to increase the fluidity of the liquid filler, or nickel and manganese to increase the strength. Another group of hard solders is the silver solders, alloys of copper, zinc and silver. These, although more expensive, are stronger and possess lower melting temperatures than brasses.

The most commonly used heat source for hard soldering is an oxy-gas torch using either propane or acetylene as the fuel. The metal parts to be joined must be clean and a flux must be used. The flux, which is molten at the brazing temperature, is an oxide solvent, and both borax and alkali metal fluorides are used as fluxes in hard soldering.

Aluminium alloys can be soldered successfully with a chloride flux and using aluminium solder alloys to effect the joint. The main group of aluminium solders are aluminium–zinc alloys, with small additions of tin or cadmium, and with melting ranges around 400°C. Aluminium–silicon alloys, melting in the range 580–600°C, may also be used as solders. Solders may be used as an alternative to welding or adhesive bonding in the fabrication of aluminium alloy windows and curtain walling.

Braze welding

Braze, or bronze, welding is a metal joining process in which the filler metal is a brass or bronze with a lower melting temperature than the metals being joined but in which, unlike brazing, the filler is not drawn into the joint by capillary action. There is a wider gap between the parts than in a brazed joint, and the filler, melted by an oxy-gas torch or an electric arc, is used to fill this gap and effect the joint. This joining process is suitable for joining cast irons or repairing cracked or broken castings. The iron is not melted in the process and, because of the relatively low temperatures involved, the tendency for thermal stress cracking of the iron is minimised and there is little change to the microstructure of the iron.

6.8 Fusion welding

There are many fusion welding processes and the heat necessary for welding may be obtained from a chemical reaction, as in oxy-gas welding, or by using electrical power.

Gas welding

Gas welding, although used much less than arc welding, is still an important process and used quite widely for maintenance and repair work. It also has some advantages for welding carbon and low alloy steels in which martensite could be formed during cooling in air after welding. The flame temperature is lower than the temperature of an electric arc and the temperature gradients in the vicinity of a weld are less than with arc welding. Also the weld area can be preheated with the flame and the flame played over the weld area after welding to reduce the cooling rate and, thus, avoid the formation of brittle martensite.

The commonly used gas for oxy-gas welding is acetylene (ethyne), and this gas is mixed with oxygen and burnt in a specially designed torch. Flames with differing characteristics can be obtained by controlling the relative amounts of oxygen and acetylene. A chemically neutral flame is obtained when the oxygen/acetylene ratio is about 1.1/1. If the oxygen/gas ratio is increased, a shorter, fiercer, oxidising flame is produced, while an oxygen/gas ratio of about 0.9/1 will give a chemically reducing flame. Flame temperatures range from about 3100°C for a reducing flame to about 3500°C for an oxidising flame. The filler metal to make the weld is obtained from a metal rod or wire fed into the flame by the welder and, generally, the filler rod is coated with a layer of flux. The flux melts and acts to dissolve surface oxides and form a protective layer over the weld.

Many metals and alloys can be welded successfully by gas welding. The chemically neutral flame is used for welding plain carbon steels and most ferrous and non-ferrous alloys, but an oxidising flame is needed for copper, brass, bronze and nickel silvers. This is to avoid any possibility of hydrogen absorption in the weld, which could lead to brittleness. For aluminium alloys and some stainless steels, it is necessary to use a reducing flame as these materials oxidise rapidly at high temperatures. The fluxes which need to be used in the gas welding of aluminium alloys are very corrosive, and care must be taken to remove all traces of flux from the completed weldment otherwise corrosion could set in rapidly.

Flame cutting

An oxy-acetylene flame may also be used for cutting steels but the design of the torch used for flame cutting (figure 6.3) differs from that for welding. The cutting torch nozzle consists of a series of small holes arranged around a large central jet. The oxy-acetylene mixture is fed through the small holes and a high-velocity jet of pure oxygen passes through the central hole. In the cutting operation the oxy-gas flame heats up the surface of the steel to a temperature of about 1000°C, when the iron will ignite and burn in the jet of pure oxygen. The heat energy generated by the combustion of iron will cause some of the surrounding steel to melt and the molten metal is blown away by the strong oxygen jet.

Figure 6.3 *Flame cutting (Courtesy of Billington Structures Ltd)*

Arc welding

Arc welding in one form or another is the most widely used method of welding. The electrical supply is low voltage and may be either alternating or direct. The arc is struck between a metal electrode and the work-piece, and the electrode may be either of tungsten or be a consumable electrode that melts, acting as a source of filler metal. This latter type is the more common. The arc temperature is of the order of 4000°C and the heating effect to the work-piece is largely confined to a very small area beneath the electrode. The tip of the electrode and the metal direct-ly under the arc melt and mix thoroughly with considerable turbulence. (In the case of a tungsten electrode, the electrode does not melt and a filler rod is fed into the arc to provide a supply of liquid metal.) The pressure produced by the arc causes a crater to form in the liquid metal pool and, as the arc progresses along the joint, liquid metal flows back into the crater (see figure 6.4).

When direct current is used, the electrode may be either positive (DCEP) or negative (DCEN) with respect to the work-piece. DCEN, being the most widely used, is known as *normal* polarity. The depth to which melting of the work-piece occurs is called the penetration of the arc, the depth of penetration being greatest with DCEN welding. With DCEP or *reversed* polarity, the electron flow is from work-piece to electrode and the greatest heat is generated at the electrode tip, giving more rapid electrode melting than with DCEN. (DCEP is not recommended for use with tungsten electrodes, as there can be some melting and erosion of the tungsten with subsequent contamination of the weld.) The extent of heat penetration into the work-piece is less with DCEP than with DCEN, but there is a smoother transfer of liquid metal from the electrode, and the emission of electrons from the work-piece has the effect of breaking up and scattering any oxide film that may be present on the metal. For this reason, DCEP is the preferred polarity for the welding of those metals that possess tenacious oxide films, namely aluminium alloys, stainless steels and some copper alloys.

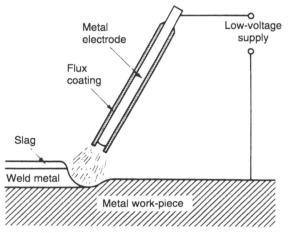

Figure 6.4 *Metallic arc welding*

With an alternating current, the arc is broken and re-established at each half cycle and this can lead to arc instability, although the use of arc-stabilising agents in the flux coating of electrodes can ameliorate the problem.

Electrodes

Bare electrode wire and welding rods are used in conjunction with gas shielded welding and can be used for unshielded arc welding, but in the latter case it is more usual to use flux-coated electrodes when steels are arc welded in air with no form of shielding, otherwise oxides and nitrides can form and remain in the weld with a consequent loss of toughness. The composition of the coatings on electrodes is complex and a variety of different coatings is used, each type designed to cater for a particular type of welding application. However, in all cases the coating is formulated to satisfy three objectives. These are: (1) to form fusible slags; (2) to stabilise the arc; and (3) to generate an inert gas shield during the welding operation.

Flux coated electrodes and welding wires are however much more expensive than bare electrode wire and a number of welding methods, other than gas shielded arc welding, have been developed in which bare electrode wire can be used successfully. These include *submerged arc welding* and *magnetic flux/gas-arc welding*, both of which are automatic flux shielded techniques, and *electroslag welding*, although this is not strictly an arc welding process.

Another welding method of interest, in which flux-cored electrodes are used, is *electrogas welding*, which is an automated shielded arc welding process with additional shielding being provided by carbon dioxide, argon, or an 80/20 argon/CO_2 mixture.

For detailed descriptions of these welding processes the reader should refer to specialist literature.

These processes are used for the welding of structural steels. Submerged arc welding is used for making shop welds in the horizontal mode. A large metal deposition rate is possible in the submerged arc process and a weld in thick plate can be completed in one welding pass. The magnetic flux/gas-arc process is also mainly for shop use but is suitable for both vertical and horizontal weld runs in structural steels. Electrogas welding is suitable for vertical, or near vertical, weld runs in structural steel plate of thicknesses ranging from 10 to 75 mm. It can also be used for joining plates with slightly curved surfaces. It can additionally be used for shop and on-site welding, and typical applications are for the construction of large gas and petroleum storage tanks and for shipbuilding. Electroslag welding can be used for making butt welds in the vertical mode in steels of thicknesses from about 10 mm upwards although, usually, it is used for sectional thicknesses greater than about 50 mm, as for smaller thicknesses other methods such as *electrogas welding* are preferable. The main applications for electroslag welding are for fabrication of thick-walled pressure vessels and the joining of castings or forgings to produce large structural assemblies.

Inert gas shielded arc welding

It is possible to produce a complete blanket of an inert gas around a weld area and, by this means, arc welding without fluxes, and using uncoated electrodes, can be accomplished readily. The most widely used inert shielding gas is argon, but helium may be used. Also, carbon dioxide can be used as a shield for the welding of steels, and nitrogen can be used with copper. The welding electrode may be of tungsten and non-consumable, in which case a separate filler rod has to be fed into the arc to provide a source of weld metal, or a consumable metal electrode can be used. The former is termed the TIG (tungsten–inert gas) or GTA (gas tungsten-arc) process, and the latter is known variously as MIG (metal electrode–inert gas), MAGS (metallic-arc gas shielded) or GMA (gas metal arc) welding (figure 6.5). In electrode holders, or welding torches, for MIG and TIG welding, the electrode is surrounded by a tube and nozzle, through which inert gas is fed and blown over the weld area. Gas shielded welding processes lend themselves to automation, and automated TIG or MIG welding is used to produce welds of consistent high quality.

When carbon dioxide, CO_2, is used as a shielding gas for the welding of steels, dissociation of the CO_2 can occur at high temperatures according to the reaction equation

$$2CO_2 \rightarrow 2CO + O_2$$

and the oxygen produced could oxidise some of the metal. The iron oxide thus formed might then react with some carbon in the weld pool

$$FeO + C \rightarrow Fe + CO$$

As the carbon monoxide generated by this reaction could produce some porosity in the finished weld, to produce consistent high-quality welds in structural steels using CO_2 as a shielding gas, it is necessary to use welding wire of a modified composition. The welding wire must contain relatively high proportions of manganese and silicon as deoxidising agents to remove FeO from the weld pool.

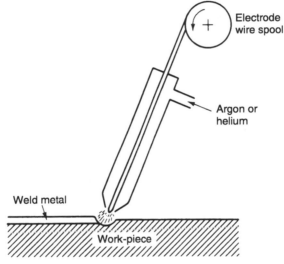

Figure 6.5 *Principle of MIG welding*

Arc stud welding

It is frequently necessary to attach studs and fixing attachments to plate or sheet material. In arc stud welding, an arc is struck between a stud and a plate for a pre-set time, usually between 150 and 500 ms. This is sufficient to melt the end of the stud and to form a small crater in the plate. The two are then brought into contact under pressure and a weld is made. The process is used for the placing of studs and fixing attachments with diameters ranging from 3 to 25 mm in steels, including stainless steels, brass and aluminium alloys for a wide range of applications in civil engineering, ship-building and general engineering. No elaborate surface preparation is necessary other than shot blasting or wire brushing.

Thermit welding

In this form of welding, the source of heat is the chemical reaction between iron oxide and aluminium:

$$8Al + 3Fe_3O_4 \rightarrow 9Fe + 4Al_2O_3 \quad (15.4 \text{ MJ/kg Al})$$

The thermit reaction can be used for the on-site welding of steel sections, such as rails (figure 6.6) in which a few kilogrammes of weld metal are required, and for the joining of very heavy sections where several tonnes of weld metal might be needed.

A sand mould, complete with feeder and riser channels, is formed around the joint area and the thermit powder is contained in a crucible positioned on top of the mould. The thermit powder, when ignited, produces pure liquid iron at a very high temperature (3500°C), and so, to produce a liquid steel of similar composition to the steel being joined and to keep temperatures down, steel scrap and alloying additions are mixed with the thermit powder. When the reaction is complete, the liquid steel is allowed to run through a tap-hole in the base of the crucible into the prepared sand mould.

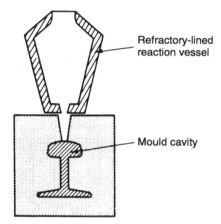

Refractory-lined reaction vessel

Mould cavity

Figure 6.6 *Thermit welding a rail*

Electric resistance spot welding

Spot welding is used mainly for the joining of thin-gauge metal sheet but it is also possible, using heavy-duty equipment, to produce satisfactory spot welds in structural steel plate in thicknesses of up to 35 mm.

The principle of spot welding (figure 6.7) is that the parts to be joined are overlapping and clamped between a pair of electrodes. A large current is then passed through the metal for a short, but accurately controlled time. The maximum resistance in the welding circuit occurs at the interface between the two parts being joined. The electrodes are made from a hard chromium–copper or beryllium–copper alloy, and are hollow and water cooled to prevent any possibility of welding occurring between the electrode and the work-piece. The spot welding cycle is in three parts. Firstly, the electrodes come together and clamp on to the work-pieces, exerting pressures in the range 70–100 N mm^{-2}. Secondly, the welding current flows for a preset time, with current densities of up to 350 A mm^{-2}. It is during this part of the cycle that the weld is made. Some plastic deformation of the hot metal occurs as a result of the electrode pressure, and this shows up as surface depressions on both sides of the joint. Finally, the electrode pressure is maintained for a short time after the welding current has ceased. This aids cooling of the weld area and ensures the formation of a fine grained microstructure within the weld nugget.

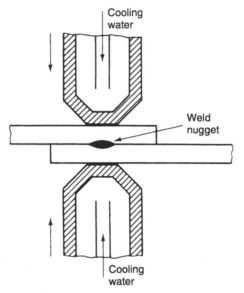

Figure 6.7 *Principle of spot welding*

6.9 Weldability and avoidance of defects

Joint design

The main types of welded joint are butt, corner and fillet welds (figure 6.8). Generally, in butt welding, square cut edges on the pieces to be joined are only feasible with thin section material, as in thicker sections there could be incomplete penetration of weld metal. In most cases the edges must be shaped to give a bevelled or chamfered edge, thus giving a Vee-butt joint or, in the case of very thick sections where there is access to both sides, a double Vee-butt joint. This edge preparation is costly, but necessary, for most welding techniques except the electroslag and electrogas processes. Usually, one of the limiting factors in manual oxy-gas or arc welding methods is the comparatively low rate of metal deposition possible. This means that several welding passes will be needed to complete a Vee-butt weld in thick section material and, frequently, this could mean 10 or 12 weld runs to complete a weld in material of 25 mm thickness (figure 6.9). One major advantage of some of the automated welding processes such as submerged arc and electrogas is that a high rate of metal deposition can be achieved and a weld can be completed in one pass only.

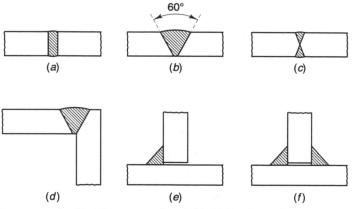

Figure 6.8 *Some types of weld: (a) square-cut butt weld; (b) Vee-butt weld; (c) double Vee-butt weld; (d) corner Vee-weld; (e) single fillet weld; (f) double fillet weld*

Figure 6.9 *Multi-pass weld in thick material*

Metallurgy of fusion welding

The filler material used for fusion welding should be of the same, or similar, composition to the metal being joined, although there are one or two exceptions to this. For the welding of steels, the welding wire or rod should contain a slightly higher carbon content than the base material, to allow for some 'burn-off' of carbon during welding and, when CO_2 is used as a shielding gas, welding wire containing relatively large amounts of manganese and silicon must be used to ensure full deoxidation of the weld pool.

When making a fusion weld, the filler material and a small portion of the metal parts being welded are melted. On removal of the heat source, the liquid weld pool will lose heat rapidly and solidify. The weld is, in effect, a small casting. The material adjacent to the weld will be subject to heating during the welding process and microstructural changes will occur in the basis metal in this area, termed the heat affected zone (HAZ), giving a variation in properties. There will, therefore, be a variation in properties across a weldment. The nature and extent of the structural and property changes will depend on several factors, including the type of metal, the rate of heat input and the rate of cooling. The heat input rate during welding is a function of both the type of welding used and the sectional size of the base metal, while the rate of cooling will depend mainly on sectional size, being slowest for heavy sections. Because of the many variable factors involved, the examples given below are broad generalisations indicating the main changes that occur.

Single-phase metals and alloys

For materials such as aluminium, copper and single-phase alloys of those metals, both the weld and heat affected zone will be of lower strength than the basis material. In the case of metals, originally in the annealed condition with a fine grained crystal structure, grain growth will take place in the heat affected zones with the greatest degree of growth occurring at the interface with the weld pool. When the liquid weld metal solidifies, the solidification will begin at this interface and coarse columnar type crystals will grow normally in the direction of the main temperature gradients (figure 6.10(a)). If the original material was in the work hardened condition, the extent of the strength loss will be even greater as the first change within the heat affected zone will be recrystallisation and annealing of the cold worked material (figure 6.10(b)).

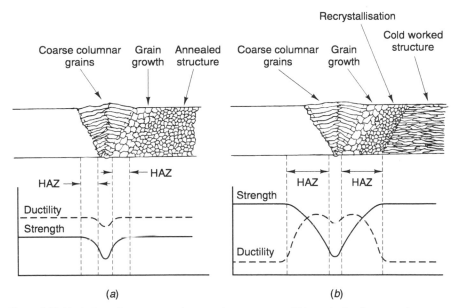

Figure 6.10 *Variation of structure and properties across a weld in a single-phase metal: (a) annealed; (b) work hardened*

Precipitation hardening alloys

Precipitation hardened alloys, for example the heat-treatable aluminium alloys, can be welded but there is a significant loss of strength in the weld and heat affected zones. The weld zone will possess a coarse two-phase structure and be comparatively weak. In the heat affected zones, the temperature will be sufficiently high to cause overageing with precipitation of the second phase (see Chapter 4, section 4.5), although in the portion closest to the weld pool, the second phase will be taken into solution and the rate of cooling after welding may be sufficiently high to form a metastable solid solution within this narrow zone, which may then age harden. These effects are illustrated in figure 6.11. The structural changes will also have another adverse effect, as there will be a reduction in the corrosion resistance in the two-phase regions. This is particularly so in aluminium–copper alloys. Generally, joints between structural members in the heat-treatable aluminium alloys are made by bolting or rivetting, because of the loss of strength which accompanies welding.

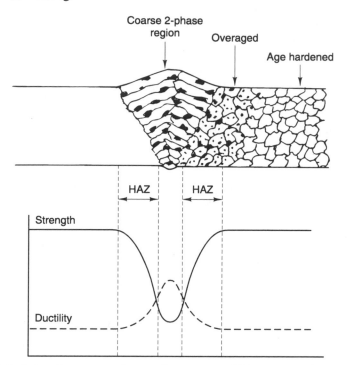

Figure 6.11 *Variation of structure and properties across a weld in a precipitation hardened alloy*

Structural steels

A weld made in a normalised low carbon steel is usually somewhat stronger than the base material. The microstructure of a steel is largely determined by the rate of cooling through the critical temperature range, and the rate of cooling of the weld and heat affected zones following welding is usually faster than that which applied during normalising. The weld pool will solidify with the usual coarse grained structure of a casting but, during rapid cooling through the critical temperature range, the coarse grained austenite will transform into a fine grained ferrite + pearlite structure. The heat affected zone will be that portion of the base metal which was heated into the austenitic region during welding, and again rapid cooling through the critical temperature range will give a fine grained structure (see figure 6.12).

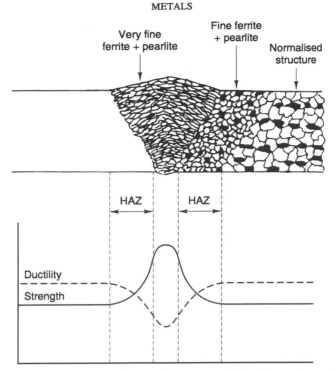

Figure 6.12 *Variation of structure and properties across a weld in a low carbon steel*

A problem with some steels, however, is that the rate of cooling may be sufficiently high to cause the formation of brittle martensite in some part of the heat affected zone. This is the case with steels of higher carbon content and for many alloy steels. Normally, martensite formation does not occur in steels with a carbon, or carbon equivalent, content of less than 0.5 per cent. For this reason, weldable structural steels have carbon equivalent values of less than 0.5 per cent.

Possible weld defects

A good weld is one in which there has been complete fusion of filler with the base metal, full penetration of filler into the joint space, and an absence of porosity, slag inclusions and cracking. With the correct choice of welding method and good welding practice, defects within welds should not occur. However, nothing in this world is perfect and defective welds are sometimes made. Some of the more common types of welding defect are illustrated in figure 6.13 and discussed below.

(a) Incomplete fusion. This can occur when there is insufficient heat input or if the surfaces have not been cleaned adequately and surface contamination prevents good bonding. The heat input and the rate of metal deposition must be matched to the work-piece thickness.

(b) Lack of penetration. This is a defect associated with welds in thick sections and, again, is due to an insufficient heat input.

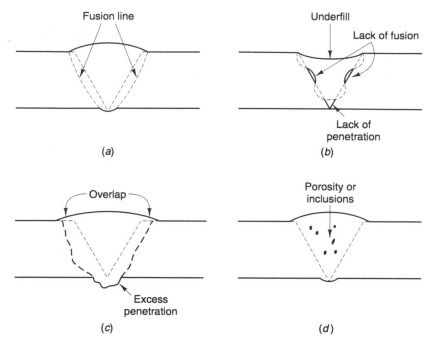

Figure 6.13 *Welding defects: (a) good weld; (b) incomplete fusion, lack of penetration and underfill; (c) excess penetration and overlap; (d) gas porosity and inclusions*

(c) Excess penetration. This defect, sometimes referred to as melt-through, is caused by too high a rate of heat input.

(d) Underfill. This is a result of too low a rate of deposition of weld metal.

(e) Overlap. This is caused by too high a weld metal deposition rate. There may be a lack of fusion between the overlap and the underlying basis material, and this defect may act as the point of initiation of fracture.

(f) Gas porosity. The presence of porosity in a weld will weaken the joint. There are several possible causes of gas porosity. Gases may be generated by chemical reactions during welding, for example, the reaction between iron oxide and carbon forming carbon monoxide: $FeO + C \rightarrow Fe + CO$. Usually, this is only a problem with high carbon steels, but it can occur in low carbon steels if any iron oxide deposits are not removed from the steel surface before welding or when using CO_2 as a shielding gas. This form of porosity can be avoided by using a welding rod or wire with relatively high manganese and silicon contents. These elements are good deoxidising agents and remove FeO from the weld pool. Another possible source of porosity is from the decomposition of contaminants such as oils and paint which have not been removed from the surface before welding. Another gas which may enter the weld pool is hydrogen, derived from the dissociation of water vapour in conditions of high humidity or if the welding flux is damp. Hydrogen may cause serious embrittlement in steels and can be a major source of gas porosity in aluminium alloys. Too high a gas pressure in inert gas shielded welding could lead to the formation of small pockets of argon in the weld pool. A melt of high fluidity

is helpful in allowing gases to rise to the surface and escape to atmosphere, and some welding wires contain small alloying additions to give a more fluid weld pool.

(g) Inclusions. In flux-shielded welding, one of the purposes of the flux is to act as a solvent for metal oxides and form a fusible slag. Care must be taken in welding to avoid the entrapment of slag particles within the weld. As with gases, a weld pool of high fluidity helps slag particles to rise rapidly to the surface of the weld pool.

(h) Distortion and cracking. The heating and cooling involved in welding may cause distortion and cracking. The solidification shrinkage of the weld pool and the shrinkage of the weld and adjacent areas during cooling can cause major internal tensile stresses, and these stresses may build up to a sufficient magnitude for cracking to occur. The problem of internal stress generation is even greater when there are physical restraints which prevent shrinkage from occurring. The use of a more intense and concentrated heat source limits the extent of the heat affected zone and reduces the amount of distortion, thus arc welding generates less distortion than oxy-gas welding. The thickness of the work-piece and the joint design affect rates of heating and cooling, and thus play an important part in the control of internal stresses and the possibility of crack formation. If it is possible, the preheating of the weld zone can help to reduce the heat input necessary for welding, and can reduce the amount of shrinkage and internal stress. In many cases a post-weld heat treatment is desirable, although this is not possible for many large and complex structures. When possible, the welded structure should be given a stress relieving treatment which involves heating at some temperature below the recrystallisation temperature of the material.

Weldability of some materials

Structural steels are readily weldable by a variety of methods. Provided that the carbon equivalent value is less than 0.5 per cent, there should be no danger of martensite formation during cooling. Steels of higher carbon content and low alloy steels are likely to form martensite on cooling and should be preheated to reduce this tendency. A post-weld heat treatment to temper any martensite which may form is also desirable.

Stainless steels are best welded using an inert gas shielded arc process, MIG, with reversed polarity. The ferritic stainless steels tend to have very coarse grain structures in the weld and heat affected zones, and this will make the joint weaker than the base material. Austenitic steels, containing both chromium and nickel, can be welded with ease but there is the possibility that there could be subsequent rapid intercrystalline corrosion within the heat affected zones, an effect known as weld decay. Chromium carbide precipitation can occur in these zones when the temperature is in the range 500–8508C, leading to a serious reduction in the chromium content of the surrounding austenite, thus permitting corrosion to occur. The risk of this problem occurring can be minimised by selecting either an 'L' grade steel, one of very low carbon content, or a stabilised steel. The stabilised steels, which are more expensive, contain small additions of titanium or niobium. These elements have very strong affinities for carbon and, if the conditions are such that carbide precipitation can occur in the heat affected zones, then the precipitate particles will be of TiC or NbC with no loss of chromium from solution in the austenite. The use of an 'L' grade steel is probably sufficient for thin sections, but the more expensive option of a stabilised steel is usual for welds in thick sections.

Cast irons can be welded but the weldability varies greatly with composition. The fairly rapid rate of cooling after welding will promote formation of a brittle white iron structure and the welding filler rod should be rich in silicon or nickel, both strong graphitisers, to counteract this tendency. In the case of spherical graphite cast irons, the filler rod also contains some magnesium to promote the formation of nodular graphite in the solidifying weld pool. In many cases, however, cast irons are braze or bronze welded.

Aluminium alloys are readily weldable with inert gas shielded arc processes, but the precipitation hardened alloys suffer a major loss of strength as a result of over-ageing effects. Consequently, fabrication of structures in these alloys involves bolting or rivetting rather than welding. There is some loss of strength in welded joints in aluminium–magnesium alloys but this is of a much lower order than with the precipitation hardening alloys.

References

BS 4 Part 1: 1993 Specification for hot rolled sections.

BS 11: 1985 Specification for railway rails.

BS 29: 1976 (1987) Specification for carbon steel forgings above 150 mm ruling section.

BS 64: 1992 Specification for normal and high strength bolts and nuts for railway rail fishplates.

BS 131: Part 1: 1961 (1989) The Izod impact test of metals.

BS 476 Fire tests on building materials and structures.

BS 499 Welding terms and symbols.

BS 970 Specification for wrought steels for mechanical and allied engineering purposes.

BS 1139 Metal scaffolding.

BS 1161: 1977 (1991) Specification for aluminium alloy sections for structural purposes.

BS 1452: 1990 Specification for flake graphite cast iron.

BS 1471–1475 Specifications for wrought aluminium and aluminium alloys for general engineering purposes.

BS 1490: 1988 Specification for aluminium and aluminium alloy ingots and castings for general engineering purposes.

BS 2789: 1985 Specification for spheroidal graphite or nodular graphite cast iron.

BS 2870–2875 Specifications for copper and copper alloys.

BS 3100: 1991 Specification for steel castings for general engineering purposes.

BS 4190: 1967 Specification for ISO metric black hexagon bolts, screws and nuts.

BS 4300 Wrought aluminium and aluminium alloys for general engineering purposes (supplementary series).

BS 4395 Specification for high strength friction grip bolts and associated nuts and washers for structural engineering.

BS 4449: 1988 Specification for carbon steel bars for the reinforcement of concrete.

BS 4482: 1985 Specification for cold reduced steel wire for the reinforcement of concrete.

BS 4486: 1980 Specification for hot rolled and hot rolled and processed high tensile alloy steel bars for the prestressing of concrete.

BS 4515: 1995 Specification for welding of steel pipe-lines on land and offshore.

BS 4604 Specification for the use of high strength friction grip bolts in structural steelwork. Metric series.

BS 4670: 1971 (1987) Specification for alloy steel forgings.

BS 4772: 1988 Specification for ductile iron pipes and fittings. [This specification will be replaced in 1996 by BS EN 545, BS EN 598 and BS EN 969.]

BS 5135: 1984 Specification for arc welding of carbon and carbon–manganese steels.

BS 5493: 1977 Code of practice for the protective coating of iron and steel structures against corrosion.

BS 5896: 1980 Specification for high-tensile steel wire and strand for the prestressing of concrete.

BS 5950 Structural use of steelwork in building.

BS 5996: 1993 Specification for acceptance levels for internal imperfections in steel plate, strip and wide flats, based on ultrasonic testing.

BS 6561: 1985 (1991) Specification for zinc alloy sheet and strip for building.

BS 6681: 1986 Specification for malleable cast iron.

BS 6744: 1986 Specification for austenitic stainless steel bars for the reinforcement of concrete.

BS 7009: 1988 Guide to application of real-time radiography to weld inspection.

BS 7123: 1989 Specification for metal-arc welding of steel for concrete reinforcement.

BS 7191: 1989 Specification for weldable structural steels for fixed offshore structures.

BS 7295 Fusion bonded epoxy coated carbon steel bars for the reinforcement of concrete.

BS 7361 Cathodic protection.

BS 7448 Fracture mechanics toughness tests.

BS 7613: 1994 Specification for hot-rolled quenched and tempered weldable structural steels.

BS 7668: 1994 Specification for weldable structural steels. Hot finished structural hollow sections in weather resistant steels.

BS 8010 Code of practice for pipelines.

BS 8118 Structural use of aluminium.

BS EN 485-2: 1995 Aluminium and aluminium alloys – sheet, strip and plate.

BS EN 10002-1: 1990 Tensile testing of metallic materials Method of test at ambient temperatures.

BS EN 10025: 1993 Hot-rolled products of non-alloy structural steels. Technical delivery conditions.

BS EN 10045-1: 1990 Charpy impact test on metallic materials.

BS EN 10088-1: 1995 List of stainless steels.

BS EN 10113: 1993 Hot-rolled products in weldable fine grain structural steels.

BS EN 10115: 1993 Structural steels with improved atmospheric corrosion resistance. Technical delivery conditions.

BS EN 10210-1: 1994 Hot finished structural hollow sections of non-alloy and fine grain structural steels.

BS EN 20898 Mechanical properties of fasteners.

BS EN 22063: 1994 Metallic and other inorganic coatings. Thermal spraying. Zinc, aluminium and their alloys.

Toner, D.E. (1969) A lesson in bronze, *Copper*, Vol. 3, pp. 8–9.

Further reading

D.R. Askeland, *The Science and Engineering of Materials*, 2nd SI edition, Chapman and Hall, London, 1990.

P.S. Bulson, BS 8118 The structural use of aluminium, *The Structural Engineer*, Vol. 70, No. 13, July, 1992.

Corrosion in Civil Engineering, Institution of Civil Engineers, London, 1979.

R.A. Higgins, *Engineering Metallurgy*, 6th edition, Edward Arnold, London, 1993.

J.B. Hull and V.B. John, *Non-Destructive Testing*, Macmillan, Basingstoke, 1988.

V.B. John, *Introduction to Engineering Materials*, 3rd edition, Macmillan, Basingstoke, 1992.

V.B. John, *Testing of Materials*, Macmillan, Basingstoke, 1992.

J.F. Knott, *Fundamentals of Fracture Mechanics*, Butterworth, London, 1973.

J.F. Knott and P. Withey, *Fracture Mechanics – Worked Examples*, Institute of Materials, London, 1993.

A.P. Mann, The structural use of stainless steel, *The Structural Engineer*, Vol. 71, No. 4, February, 1993.

E. Mattson, *Basic Corrosion Technology for Scientists and Engineers*, Ellis Horwood, Chichester, 1989.

E.C. Rollason, *Metallurgy for Engineers*, 4th edition, Edward Arnold, London, 1973.

II TIMBER

Sequoiadendron giganteum

Introduction

Trees cover approximately 30 per cent of the earth's land surface (Hibberd, 1991) and include, among the many tens of thousands of known species, the most massive (the Wellingtonia, over 1000 tonnes), the tallest (the giant Redwood, up to 100 m) and the oldest (the Bristlecone pine, over 5000 years) living organisms in the world (Hart, 1973; Attenborough, 1979). While not reaching such extremes, even in the UK, where the total area under forest is only a third of the world figure, trees are an expected part of the urban and rural landscape.

Timber as a construction material has a long history, probably tying with stone as the oldest structural material still in common use, and easily first in terms of annual world consumption (Gordon, 1988). The versatility of timber is part of its continuing attraction. It is amenable to factory mass production of, for example, doors and windows, or one-off production of technically demanding glued-laminated structural components, yet is equally easily worked using simple hand tools. It is light but strong and tough, it has excellent thermal insulation properties and is more fire resistant than steel or concrete, and it has a warm and attractive appearance.

The variability of timber, which derives from its natural origin, is both a benefit and a drawback. It is beneficial in that no matter what the requirement there will probably be a timber to meet it, but a drawback in that even for a single species the properties vary from tree to tree. For timber to be used successfully it is essential to understand its growth pattern, structure, defects and processing, which are the main topics of this Part of the book.

In the previous edition of this book it was suggested that timber was a material which was inexhaustible in supply. Current environmental concern has prompted a reappraisal of this position as a result of the imbalance between deforestation and afforestation in some parts of the world. However, it has also been suggested that a real shortage of timber will represent a triumph of mismanagement (Gordon, 1988), an outcome which should be avoided by the intelligent use of timber, with an understanding of its natural characteristics and an appreciation of its value.

104

7

Structure of Wood

To utilise any structural material to optimum effect it is necessary to understand its composition and structure, since these can have a major influence on a material's properties. For timber this necessitates a knowledge of the nature and growth patterns of trees, since the composition and structure of wood derive from the requirements of the growing tree rather than the requirements of the timber user. With this knowledge and an appreciation of the variations which occur between and within species, it should then be possible to specify timber correctly for any performance requirement.

7.1 Classification of trees

All commercial timbers can be classified into two broad groups: softwoods and hardwoods. When first used in the middle ages these terms would have been indicative of the relative hardness, density or case of working of the types of timber then in common use, possibly comparing native oak with imported spruce for example (Ridout, 1992). Nowadays, however, the terms hardwood and softwood are quite misleading: balsa is a hardwood but is softer and less dense than any softwood, while pitch pine is a softwood which is harder and more dense than many hardwoods.

The true distinction between the two groups of timber is botanical, as follows:

1. Softwood is produced from the gymnosperms, the coniferous trees such as pines and spruces, which have characteristic needle-like leaves. These trees are generally evergreen, but the group does include some species, for example the larch, which lose all their needles in autumn.
2. Hardwood is produced from one group of the angiosperms, known as dicotyledons, which are the broad leafed trees, such as oak, beech and ash. The temperature zone hardwoods are generally deciduous, while most tropical hardwoods retain their leaves all year round.

The ease with which conifers are grown, their speed of growth and the abundant supply of timber available from the northern temperate zone forests make commercial softwood much cheaper than hardwood. Softwood is also much easier to work with than hardwood, and these factors taken together ensure that the bulk of all timber used in construction is softwood. The higher cost of hardwood timbers is related to various factors, including perceived higher quality, increasing scarcity and greater transport costs. In general, hardwoods are only used in situations where their superior appearance, better natural durability or higher strength can justify their greater cost, as for example in the recent re-laying of the track beds of the Forth Rail Bridge using opepe because of its superior durability (New Civil Engineer, 1995).

7.2 Growth structures

Basically trees can be seen as being composed of three discrete parts: the roots, the trunk and the crown (figure 7.1), each of which has a function to perform in supporting the growth of the tree. The functions are distinct but interdependent and together contribute to the quality of the tree, which in turn influences the quality of the marketable timber derived from it.

The function of the root system is twofold. Firstly, it absorbs moisture and minerals from the soil which are transferred via the trunk to the crown. Secondly, the spread of the roots through the soil acts as an efficient anchorage to enable the growing tree to withstand wind forces.

The trunk provides rigidity and mechanical strength to maintain the crown of leaves at the height necessary for it to compete successfully for light and air, and it is this competition which produces the best commercial timber. In an open parkland setting, the lack of near neighbours results in a tree form with little height growth in the trunk and multiple branching from quite close to the ground (Hart, 1973). The competition for light in a forest or plantation setting forces a greater height growth and ensures that there are relatively few large branches at lower levels, producing more good-quality timber.

The trunk also provides a two-way transport system for moisture travelling up from the roots and sap travelling down from the crown. The third function of the trunk is to store the sap and to convert it to the form needed for growth.

The function of the crown of leaves is to capture as much energy from sunlight as possible, to enable the photosynthetic process to break down the carbon dioxide absorbed through the leaves. This releases oxygen to the atmosphere and causes the carbon to combine with the water and minerals from the roots to form the glucose-based sap which is essential for growth.

7.3 Cross-sectional features of the trunk

Although to the tree all of the sub-systems described above are equally important, to the timber user only the trunk is of concern since it is only from the trunk that marketable solid timber sections can be produced. A cross-section of a trunk is shown in figure 7.2, the main features of such a cross-section being common to both hardwoods and softwoods.

On the extreme outside of the trunk is the outer bark, a rough textured, dense material which acts as a protective coating to the vital food transfer and growth

Day time: *Night time:*

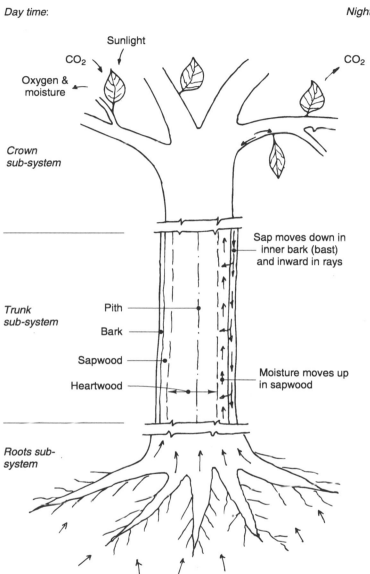

Figure 7.1 *Growth structure of tree*

layers just inside. The bark protects the living tree from extremes of temperature, drought and mechanical damage. The inner bark, or bast, is softer and moister than the outer bark and functions to conduct the sap produced in the leaves to all the areas of active growth or into storage until required for growth.

Trees are members of the group of plants known as exogens which grow by the addition of an outside layer of new tissue. In trees this takes place in the cambi-

Figure 7.2 *Cross-section of tree*

um, which is an extremely thin, delicate layer, varying from one to ten cells thick, depending on the season. Growth takes place in the cambium by cell division, some cells effectively dividing tangentially to maintain the continuous cambium on the increasing circumference of the trunk, others dividing radially to produce bark cells (phloem) to the outside and wood cells (xylem) to the inside. The wood cells are hollow tube-like structures, ranging in diameter from about 0.02 to 0.5 mm and in length from 1 to 6 mm, depending on species, and in general some 90–95 per cent of all the cells grow in a vertical direction.

The nature of the growth pattern of trees is such that a new layer of tissue (cells) is laid down each growing season, forming a conical sheath from the tips of the leading shoots right down to the roots. In trees growing in the temperate regions, which have clearly defined growing and dormant seasons, the growth rings are generally distinct, usually more so in softwoods than hardwoods, and are commonly referred to as annual rings. However, in some hardwoods there may be more than one growth ring per year, depending on climate, while in some trees the rings may be quite indistinct if growth is more or less continuous throughout the year. When the growth rings are clearly visible it is due to differences in the character of the wood grown in the early part of the growing season compared with that later in the season. In softwoods the difference is due to differences in the size of cells, while in hardwoods it is due to variations in the amount of different types of cell produced. In both cases the earlywood is generally softer, weaker and more porous than the latewood. As a result of this difference, the density, and hence the strength, of timber is affected by growth rate or ring-width, each species having a particular growth rate which gives the greatest strength. However, for all species, very wide or very narrow rings are generally an indication of weak timber (Desch and Dinwoodie, 1996).

In all timbers the outer, or youngest, growth layers are called the sapwood. In these layers the moisture absorbed in the roots is conducted up to the crown in some cells while in others the sap from the crown is stored until required by the cambium. Some of the cells in the sapwood remain alive for a long period to provide the conversion function required to turn the sap into the form necessary for use in the cambium. The sapwood is typically quite narrow, only 12.5 to 50 mm wide. Its main characteristic is its high moisture content when newly felled and its often poor natural durability, even once seasoned, because of the generally high starch content in the cells which may attract insects to the timber.

The main central part of the tree is called the heartwood. This section of the trunk is concerned solely with providing the required mechanical rigidity to the tree. The heartwood is no longer involved in the food transfer or storage function, and in many timbers is darker in colour than the sapwood as a result of the deposition of a whole range of complex organic materials in the cells as they reach the end of their time as sapwood. These extractives, as they are called, are responsible for the better natural durability which is often observed in the heartwood of many timbers.

The relatively small percentage of cells which grow in the horizontal direction generally occur as groups or bundles of cells known as rays, or medullary rays if they originate right at the centre of the stem. They are used by the tree for the transfer of food from the inner bark, its storage and subsequent conversion for use in producing cell tissue. In softwoods these rays are generally quite indistinct since there may be only 2 or 3 cells in the bundle. In hardwoods there are usually many more cells in each ray and they are thus much more easily seen, and they often contribute to the attractive surface appearance (or figure) of many hardwood timbers.

Right at the centre of the trunk is the pith (or medulla) which varies in size up to 12.5 mm in diameter. The pith is the relatively soft original tissue of any stem and it is prone to decay.

7.4 Cell types and function

The cells in timber are required to perform a variety of functions: conduction of moisture, mechanical support and food storage and conversion. In softwoods, the first two functions are performed by a single cell type, the tracheid, which varies in size and section during the growing season to favour one or other function. In hardwoods, which in evolutionary terms are a more recent development and are much more complex in terms of cell type, these two functions are performed by two different cell types, fibres and pores. The third function, that of sap storage and conversion, is performed by another type of cell, parenchyma, which occurs in both hardwoods and softwoods.

In softwoods there are essentially only the two cell types present, tracheids in the longitudinal direction and parenchyma in the transverse direction. In the early part of the growing season the tracheid cells have thin walls and large cell cavities which readily facilitate the function of conduction. The earlywood cells are also somewhat larger overall than the latewood tracheids, which, with their small cell cavities and thick walls, are optimised for mechanical support. The cells may be up to 6 mm long by 0.06 mm in diameter and are quite pointed at their ends, overlapping each other rather than butting together. Conduction of water from cell to cell thus occurs not from the ends of each cell but through openings, or pits, in the overlapping and adjacent cell walls. While all cells remain alive for some period after their formation in the cambium, the horizontal storage tissues known as parenchyma remain alive for much longer than the tracheids. This is a necessary feature of the growth pattern of the tree because plant food is usually stored in a form different from that required in the cambium. There is thus a necessity for satisfactory pre-use conversion which can only take place in a living cell.

The cells called pores or vessels which perform the function of conduction in hardwoods are generally shorter and more nearly equal in length to diameter than the long thin tracheids of softwoods. These cells differ also in being arranged end-to-end to form continuous conduits for moisture movement, which may extend for the full height of the tree or be limited to only a few hundred millimetres in length depending on species. The cells called fibres which provide mechanical support in hardwoods are relatively short, only 0.25–1.5 mm long, and thin, with a diameter about one-hundredth of their length. These cells have relatively thick walls and small cavities and play virtually no part in the movement of moisture to the crown. Hardwoods also have axial tracheids, considered to be an evolutionary intermediary between the softwood tracheids and the hardwood fibres, but in very small amounts. Hardwoods contain radial parenchyma or ray cells which perform a similar function to the same cells in softwoods, but also contain axial parenchyma in varying amounts depending on species. As with softwoods, some hardwoods also show a differentiation in the distribution of the specialist conduction and support tissue at different periods in the growth season. A concentration of relatively large vessels in the early stages of the growth season gives rise to the descriptive term ring-porous hardwood. Conversely, in many other hardwoods the vessels occur uniformly across the section, giving a diffuse-porous timber.

7.5 Cell structure and chemistry

The cell walls (see figure 7.3) comprise a primary wall and a secondary wall surrounding the central cavity or lumen. The primary walls of adjacent cells are separated, and bonded together, by the middle lamella which forms an integral part of the cell structure. The primary cell wall is formed in the cambium as the cells divide, and is initially quite flexible in size and shape to conform to the growing cell. Once the cell has been differentiated as a wood cell (of whatever type), the secondary wall is formed in three layers, as shown in figure 7.3. Both primary and secondary walls are composed of microfibrils, which are thin threads of cellulose, formed by the polymerisation of glucose, covered in a layer of hemicellulose and bound together by lignin. Formation of the secondary wall layers generally occurs quite rapidly in the original growth season. The lignin 'glue' develops more slowly, however, in the middle lamella and within the cell wall around each cellulose 'fibre'.

The chemical composition of the dry cell wall material varies widely with species, ranging from 45 to 60 per cent cellulose, 10 to 25 per cent hemicellulose, and 20 to 35 per cent lignin. In addition, the chemical analysis of wood shows up a range of materials collectively referred to as 'extractives' which exist within the cell cavities, particularly in the heartwood. Since many of the substances present as extractives are toxic, their presence helps to explain the better durability of the heartwood of most timbers.

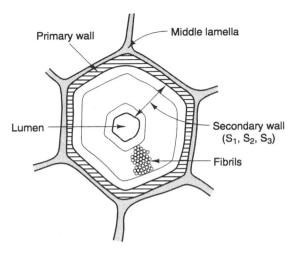

Figure 7.3 *Cell structure*

7.6 Structure, properties and use

The cellular structure of timber explains many aspects of the use of timber. For example, timber holds glue better than other non-porous materials because the glue is able to penetrate the cell cavities to form a 'key'; similarly, the preservation of wood can be achieved by forcing a liquid preservative into the open cellular structure. Widely varying surface finishes occur because timbers with thick cell walls and small cell cavities are hard and difficult to work, whereas those with thin cell walls and large cell cavities are easy to work.

The most important effect of the structure of timber, however, is apparent in the anisotropy of timber, which means that the properties of timber vary significantly with direction. For example, the strength of timber across the grain (i.e. at right angles to the long axis of the cells) may be less than one-tenth of the strength along the grain. This is due not only to the predominance of cells in the axial direction and to their elongated shape in that direction, but also to the fact that in the thickest layer in the cell wall, the S_2 layer, the basic cellulose fibres (the microfibrils) are aligned almost in the direction of the long axis of the cell. The S_2 layer microfibrillar angle is, in effect, one of the main determinants of the strength of wood substance.

8

Production of Solid Timber Sections

The transformation from living forest tree to finished construction section involves four basic processing stages: felling, conversion, seasoning and grading. The first of these, felling, is part of the whole forestry process (see Hibberd, 1991) and is essentially beyond the scope of this book. Whether a log comes from selected felling by chain saw in an old growth forest or from the clear felling of a plantation crop by mechanical harvester, as in figure 8.1, does not affect the quality of timber. The quality may vary but this is to do with variation in growing conditions and age at felling rather than the method of felling. The one aspect of forestry which is relevant to all construction users of timber, however, is the importance of ensuring that their timber is sourced from well-managed, sustainable supplies.

Figure 8.1 *Mechanical harvester felling, de-limbing and cutting to length in one operation (Photograph courtesy of Fountain Forestry Ltd)*

8.1 Conversion of timber

Conversion is the term used to describe the processes in which the felled trunk is sawn into marketable sizes of timber. While the move to sustainable forestry has led to the development of portable sawmills which are used in Papua New Guinea, for example, to convert single hardwood trees in the forest (Times Higher Education Supplement, 1994) the vast majority of construction timber is converted in fixed sawmills normally located close to the source of timber and processing up to 10 000 m^3 per annum in the larger mills. The log is first sawn into manageable lengths, either at stump in the forest or at the mill, and is then de-barked, giving a cleaner log for sawing and increasing the value of the off-cuts which can be used in particle- or fibre-board manufacture. In large mills de-barking is done mechanically but in smaller mills it may still be done by hand, as illustrated in figure 8.2. The logs are then sawn lengthwise into the required sections, see figures 8.3 and 8.4, generally using variations on the two basic methods of plain sawing and quarter sawing, see figure 8.5.

Figure 8.2 *De-barking by hand prior to conversion*

Figure 8.3 *Combination band saw and circular saw for ripping to size and cross-cutting to length*

Figure 8.4 *Eight-blade frame saw*

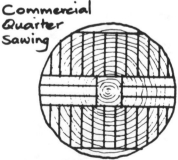

Figure 8.5 *Sawing methods*

The log in figure 8.3 has been roughly squared by four passes through the large band saw and is now being sawn to size. The operation is computer controlled and the set-up includes a small circular saw (at the right of the figure) which cross-cuts the sections to length as they pass through the band saw. Figure 8.4 shows a frame saw being used to cut small-diameter logs into planks in a single pass, the logs having already been through another two blade frame saw to remove the two half-round flitches and give a log with two square edges.

The method of sawing generally has little effect on the mechanical properties of the timber. The most economical method for the mill is plain sawing since the whole log is processed in one pass through the saw, while any variation on quarter sawing requires the log to be taken back and repositioned on the log carriage for each cut through the saw. In plain sawing, however, the sections become progressively more tangential to the growth rings as they get further from the centre cut, and this can make these sections more prone to distortion in seasoning (see section 8.2). In quarter sawn timber, where the growth rings cut the section at more than 45°, the timber is generally more stable in seasoning and may give better wear resistance in flooring applications, for example. The major difference from the method of sawing, however, is seen in the appearance of the cut surface, see figure 8.6. Many hardwood timbers are cut in particular ways to give the most attractive figure, a combination of growth ring and ray structure, for aesthetic reasons. If this is required, it is generally necessary to specify it well in advance of requirements since special saw runs may be necessary and this can result in much higher costs. A more complete description of the process of conversion is given by Walker (1993).

Figure 8.6 *Appearance of cut surface: (left to right) flat sawn, commercial quarter sawn and radial sawn*

8.2 Seasoning

When just felled, timber may have a moisture content in excess of 100 per cent. Before it can be used, this moisture content must be reduced to a level at which the timber will be generally in equilibrium with its end-use environment if problems are to be avoided. Typical moisture contents for particular environments are shown in table 8.1.

Seasoning is the term used to describe the process of drying timber in a controlled manner to reduce its moisture content without introducing unwanted defects, such as splits or distortion. Moisture is present in unseasoned timber both within the cell cavities and the cell walls. When the timber is seasoned, the water

TABLE 8.1
Approximate moisture contents of timber for particular end uses

End use	Average moisture content
External joinery	18% plus
Internal joinery with intermittent heating	15%
Internal joinery in continuously heated areas	12%
Internal joinery near heat source	8%
Flooring with underfloor heating	6%

within the cell cavities may be removed without any effect on the timber other than a reduction in its effective bulk density. This is the case down to a moisture content of approximately 27 per cent, known as the fibre saturation point, at which all free water has been removed from the cell cavities. Drying below this point, which is necessary for virtually all constructional timber, requires removal of water from the cell walls, which results in shrinkage. The art in seasoning is to remove this water with as little detrimental effect on the timber section as possible. The difficulty in seasoning is that, as a result of its inherent anisotropy, timber shrinks by different amounts in the tangential, radial and longitudinal directions, so that depending on the conversion method, pieces will be more or less prone to distortion, see figure 8.7. The quarter sawn board (A) will shrink a little in thickness but is relatively stable in width and will thus maintain its rectangular shape, while the plain saw board (B) shrinks in width by a greater amount on the outer face compared with the inner (i.e. closer to the pith) face and will tend to curl up or cup. Good seasoning practice requires adequate support, restraint and control of drying to minimise degrade of timber from splitting or warping (Walker, 1993).

Figure 8.7 *Effect of conversion method on tendency to distortion on seasoning*

There are two methods of seasoning timber: natural air seasoning and kiln drying. In natural air seasoning, the timber is stacked in the open air or in open-sided sheds in such a way as to promote drying without artificial assistance. The timber stack is supported clear of the ground to prevent rain splash, and adjacent pieces in each layer are kept separate to provide air circulation by means of spacers or sticks which are generally about 25 mm square. If the timber is stacked in the open a top cover should be used to keep off rain, or snow, and protect the stack from direct sunshine. The end pieces may also be treated in some way to reduce the tendency to splitting caused by the more rapid rate of drying through the end grain. Although hardwoods and those softwoods which dry slowly are best stacked in winter so that the timber will not be affected by summer heat until its moisture content has been reduced to some extent, this may not be possible in commercial practice. The advantage of air seasoning is that it is a cheap method with very little loss in quality of timber if done properly. The disadvantages are that it is a relatively slow process so that both timber and space may be immobilised for long periods for the slow drying hardwoods, while very little control is possible, and then only by varying the size of spacers or by the use of mobile open-slatted sides to the sheds. A typical air seasoning set-up is shown in figure 8.8, which illustrates both good practice, in the regular alignment of inter-layer sticks, and bad practice in the poor base support to the stacks which in the extreme right hand stack has resulted in serious bowing in the bottom parcel of timber which will be set into these pieces.

Figure 8.8 *Air-seasoning of quick drying softwood*

Air seasoning in the UK cannot generally reduce the moisture content below about 18–20 per cent. To reduce the moisture content below this value requires kiln drying, which employs a heated, ventilated and humidified chamber. In kiln drying, temperatures up to 80°C are used, which cause the moisture in the wood to move more rapidly to the surface, from where it is removed by the circulating air. If this air were merely heated, excessive evaporation of moisture from the surface would take place – faster than moisture could move out to the surface. The outer parts of the timber, and especially the ends of sections, would then tend to shrink peripherally so that splitting might occur. In kiln drying it is therefore necessary to humidify the circulating air in order to control the rate of evaporation.

As drying proceeds, the temperature in the chamber is raised as the humidity is reduced to promote drying. Air circulation should be uniform over the face of the timber stack at which it enters, and the velocity of the air through the stack should be sufficiently high to be consistent with economic operation. In order to assist in controlling the humidity of the circulating air, it is common practice to exhaust some of the air from the system from time to time and to replace it with fresh air from outside, any deficiency in water vapour being supplied by means of carefully regulated spray.

Kiln drying is a much more rapid method of drying timber, air seasoning periods of weeks or months being reduced to days or weeks, and is now stipulated for any timber which is to be graded (as described in section 8.3) for use as structural timber. Once dried to a low moisture content, it is necessary to ensure that the timber is transported and stored on site in a manner which maintains its moisture content. Although it is true to say that it is difficult to completely re-wet timber that has been kiln dried, it is not impossible, and site storage should be arranged to ensure that dried timber is kept dry.

8.3 Stress grading

In order to design any structure it is necessary to know the strength properties of the material being used. With steel or concrete it is possible to assume a particular characteristic strength in design, knowing that material meeting this specification can then be manufactured by the steel mill or concrete batching plant. Solid timber, however, is a natural product whose 'manufacturing' stage is the process of tree growth and which is thus effectively controlled by climate, soil conditions and forestry management. For timber, therefore, it is necessary to assess the strength properties of the sections available rather than to specify a particular strength, the assessment being referred to as grading. In the UK, grading is carried out in accordance with the rules of BS 4978 for softwoods and BS 5756 for tropical hardwoods, and may be done visually or by machine (for softwoods only). Good descriptions of the development of stress grading are given in Mettem (1986) and Mettem and Richens (1991), and of the mechanics of the process for softwoods in TRADA publication TBL35 (1991b).

In visual stress grading, all four faces over the whole length of a section are inspected to assess the various strength-reducing features which may be present, see section 8.4. Each section of timber must comply with the limits set out for the appropriate grade, full details of which may be obtained from the relevant standard. For softwoods there are two grades, General Structural (GS) and Special Structural (SS), for normal structural use, together with another three grades, LA, LB and LC, for timber to be used in glued-laminated timber manufacture (see Chapter 11, section 11.2). For tropical hardwoods there is only one grade: Hardwood Structural (HS). Visual stress grading is a relatively slow process, even with an experienced grader, and it has the major disadvantage that it does not take account of density, since this cannot be measured accurately during visual grading.

Machine stress grading makes use of the fact that for structural sizes of timber there is an acceptably accurate correlation between the stiffness (modulus of elasticity) and the bending strength (modulus of rupture) of a section when tested over a short span, the relationship being unique for each species of timber. Once tests have established this relationship for a given timber, the grading machine (see figure 8.9) can be programmed to identify any of the four grades covered by BS 4978, i.e. MGS, MSS, M50 and M75. The grading machine is continuously fed with timber to which it applies a small transverse load on a span of about 1 metre. The tim-

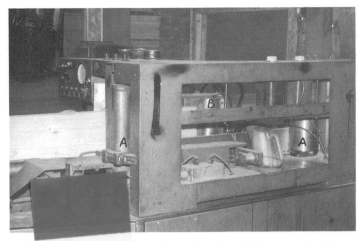

Figure 8.9 *Machine stress grading of timber: (A) reaction rollers; (B) loading roller (Photograph courtesy of James Donaldson & Sons Ltd)*

ber is moved through in increments of, typically, 150 mm and the measurements of deflection for each increment are stored until the whole piece has passed through. The maximum deflection recorded for the piece of timber, that is, the deflection associated with the minimum modulus of elasticity and thus the minimum strength, is compared with the maximum deflection associated with the stress grade being sought, and if it passes, the piece is stamped with the relevant grade mark.

Machine grading has proved to be more reliable than visual grading, and generally will allocate a larger percentage of timber in a given batch to the higher grades than will be the case for visual grading, which tends to be more conservative. An example of how this may occur would be where a higher density in a piece, which cannot be assessed visually, compensates for other defects. Machine grading is also more rapid than visual grading, with throughputs of up to 90 metres per minute being possible in some machines (Baird, 1990). Before or after grading by machine, however, the timber must also be visually assessed for certain defects, the limits again being set for each grade in BS 4978.

8.4 Defects in timber

Defects can develop in timber at various stages, some occurring naturally during growth while others result from the operations involved in the conversion and seasoning processes. These defects can affect the quality of timber produced, either by reducing its strength or marring its appearance. This section is mainly concerned with those defects which affect the strength properties of structural timber.

Knots

These are perhaps the most obvious natural defect. A knot is the part of a branch which becomes enclosed within the expanding trunk as the tree grows, and a variety of terms are used to describe the condition of the knot and its position and shape in the converted timber cross-section. With a *live* knot there is complete continuity between the fibres of the branch and the trunk. *Intergrown* describes a knot in which at least 75 per cent of the cross-section of the knot perimeter fibres are continuous with those of the trunk, while a *dead* knot is one in which continuity is less than 25 per cent. With an *encased* knot the entire cross-section of the knot is surrounded by bark or resin. Depending on its position as visible in the converted timber, a knot may also be described as an *edge, margin, splay, spike, round or arris* knot. The knot shown in figure 8.10 illustrates both intergrowth on the lower portion of the perimeter and encasement on the upper side, while on the radial face at left it appears as a splay knot whereas on the tangential face at right it is clearly a round knot.

Figure 8.10 *Development of knots in timber*

In terms of their effect on the appearance of timber, knots may be considered detrimental or ornamental, 'knotty pine', which is popularly used for panelling and furniture being an example of the latter. The effect of knots on strength arises not from any particular weakness of the wood in the knot but rather from the effect of the knot in distorting the direction and continuity of the grain of the section in which it occurs. In grading hardwoods, knots are simply assessed in terms of their dimension measured on the surface of a piece in relation to the dimension of that surface. In grading softwood, where a ring of branches and thus knots commonly occur at intervals up the trunk, knots are assessed by the Knot Area Ratio (KAR) which is a measure of the amount of knot tissue as a proportion of the gross area of the section of timber. The method of assessment involves projecting the surface occurrence of knots at a section as cones with their apices at the pith, and is described clearly in BS 4978. The technique may be visualised from figure 8.11, which shows a cross-section through a monkey puzzle tree (*Araucaria araucana*) at a branch whorl. If the rectangular section shown as a dashed line on the figure had been sawn from this tree, the knot area ratio at this section would be the sum of the shaded areas as a proportion of the whole rectangle. Since the effect of knots relates not only to their size but also to their position on a section, particularly on a section loaded in flexure, the softwood grading rules also define a margin condition, the margins being the outer quarters of the depth of a section. An indication of the effects of knots can be obtained from the experimental data given in table 8.2, which was obtained on softwood specimens tested in flexure.

Figure 8.11 *Illustration of Knot Area Ratio calculation*

TABLE 8.2
Effect of knot on flexural behaviour

Specimen	Strength $(N\,mm^{-2})$	Elastic modulus $(kN\,mm^{-2})$
Clear	48.5	8.0
Margin knot, at centre span, tension face	18.5	4.5

Sloping grain

The grain direction in a timber section is the direction of the main wood cells, tracheids or fibres and vessels, in relation to the long axis of the section. In the living tree these cells do not always grow perfectly vertical, straight and parallel to the long axis of the trunk. The natural taper from bottom to top of the trunk may be enough to cause sloping grain in a converted piece of timber which has been sawn parallel to the pith, but more generally sloping grain derives from the tendency in some trees for the cells to grow in a spiral pattern around the trunk (see figure 8.12), the direction of which may change from time to time. In assessing grade, it is the general direction of the grain which must be determined and local variations in grain direction caused by knots are ignored. The grain in timber is not easily visible (it is not the same as the growth ring pattern) and is most readily found with the use of a special scribing tool, see TBL 35 (TRADA, 1991b) for details. The effect of sloping grain is to reduce the strength of a timber section as the slope, i.e. the angle between grain direction and the long axis of the section, increases. Limits vary from 1 in 6 for the GS grade to 1 in 18 for LA, the best laminating grade.

Figure 8.12 *Trees do not always grow straight and parallel to the pith*

Rate of growth or ring width

Assessment of the rate of growth of a tree is not an assessment of its density, except insofar as the density of a particular species is affected by the relative proportion of earlywood and latewood which is influenced by growth rate. Generally in very fast grown timber there is a greater proportion of earlywood which is less dense and so gives reduced strength. The visual grading of softwood therefore includes upper limits on average annual ring width of not more than 10 mm for GS grade and not more than 6 mm for SS. In machine grading this is automatically incorporated in the assessment of grade.

Cracks, fissures, resin pockets and bark pockets

These may occur during growth or may result from shrinkage stresses in seasoning, and are generally given specific names to identify them more readily. Checks, shakes and splits are terms which refer to particular fibre disruptions which affect the strength of the timber by reducing the cross-sectional area of complete fibrous section. Heart-shakes, which occur in the centre of the trunk, may even indicate the presence of decay, or the beginnings of decay. Resin pockets are fissures containing resin which constitute a strength defect and may also interfere with decorative treatments in the finished use. Resin pockets and fissures are treated effectively equally in grading softwood but have different rules for hardwood grading. Bark pockets arise when the cambium at a particular point is damaged so that growth at that point stops. The adjacent timber continues to grow beyond the damage and eventually the cambium re-establishes continuity, thus enclosing the bark at the originally damaged area, as can be seen in Chapter 7, figure 7.2. For softwoods, bark pockets and resin pockets are treated the same in grading, while for hardwoods there are different requirements.

Wane

This occurs during conversion and is the result of undue economy in attempting to use every possible piece of timber converted from the trunk. Wane is the term which describes the appearance of the rounded periphery of the trunk on one or more faces of a sawn section, wane being apparent on both the top and bottom edge of the visible face on the section being sawn in figure 8.3. The effect of wane is quite simply that, since the cross-sectional area of the section is reduced, there is a consequent reduction in strength in the parts affected. Wane is measured as a proportion of the overall width or thickness of a section, and allowable proportion varies from one-third (GS) to one-quarter (SS and HS) to one-eighth (M75). Wane is not acceptable in any of the laminating grades.

Distortions

These are directly related to the movement which occurs in timber as a result of changes in moisture content. Excessive or uneven drying, poor stacking and inadequate spacing and support during seasoning can all produce defects or distortions in timber. As described previously, cupping occurs when a sawn board curves in section away from the heart of the tree, owing to differences in the amount of shrinkage on each face. Twisting occurs in a section owing to differential drying out of distorted grain, typically spiral grain. Springing and bowing are due to errors in placing of the spacing sticks between timbers in a drying stack, resulting in uneven support or loading while the seasoning timber is still wet and more readily deformed even under its own weight. Once dried, the deformation is set into the timber. All such defects have an effect on structural strength as well as on fixing stability. Limiting dimensions for each of these distortions are generally the same for all grades of softwood and hardwood, see BS 4978 and BS 5756.

Insect damage

This may occur in the standing tree, the newly felled log or the timber yard, as a result of the activity of a variety of insects, including pinhole borers, bark borer and powderpost beetles and wood wasps. Damage is caused by tunnelling, which with some species of insect is done by the larvae while with others it is the adult beetle. Most of these are killed as the timber is dried, or they are unable to re-infest dry timber, although the powderpest beetle may remain active unless the timber is kiln sterilised. Active infestation is not permitted (for obvious reasons) in any graded timber, the large holes (6–7 mm) associated with wood wasps in softwoods are also not permitted, but the smaller holes of insects such as the pinhole borer are allowed provided they are not present in such profusion as to reduce the strength of the timber. Aside from the strength considerations, the presence of worm holes, and the staining of surrounding timber which occurs with the pinhole borer, for example, may seriously reduce the value of timber in terms of appearance.

Fungal decay

As with insect damage, there are a number of types of fungi which attack standing trees, logs and even recently converted timber. Although most such rots do not develop further once the timber is converted and seasoned, in general it is good practice to reject such timber. The exception to this is the presence of sapstain or bluestain which are non-rotting fungi, caused by a mould fungus, which some-times develop on the surface of newly converted timber, European redwood, obeche and ramin being examples of susceptible species. Provided seasoning is started as soon as the timber is converted, and the surface moisture content is rapidly reduced, sapstain will not occur. If it does appear it will be quickly inhibited once the timber dries below about 20 per cent moisture content, and generally it has no effect on strength properties, although as with insect damage the appearance of the timber may be spoiled.

Reaction wood

This is an occasional feature of growing trees where a special type of cell tissue develops to resist relatively constant stresses imposed by wind or other forces which tend to bend the trunk (or more commonly the branches). In hardwoods, reaction wood occurs on the side of the trunk which is in tension and is referred to as tension wood, while in softwoods it forms on the side of the trunk which is in compression, and is then known as compression wood (Desch and Dinwoodie, 1996). Tension wood is generally less dense than normal wood and, although it is slightly stronger in tension and a little tougher (i.e. more impact resistant), it is very weak in compression. Compression wood is denser than normal wood, owing to a higher than normal lignin content in the cell walls, but is little stronger in compression and is very weak in bending. Reaction wood is generally more difficult to work than normal wood, but its main problem is in fact the abnormally high shrinkage which it suffers on drying, particularly in the normally stable longitudinal direction. In sections which contain both normal and reaction wood, therefore, the probability of distortion on drying or warping in service is much increased, so that timber containing reaction wood is normally rejected in grading.

Sapwood

As was explained in the previous chapter, sapwood is a natural and necessary feature of the growing tree. It is not in itself a defect, but may cause problems, for example, in seasoning since it tends to exhibit higher shrinkage than heartwood. There are also general concerns about durability, since the sapwood of most species is classified as non-durable, which may lead to sapwood being excluded from structural sections although there are no differences in strength. Since much of the softwood used in the construction industry is obtained from immature trees and therefore contains a high proportion of sapwood, exclusion of the sapwood from structural sections would be uneconomic. If timber is at risk in service, see Chapter 9, effective preservative treatment can eliminate any differences in durability between sapwood and heartwood.

8.5 Grade stresses and strength classes

Once the defects in a timber section have been assessed and the section has been visually or mechanically assigned to a grade, there is a further step necessary before design stress values can be obtained. The various grades do not have unique stress values associated with them. General Structural or M75, for example, are categories which have different design stress values depending on which particular timber species is being graded. To obtain permissiible design stresses for structural calculations, it is therefore necessary to identify species as well as grade.

The design stresses for each species in the original UK grading system were based on an assessment of the results of large numbers of comparatively small clear specimens, tested using the methods of BS 373. Owing to the significant variation in strength which can arise from apparently similar specimens of a given species, statistical analysis of these results was used to estimate the strength below which only 1 per cent of results would fall. A reduction factor was then applied to this 1 per cent strength value, made up of factors to reflect the effects of, for example, size of specimen, rate and duration of loading, together with a general factor of safety, to produce a basic stress for clear timber. Further reductions were applied to the basic stress to produce a system of grade stresses which varied from 40 per cent to 75 per cent of the basic stress, depending on the severity of the defects present.

The current grade stresses given in BS 5268 are calculated somewhat differently since they are based on data from tests on structural sizes of timber sections, i.e. sections containing defects, and a lower fifth percentile cut-off is used in the statistical analysis of the data. Basic stresses are thus not relevant, and are not quoted in BS 5268, but the stress values for the various grades of GS, SS, MSS, etc. are of the same order as those derived for a particular species in the original method.

The major difficulty for the non-specialist structural timber designer is that, as stated above, the stress values are not constant with grade but vary with species. The range for allowable stress in bending, for example, for grade GS is from 4.1 N mm^{-2} for home grown European spruce to 7.4 N/mm^{-2} for Caribbean pitch pine. If there were only four grades this might not be an insurmountable problem, but BS 5268 also accepts timber graded to appropriate standards by national bodies in the USA (National Grading Rules for Softwood Dimension Lumber, NGRDL) and Canada (National Lumber Grading Association, NLGA, and Machine Stress Rated Lumber, MSR), which can have up to twelve grades for each species.

To simplify the system, BS 5268 introduced a system of nine strength classes, SC1 to SC9, which have single stress values for each strength property, i.e. bending, tension, compression parallel and perpendicular to grain and shear, together with a modulus of elasticity value (the reader will recall the anisotropy of timber, which necessitates the many strength values). As an example, the grade stress values for dry timber for bending parallel to the grain for the nine strength classes are shown in table 8.3.

TABLE 8.3
Permissible stresses in bending for strength classes of
BS 5268

Strength class	Bending strength (N mm^{-2})
SC1	2.8
SC2	4.1
SC3	5.3
SC4	7.5
SC5	10.0
SC6	12.5
SC7	15.0
SC8	17.5
SC9	20.5

In general, strength classes SC1 to SC5 cover the common construction softwoods while SC6 to SC9 comprise the denser hardwoods. The various species/grade combinations fit into the strength class system as described in Tables 3 to 8 of BS 5268: Part 2. For example, SC3 timber, which is normally the lowest strength class to be used structurally, may be GS grade imported whitewood or SS grade Canadian sitka spruce. For simplicity, machine graded timber may be graded directly to strength classes rather than grades. The advantage of this system lies in the simplicity of a single set of nine classes of timber rather than the multiplicity of grades for the numerous species in commercial use. There is a drawback, however, in that all species of timber assigned to a particular strength class may not have similar properties in respect of appearance, nailing or gluing characteristics, durability or acceptance of preservative treatments. While it may thus be possible to design members using the general strength class stress values, it may be necessary to specify particular species from that strength class, depending on any particular requirements.

Although BS 5268 is the current code of practice for the design of timber structures in the UK, it is soon to be replaced by EuroCode 5, the proposed pan-European Standard. This code represents a fundamental shift in approach to the design of structural timber in that it is a limit state design method in contrast to the permissible stress design embodied in BS 5268. For this reason, in the EuroCode the design stress values for different categories of timber (from C14 to C40 for softwoods and C30 to C70 for hardwoods) are characteristic values based on the strength below which not more than 5 per cent of test values should fall. The values for bending strength in the EuroCode thus range from 14 N mm^{-2} to 70 N mm^{-2}. The differences in strength class values between BS 5268 and EuroCode 5 should not be confused as representing a change in the quality of timber – they simply reflect a different design philosophy between the two codes. The final, grading, stage of producing solid timber structural sections may in future be carried out directly to EuroCode 5 strength classes, but when used to design a structure the same section size should result if, say, whitewood is used, irrespective of whether it is described as SS to BS 4978, SC4 to BS 5268 or C24 to EuroCode 5.

9

Durability of Timber

In common with all other biological materials, timber is bio-degradable and the concept of durability of timber might thus be seen as a contradiction in terms. However, it must be recognised that, as with any structural material, deterioration requires certain ambient conditions, and that even in such conditions different grades of material, which for timber means different species, will deteriorate at different rates. Some timbers are naturally durable and may be long lasting in even quite adverse conditions. Others are very susceptible to deterioration but may still be successfully used provided they are treated in some way to improve their resistance to fungal decay and insect attack, these being the two most serious problems for timber in service. As with all materials, if a timber structure is to achieve its design life, a proper assessment of the service environment must be made to select the most appropriate species or treatment. The design, and more importantly the detailing, must be carefully considered and executed to avoid building in problems such as moisture traps, and a programme of planned maintenance must be continued throughout the life of the structure.

9.1 Natural durability

The reader will recall from Chapter 7 that the chemical composition of timber includes not only the basic cell-building blocks of cellulose and lignin, but also varying amounts of extractives. These are substances such as oils, gums, resins, tannin and other phenolic compounds, which are deposited in the heartwood of the growing tree, and contribute to the colour, odour and, because of their toxicity, durability of the timber. Since the content and composition of extractives vary with species, so does the natural durability. In the UK, the durability of timbers is assessed from heartwood stakes of 50 mm square section in ground contact. As with all timber properties, there is great variability even within species and so the data from soil burial tests is used to classify commonly used timbers in five broad categories ranging from perishable (stakes survive less than 5 years) to very durable (survival for more than 25 years), see BS 5268: Part 5. The very durable timbers are all hardwoods and include greenheart, iroko, jarrah, opepe and teak. The best softwoods are western red cedar and pitch pine, both of which achieve the classification of durable (survive in ground contact for 20–25 years). The majority of the softwoods which make up the bulk of all timbers used in construction are classified as non-durable (survival averaging 5–10 years). Since these classifications relate to heartwood only, sapwood being classified as perishable or non-durable at best, the growing trend for the use of short rotation, fast-grown plantation softwood, which has a greater proportion of sapwood, means that most construction timber must be considered to be potentially susceptible to insect attack and fungal decay.

9.2 Insect damage

Timber in service in buildings may be attacked by a variety of insects, the two most important being termites and beetles. On a world scale, termites are the most damaging of all insects (TRADA, 1987), and although they are generally restricted to tropical and sub-tropical regions, they are also found in Southern Europe and limited areas of France and Germany. In the UK the main cause of insect damage is beetles. A number of species infest timber (Bravery *et al.,* 1992) but only a limited number are identified as primary pests normally requiring insecticidal treatment if discovered.

Beetles infest timber because the organic nature of the material is favourable to their life cycle: from egg to larva to pupa to adult beetle. The adult beetles lay eggs in surface cracks, crevices, old flight holes or, in one case, hardwood pores. The hatching larvae or worms tunnel through the timber for the whole of their growth period, which may extend up to ten years or more depending on insect species and conditions of temperature and moisture content in the timber. Eventually the larvae pupate and then hatch into the adult beetle, which emerges through a flight hole to fly off and perpetuate the cycle. Some beetle species are cell content feeders, living on the residual starch content of cells, while others can consume the cellulose of the cell walls. In either case the net effect of this tunnelling is to reduce the cross-sectional area of the timber and so reduce its strength. The damage can be quite extensive, as illustrated in figure 9.1, which shows worm holes in a section of 75 mm wide imported hardwood strip flooring.

Figure 9.1 *Worm damage in imported maple flooring*

The common furniture beetle (Anobium punctatum) is found throughout the UK and infests the sapwood of both softwoods and hardwoods, even in dry timber (moisture content 12–14 per cent). If the timber has been subject to some fungal decay, the furniture beetle may also infest heartwood. Death watch beetles (Xestobium rufovillosum) generally infest only partially decayed hardwood, principally oak, and are found mainly in England, seldom in Scotland. Dampness is a prerequisite for infestation but, once established, attack may continue in drier timber. Attack may extend into any softwoods adjacent to infested hardwood but only if the wood is damp or decayed. House longhorn beetles (Hylotrupes bajulus) can cause serious damage to the sapwood of softwood timbers in structures. The larvae may grow up to 30 mm in length, and since they may remain in the larval stage for up to 11 years (Richardson, 1993) they can cause quite severe damage before the adult finally emerges through an oval flight hole about 10 mm across. The house longhorn is common throughout Europe but in the UK is confined to areas of south-east England, where specific treatments are required for timbers to comply with the Building Regulations. The Lyctus powder post beetles (Lyctus brunneus or lyctus linearis) can only infest hardwoods since they lay their eggs in the pores of the timber. Only those hardwoods with a sufficiently large pore size are susceptible and, obviously, softwoods are immune. These beetles are starch feeders and so only the sapwood of suitable hardwoods is infested. While initial infestation may occur in timber yards, chiefly in England, degeneration of the starch in cells over a period of years makes re-infestation of timber in service by emerging adult beetles very unlikely.

In assessing insect damage to timber there are two requirements: the first is to decide whether attack is current; the second is to determine the extent of damage, possibly requiring removal of the surface layers of a timber member to expose the extent of tunnelling, which may not be accurately indicated by individual flight holes. If necessary, treatment may require cutting out damaged timber and splicing in new, treated sections. If infestation is still current, all timber should be treated with preservative or insecticide, see section 9.5. In many cases, insect attack may have long since ceased, or can be eradicated by removing any source of moisture and allowing the timber to dry out. In these circumstances, and provided the existing reduction in member section is not too severe, no additional treatment may be required.

9.3 Marine borers

There are a number of marine organisms which can seriously degrade timber used in salt water. Attack varies with water temperature, being greatest in tropical regions and least in temperate climates, although even in the UK marine timbers may be at risk, particularly those in coastal areas swept by the gulf stream. These organisms are also affected by water salinity, too high or too low a salt content preventing optimum development of the organisms. The two most common organisms are the shipworm (teredo navalis), a mollusc, and the gribble (limnoria lignorum), a crustacean, which is generally the more serious problem.

The shipworm bores into timber through a very small entry hole which is almost impossible to detect. The end of the shipworm remains at the surface to enable it to maintain water contact for breathing while it tunnels into the timber, growing in length and diameter, possibly reaching up to 300 mm in length and 15–20 mm in diameter. Its tunnels are characteristically lined with a calcareous deposit and it may tunnel extensively within affected timber, possibly completely destroying the strength of the timber with little external evidence of attack. The gribble creates small tunnels in the near surface layers of exposed timber, weakening the timber which may then be abraded by water action, exposing fresh surfaces for attack. Attack is often most noticeable at the water line, as wave action and continued tunnelling produce a waisting effect on timber members. Although a few timbers have some natural durability against marine borers, most notably greenheart, all timbers are liable to attack by marine borers and any timber which is to be used in marine works where infestation is possible should be treated with a suitable (i.e. environmentally acceptable) preservative using one of the pressure methods of application.

9.4 Fungal growths

These may be destructive or non-destructive, and the following notes give outline descriptions of both types. Fungi are parasitic organisms which contain no chlorophyll, so cannot synthesise food for themselves, but are able to consume ready-made organic matter such as starch, cellulose or lignin from timber. Fuller descriptions of these organisms may be found in the literature, particularly Bravery *et al.* (1992), Richardson (1993) and Desch and Dinwoodie (1996).

Non-destructive fungi

These may be of the mould type, which feed on free sugars in the timber surface and show as a black, green or brown powder which is easily brushed away when dry, or of the staining type, which feed mainly on starch in sapwood cells, the cell walls not being weakened. While neither type is damaging to the timber, they require the same damp conditions as enable destructive fungi to prosper and should be looked on as an early warning of adverse conditions developing. Proprietary fungicide can usually eradicate such fungi, but this will only be effective if the moisture content of the timber is reduced and kept low.

Destructive fungi

There are three broad categories of fungi which are destructive to timber, namely soft rots, brown rots and white rots. Soft rots are associated with very wet timber and were first identified in the internal timber slats of water cooling towers (Richardson, 1993). They also occur in water-logged timber in ground contact, provided there is some availability of oxygen. These rots remain within the cell walls where they consume both cellulose and hemicellulose, causing the surface of the timber to soften and become darker in colour. The depth of penetration may be only a few millimetres, with sound timber below, although in ground contact the rot may progress more deeply into the timber. Brown rots are so called from the colour of the lignin matrix, which is left behind after the cellulose and hemicellulose from the cell walls have been consumed. White rots are able to consume lignin as well as cellulose and hemicellulose, leaving behind a white stringy mass of material. Both brown and white rots exist in the cell cavities in timber and, while they may attack softwoods and hardwoods, brown rots are commoner in softwoods while white rots tend to predominate in hardwoods (Ridout, 1992). Like the soft rots, they require moisture and oxygen in order to thrive.

Within the categories of brown and white rot many types of fungus exist, the most serious being the dry rot fungus Serpula lacrymans, a form of brown rot, which can spread vigorously and extensively. Despite its common name, this fungus cannot develop in dry timber, although once established it can tolerate drier conditions than most other fungi, down to a minimum of 20 per cent moisture content. Optimum conditions for its growth are a moisture content of 30–40 per cent, a temperature of 23°C and a lack of ventilation. After germination, the fungus develops as minute hyphae which coalesce, covering the surface of the affected timber in what looks like dirty cotton wool. The hyphae are capable of travelling across brickwork and concrete, beneath plaster, through walls and along steelwork in search of nutrient material, including moisture, which is conducted throughout the fungus. Since dry rot requires stable conditions, it usually develops in concealed areas and the spread of dry rot fungus within a building may be extensive before the first visible signs appear. Such signs may include reddish spore dust from the developing fruit body, distortion of the surface of wood panelling, excessive deflection of affected timber floors and softening of the timber. There is usually also a characteristic smell. In the final stages of attack, wood becomes dry and friable and the surface of the wood breaks into cubical pieces.

The common name wet rot is used to describe a range of different fungi of both brown and white rot types, all of which generally require very damp conditions. Of the many types of wet rot which exist, some of which may show external growths and others of which may only occur within the timber, cellar fungus (Coniophora puteana) is perhaps the most widely encountered. This is also a brown rot, and the decay and growth are almost completely internal within timber, with very little external evidence apart from a dark discoloration and some longitudinal surface splits or cubical cracking. Development of the fruit body is not common. In the final stages of attack, the wood becomes very brittle and is readily powdered. Cellar fungus spreads less vigorously than dry rot and requires wetter conditions such as may develop from persistent leaks or condensation.

Eradication of both dry and wet rot is essentially the same. The first requirement is to eliminate *all* sources of moisture supporting the rot, and then to dry out both the timber and all adjacent materials, either naturally or with the aid of dehumidifiers. The next stage is a thorough survey to identify all infected timber, and all such timber, and up to 500 mm of adjoining clean timber, should be cut out and burnt. This also requires the removal of any attached materials, such as plasterwork. Further treatment of brick and plaster adjacent to infected timber may be carried out by heat sterilisation followed by the application of a masonry biocide such as dodecylamine salicylate (Ridout, 1992). Any timber not removed should be treated with a suitable preservative. All replacement timber should be fully treated with preservative using one of the vacuum/pressure methods.

It will be appreciated from the above that the eradication of rot may be as damaging as the rot itself, and it may be that such drastic treatment is not always necessary. If the fungus is discovered early enough so that little damage has been done to the timber, or if the infected timber is non-structural, and rapid and complete drying of the timber and adjacent materials can be effected, the fungus will die as it dries and no further treatment may be required other than the judicious application of a fungicide, to prevent any re-infection while drying proceeds, and regular inspection to ensure no re-occurrence of the original dampness. Case studies describing such an approach are given by Ridout (1992).

9.5 Preservative treatment

Need for preservative treatments

As indicated in the preceding sections, timber in service is at risk from a number of organisms. One solution to this problem might be to use only those timbers which are naturally very durable, but this would only rarely be cost-effective and would be extremely wasteful of the vast bulk of the forest resource. A better solution is to make use of the range of preservative treatments which are available and which can ensure an acceptable service life even for non-durable species.

As indicated at the start of this chapter, the first step in ensuring durability is a careful assessment of the environment in which the timber is to be used, to identify possible risks. At its most basic this may be a simple question of whether the timber can be kept dry, in which case the risk of fungal decay is minimal, or is liable to intermittent or continual wetting which would give a high risk of fungal decay. In BS 5268: Part 5, five risk categories are used, four to define increasing risk of fungal decay and the fifth relating to exposure to marine borers. A similar system of hazard classes is defined in new European standards (see Bravery, 1992 and BS EN 335). Having defined the risk of attack, BS 5268 then also considers the likely safety and economic risk, essentially assessing consequences of failure and relative ease (cost) of remedial work or replacement. Putting both of these together it is then possible to decide on the necessity for preservative treatment. Once this has been done, the appropriate type of preservative and application method can be selected. For non-structural timber, the requirements for preservative treatment are covered in BS 5589.

Types of preservative

Types are many and varied within the definition that any preservative must be poisonous to agents of decay. A full list of all approved wood preservatives is given in the HMSO publication *Pesticides* (updated annually). Of the many available, those which are in common use are selected because they meet all or most of the following requirements: toxicity to wood-destroying insects or fungi; permanency; economy and availability; penetrability; least toxicity to humans, plants and animals; non-corrosive nature to other materials; and non-promotion of flammability. Wood preservatives are classified in BS 1282 as follows.

Tar oil preservatives. These are very widely used, particularly in the form of coal-tar creosote to BS 144 and are excellent for use on timbers which are to be located on the exterior of buildings or otherwise to be exposed to the elements. The water repellency afforded to treated timber is an advantage in these external applications. Drawbacks with this type include encouragement of flammability without several months' weathering, the unpleasant odour and the difficulty in painting treated timber. Creosote is unsuitable for internal use in buildings.

Water soluble preservatives.   These are generally odourless and non-staining with no restrictions on decorative treatment, making timber so treated suitable for interior finishing joinery work. Because of the waterborne nature of this type of preservative, it may be necessary to 're-dry' the timber to an acceptable moisture content. Normally, such preservatives do not increase the flammability of timber, and in some cases actually reduce it. The most commonly used types are those where the preservative is copper/chrome/arsenic (see BS 4072) which 'fixes' in the timber after a few days and is then very resistant to leaching, or various boron compounds (BWPA, 1986) which are prone to leaching in wet conditions.

Organic solvent preservatives.   These are generally poor in respect of initial non-flammability because of the volatile nature of the liquid solvent (as opposed to the preservative). Such preservatives have good penetration, are generally non-corrosive, treated timber can be decorated with a variety of paints, and there is no necessity to 're-dry' timber after treatment. After the solvent has evaporated, the treated timber is no more flammable than untreated timber. A variety of types are available (see BS 5707). Because of increasing concern over solvent emissions to the atmosphere, a number of organic solvent preservatives, particularly those used for remedial treatments in buildings, are now formulated as emulsions with water. Solvent is still required, since the preservative chemicals are immiscible with water, but in smaller quantity, the water:solvent ratio in the emulsion generally being in the order of 10:1. Penetration is generally less effective for these emulsion formulations. Some materials are also prepared as thick pastes for remedial treatments. These are spread on to the timber surface and can achieve good penetration of preservative into the timber.

Preparation for preservative treatment

This consists generally of ensuring that the surface of the timber is clean and that it has been seasoned to a moisture content at least below the fibre saturation point. For the organic solvent type preservatives, which do not wet the timber, it can be seasoned to its end-use moisture content. For the boron compounds, on the other hand, which enter the timber by diffusion, it is essential to have a moisture content above 50 per cent. Ideally, all cutting and machining should have been completed. Where it is necessary to cut timber after it has been treated with preservative, the exposed surfaces should be re-treated. Any waste generated from processing treated timber must be disposed of safely in accordance with statutory requirements under the Environmental Protection Act, 1990. Disposal may be either by incineration in an approved plant or disposal to land fill sites having the appropriate licence to handle hazardous waste.

Methods of preservative application

These vary from a surface treatment of notional protective value to full pressure impregnation. The method of treatment used is normally based upon the end-use of the timber, the risk category and the amenability of the species of timber to treatment, see BS 5268: Part 5. The following outline descriptions of methods are arranged in ascending order of efficiency and effectiveness; more detailed descriptions can be obtained from BRE Digest 378 (1993).

Brush or spray.   These are the least effective methods but are better than none and, provided the preservative is flooded over the surfaces to encourage absorption, reasonable penetration is possible in permeable timbers. These are generally the main methods of application for remedial treatments to *in situ* timber using solvent or emulsion based materials.

Immersion.   This method may be specified for a variety of preservatives, although the organic solvent type is most frequently used in this method. Depending on the preservative used, some pre-heating may assist penetration. The period of immersion may vary from only a few seconds, to a few minutes or up to one hour. Tanks should be covered and solvent filters/recovery plants can help to reduce emissions.

Hot and cold open tank.   This was an extremely efficient method for achieving penetration in permeable timbers. Nowadays the process is generally only used for treating fence posts with creosote, see BRE Digest 378 (1993).

Pressure vacuum treatments.   By far the most efficient and controllable methods of preservation, pressure or vacuum treatments are widely and economically practised. Basically, there are two different pressure processes: the full cell process and the empty cell process. In the full cell process, the timber is placed in an enclosed pressure vessel and is subjected to a low vacuum for periods varying up to about an hour. While the vacuum is maintained, the preservative, usually preheated, is introduced into the vessel until it is filled. Pressure (up to not more than 14 bar) is then gradually increased and maintained for periods up to several hours until the required amount of preservative has been introduced into the timber, after which pressure is reduced and the vessel is drained of preservative. A further vacuum is applied for a brief period just long enough to clean the surface of the timber. In the empty cell process there are several proprietary methods, each varying slightly, but basically differing from the full cell process only by the absence of a preliminary vacuum period. The double vacuum method is essentially a variation of the full cell process. The initial vacuum drawn in the cylinder before introducing the preservative is similar but is generally held for a shorter period. Once the preservative is introduced, the vacuum is released and the timber is left at atmospheric pressure to absorb the preservative. At the end of this period, the vessel is drained and a final vacuum is applied to draw out excess preservative and leave a cleaner product. The process may be varied, depending on the permeability of the timber being treated, to include a low pressure (1–2 bar) during the impregnation period. Figure 9.2 shows a modern pressure treatment plant, the area of the pressure vessel being bunded and the whole treatment and storage area being covered.

Figure 9.2 *Pressure vessel for preservative treatment. Complete treatment area is roofed over and bunded to comply with Health and Safety Executive requirements (Photograph courtesy of James Donaldson & Sons Ltd)*

Diffusion process.   Unlike the previous treatment methods, this is carried out on green timber with a moisture content in excess of 50 per cent, normally as soon as the timber is converted. Very soluble boron compounds are applied to the surface of the timber which is then close stacked under cover to prevent evaporation of moisture. Over a period of weeks or months, the preservative diffuses into the wet timber, achieving good penetration even in impermeable timbers. The one drawback of these preservatives is that their very solubility makes them leachable and they are not suitable if the timber is to be used in wet conditions. A variation of this process which can be used to re-treat *in situ* timber is the introduction of solid rods of preservative into pre-drilled holes. Any moisture in the timber dissolves the compound, which then spreads through the damp timber.

Health and safety considerations

By their very nature preservatives are toxic, and this toxicity may extend to plants, animals and humans as well as the intended targets of wood-boring insects and fungi. All wood preservatives used in the UK must be approved for use by the Health and Safety Executive under the Control of Pesticides Regulations (1986). Non-approved formulations must not be used and the approved materials must be used in accordance with the manufacturers' stated instructions. A full list of currently approved formulations, indicating any restrictions on use (industrial, professional or amateur) is given in *Pesticides* (HMSO, updated annually). The COSHH Regulations (1988) must also be complied with in the use of preservatives to ensure the safety of operatives. If remedial treatment of timber is required in buildings where bats roost, the Nature Conservancy Council must be consulted before any work is started, since bats are very susceptible to many common preservatives and are protected under the Wildlife and Countryside Act (1981).

Aside from the legal requirements outlined above, the growing public concern over the general effects of toxic chemicals in the environment has led to a situation where some preservatives are no longer used in some countries on a voluntary basis, although they are still used elsewhere. The preservative pentachlorophenol (PCP) which is still used in the USA is no longer used in the UK, although not banned completely (see latest edition of *Pesticides* (HMSO, updated annually)). Tri-*n*-butyltin oxide (TBTO) is still used in the UK, e.g. for window joinery, but is not acceptable in Continental Europe. Its use in the UK is restricted to industrial processes or in paste formulations applied by professionals. Similarly, the common UK preservative copper/chrome/arsenic (CCA) is under scrutiny in Continental Europe because of its arsenic content, even though this is fixed in the treated timber. Research to develop effective alternatives is continuing, with attention being paid to replacing existing chemicals in preservatives, such as the use of boron or permethrin as a safer insecticide than arsenic, or to alternative strategies for treatment, such as the use of biological control organisms (Score and Palfreyman, 1994) which are target-specific.

9.6 Timber in fire

Of the three main structural materials, steel, concrete and timber, timber alone is combustible. When the temperature exceeds 250°C, timber starts to decompose, giving off flammable gases and turning to charcoal. Small sections in a fire for a sufficient period may be completely burned through. In large sections, however, timber is a safer material than steel or reinforced concrete, since its unique properties may leave it still in place after a fire which would cause steel or concrete sections to collapse. The reasons for this are, firstly, that timber chars at predictable rates, BS 5268: Part 4 quotes values of 40 mm per hour for structural softwoods and 30 mm per hour for the more dense hardwoods. Structural members can, therefore, be designed to allow for this rate of loss of section. Secondly, timber is a very good thermal insulator, and thus conversely a poor conductor. Very high external fire temperatures are not transmitted into a section and thus the interior maintains a low temperature and retains load-carrying properties. Thirdly, the char which develops on the exterior not only has an even lower thermal conductivity and acts as an effective insulation layer as it builds up, but also blocks the production of flammable gases from the interior, so reducing their contribution to maintaining burning. The effect of fire is illustrated in figure 9.3, which shows a section of a pine lintel salvaged from an old farm cottage and burned for about half an hour. The original section on the left is 178 × 76 mm. The burnt section on the right has lost approximately 15 mm all round, but the residual section shows no change in appearance from the original except in the pyrolisis zone just inside the charcoal layer. The slightly darker general colour of the burned section is from the charcoal being smeared across the surface as it was sawn through.

While timber itself is thus effective in resisting fire, any metal fixings, screws, nails or bolts are not and, by transmitting fire temperatures into the interior of a section, can result in localised charring which will reduce the effectiveness of the fixing. To counteract this and maintain structural integrity, either the fixings must be protected at the surface or they should be recessed into the timber to a depth

Figure 9.3 *Timber section before and after fire*

equal to the expected depth of charring for the required period of fire resistance, and the hole plugged with timber glued into position.

Another indicator of performance in fire is the surface spread of flame. Timber is normally rated 3 for surface spread of flame according to BS 476: Part 7, but its performance can be improved by the use of flame-retardant treatments applied by impregnation (in a similar manner to preservative treatments) or as a surface coating such as intumescent paint or varnish (see TRADA, 1993).

10

Properties and Uses of Solid Timber

As with all materials, the properties of timber which are important depend on the particular application for which it is being considered. For some it may be durability, others may emphasise appearance, but for structural applications the basic requirements are normally strength and deformation. Durability has already been considered, in Chapter 9, while what constitutes the most attractive timber for decorative uses is a subjective aesthetic judgement (see TRADA, 1985). In this chapter, the properties of density, strength and deformation are considered in relation to the various characteristics of timber which have an influence upon them, and then the applications of various timbers in construction are outlined.

10.1 Density

The density of the solid cell wall material in wood is effectively constant, irrespective of species, the value generally quoted being 1500 kg m^{-3} (Desch and Dinwoodie, 1996). The difference between this value and the densities of commercial timber species, which for dry timber range from under 400 kg m^{-3} for the softwood Sitka spruce to in excess of 1000 kg m^{-3} for the hardwood Ekki (Lavers, 1983), is due to the cellular structure of timber, the variation from species to species resulting basically from differences in their ratio of cell space (air) to cell wall. Even within a given species, however, there can be quite large variations in density between samples from different trees and even from different parts of the same tree.

Variations in density from tree to tree are often the result of differences in growth rate under the influence of latitude, climate, soil conditions or even simply a tree's position on the edge or in the interior of a forest. The effect of increased growth rate is to reduce density in softwoods, owing to increasing amount of lighter earlywood tracheids, and increase density in ring-porous hardwoods, resulting from increase in fibre content, with the reverse effects for both types with reduced growth rate. These relationships break down, however, if the rate of

139

growth is excessively fast or slow (Desch and Dinwoodie, 1996). Within individual trees there is generally a predictable variation in density, which decreases with height up the trunk and increases with distance outwards from the pith, most noticeably in the corewood closest to the pith. This expected variation can be upset by variations in the growth rate during the life of the tree. Density values calculated for timber samples are also affected by moisture content (any comparisons should be made at constant moisture content) and by the amount of any extractives present in cells.

The major practical importance of density in timber is its role as an indicator of strength, since both properties largely depend on the amount of cell wall material present. A plot of dry density (12 per cent moisture content) against average wet and dry bending strengths, using data from Lavers (1983) for the more common species named in BS 5268: Part 2, is presented in figure 10.1. It is clear from this that while there is a relationship between the two, which is better for the softwoods at the lower end of the density range and becomes quite poor for the hardwoods, precise prediction of strength from density is not possible because of all the many other factors which affect strength. However, there may be some merit in specifying timber with a minimum density for a particular species as a cut-off to ensure strength values above a required minimum.

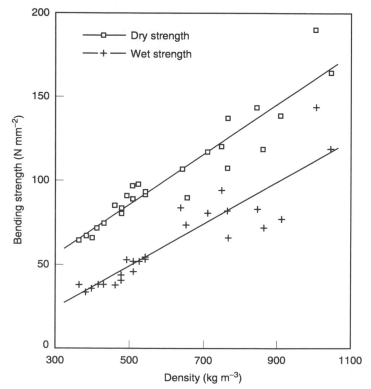

Figure 10.1 *Strength–density relationship (after Lavers, 1983)*

10.2 Strength properties

As outlined above and in Chapter 8, the strength of timber is affected by density and a wide range of naturally occurring defects, consideration of which form the basis of the grading system. It has also been explained in Chapter 7 that timber is anisotropic because of its cellular structure, and thus strength properties must be considered in relation to the direction of loading relative to the grain direction. This is illustrated in table 10.1, which shows bending test data for samples from two timbers cut from wide boards such that their grain direction was either parallel to the span or perpendicular to the span at test, an artificial condition but one which clearly emphasises anisotropy.

TABLE 10.1
Variation of bending strength with grain direction

Material	Strength in bending ($N\ mm^{-2}$)	
	Parallel to span	*Perpendicular to span*
Hardwood A	85	5
Hardwood B	69	3

In addition to all the parameters previously identified, timber strength is also affected by moisture content, rate of application and duration of loading and temperature. The effect of moisture content has already been indicated in figure 10.1, where the wet strengths in bending are up to 50 per cent less than the dry strengths. The effect on strength in compression can be even greater, as shown in table 10.2, data for which were obtained from $50 \times 50 \times 200$ mm clear specimens.

TABLE 10.2
Effects of moisture content on compressive strength
(compression parallel to grain, imported whitewood)

Moisture content (%)	Strength ($N\ mm^{-2}$)	Elastic modulus ($kN\ mm^{-2}$)
9	37.0	8.5
33	16.5	4.5

The reason for the effect of moisture content on strength is generally considered to be related to the strengthening of the secondary bonds between microfibrils in the cell walls as they come closer together, owing to the removal of intervening water on drying (Desch and Dinwoodie, 1996). This is supported by the general form of the relationship between strength and moisture content, see figure 10.2, which exhibits a gradual reduction in strength with increase in moisture content up to the fibre saturation point, after which the strength remains constant. A quantitative example of this for a particular timber is given by Lavers (1983). Allowance for the effect of moisture content in the design of timber structures which will be

CIVIL ENGINEERING MATERIALS

exposed to wet conditions is made by multiplying the grade stresses of BS 5268 (which are derived for dry timber) by reduction factors which vary from 0.6 for compressive strength to 0.8 for bending and tension and 0.9 for shear.

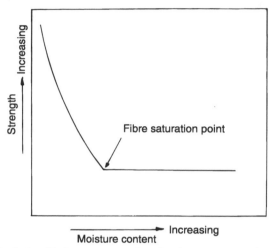

Figure 10.2 *General relationship between strength and moisture content (after Lavers, 1983)*

The strength properties of timber are normally assessed by relatively short duration tests, typically of the order of five minutes to failure. In reality, timber structures may experience load applications ranging from very short term (wind gusts) to long term (self-weight plus permanently imposed load). Since timber is essentially a viscoelastic material, suffering increasing deformation under load with increasing duration of load application, the effect of this is that the apparent strength varies with the duration of load, as is illustrated diagrammatically in fig-

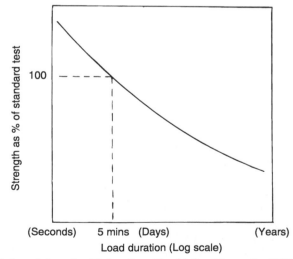

Figure 10.3 *Variation of strength with duration of load (after Dinwoodie, 1994)*

ure 10.3. While the form of this curve has been criticised by Madsen (1992), who suggested that for structural sizes (rather than the usual small clear specimens) it should be concave downwards rather than as shown concave upwards, the general agreement is that timber strength is time-dependent. The reasoning for this is generally based on the concept of limiting strain failure, increasing creep deformation under load eventually achieving the critical strain, in much the same way as has been suggested for concrete (Neville, 1986). In BS 5268, the grade stresses are based on long-term loading, and modification factors are used to adjust these stresses if the design loadings are more temporary, the factors varying from 1.25 for medium-term loading (e.g. snow), up to 1.75 for very short-term loading (short-duration wind gusts).

The strength of timber is also affected by temperature, the general effect being a linear decrease in strength with increase in temperature. This effect is very dependent on moisture content, dry timber suffering much less decrease in strength per °C rise in temperature than wet timber (Dinwoodie, 1994), and the effect also varies with the particular strength property being considered. If the increased temperature is below approximately 100°C and is sustained for only short periods, the reduction in strength is recovered on cooling, so that kiln drying temperatures, for example, have no permanent effect.

10.3 Movements in service

Timber in service experiences movement or strain due to applied loads, changes in moisture content or changes in temperature. Like all materials, timber expands and contracts on heating and cooling. As would be expected, the coefficient of thermal expansion for timber exhibits the usual differences due to anisotropy, values of $30-70 \times 10^{-6} \text{ K}^{-1}$ across the grain contrasting with values of $3-6 \times 10^{-6} \text{ K}^{-1}$ along the grain. For the range of temperatures to which timber is typically exposed in service, thermal movement can normally be ignored.

When timber is seasoned it shrinks on drying below the fibre saturation point. If it is subjected to changes in moisture content in service, it will shrink further if dried or expand on taking up moisture from the environment. The amount of movement varies with species and also slightly between the tangential and radial directions. Values quoted for a change in environment from 90 per cent to 60 per cent relative humidity are 1–3.5 per cent and 0.5–2 per cent for tangential and radial movements respectively (Everett and Barritt, 1994). Movement in the longitudinal direction is normally less than 0.1 per cent and is considered negligible. The usual strong anisotropy in moisture movement is suggested to result from the arrangement of microfibrils in the S layer in the cell wall, the near parallel orientation of these to the cell axis meaning that as water is interposed between them on wetting, expansion will be across the cell, not along it (Desch and Dinwoodie, 1996).

The amount of movement actually experienced in service is generally quite small in protected timber, since the changes in environmental humidity are normally quite limited. For timber in practice, a more useful description of tendency to moisture movement is the arbitrary classification of timbers into groups exhibiting small (less than 3.0 per cent total tangential and radial movement), medium (3.0–4.5 per cent) and large (over 4.5 per cent) movement between 90 and 60 per cent relative humidity (BRE, 1969). In many applications, the movement category is not important. If the timber is likely to experience variations in humidity in an application where dimensional stability is vital, however (for example, in internal joinery), the movement classification can be helpful in selecting the most appropriate timber, all other considerations being equal.

The deformation response of timber to load is quite complex. Under low levels of load applied for short duration, timber deforms essentially elastically, induced strain being fully recovered on removal of the applied stress. The slope of the effectively straight line stress–strain curve which can be obtained from this type of test gives a value for the modulus of elasticity (ratio of stress to strain, often referred to as stiffness) for the timber. If the load applied is increased beyond a certain point, termed the limit of proportionality, deformation increases more rapidly and the stress–strain relationship becomes increasingly more curvilinear before strength failure finally occurs. The exact form of the curve depends on the particular load regime, tension, compression or bending; an example of an experimentally determined load–deflection curve for specimens in bending is shown in figure 10.4. The plot for the specimen tested with grain parallel to span does not show the smooth curve to failure which would normally be given by a tension or compression test (Desch and Dinwoodie, 1996), the jagged effect towards failure being caused by the progressive splitting of the extreme tension fibres.

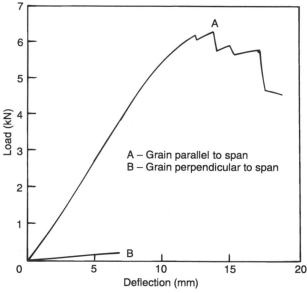

Figure 10.4 *Load–deflection behaviour in bending*

In practice, timber is not loaded monotonically to failure, but rather a certain level of load is applied for a long period of time. Under these circumstances there is an initial 'elastic' strain on application of load, followed by a continuing increase in strain at a constantly decreasing rate throughout the duration for which load is applied. The time-dependent deformation is referred to as creep and is also a common feature of the deformation behaviour of materials such as concrete, bitumen and the thermoplastics.

The deformation behaviour of timber is influenced by the same range of intrinsic and external factors which affect strength, and is equally variable with species. The stiffness of timber shows a similar correlation with density as does strength, and is equally adversely affected by the direction of the grain in a piece in relation to the direction of stressing, see figure 10.4. Stiffness is reduced with increasing moisture content up to fibre saturation point, a reduction factor of 0.8 for modulus of elasticity being given in BS 5268 to convert from dry to wet exposure conditions. Creep deformation is also greater for the same load in wet timber as in dry, and both the rate of creep and the total creep movement are increased with increase in temperature. The modulus of elasticity is reduced with increase in temperature – like strength, the reduction is greater for wet timber than for dry but, unlike strength, the effect appears to be non-linear (Dinwoodie, 1994).

10.4 Uses of timber

Timber has been used in construction for thousands of years, from a time well before it was possible to describe its properties scientifically. Nowadays, the advantages of timber as a construction material, in terms of high strength/weight ratio, ease of working, good thermal insulation, toughness and attractive appearance, are clearly recognised and the uses of timber are so numerous and varied that a comprehensive coverage is beyond the scope of this section, which provides only a general outline. An excellent summary of uses is given in TRADA (1991a), while more detailed information can be obtained from Baird (1990) and Harding (1988) for softwoods, and Mettem and Richens (1991) for hardwoods.

Structural applications

This term is considered to include all uses of timber in permanent carcassing or structure, and a range of such uses is given below.

Marine work

Much of the marine work traditionally requiring timber for wharves, piers, sheet piling and cofferdams is now undertaken using concrete or steel products, or both. There are several reasons for this, including the increasing scarcity and cost of good-quality timber of the types traditionally used, such as oak or greenheart. Nevertheless, there are many applications where it is still an economic proposition to use timber for marine construction purposes (for example for groynes, figure 10.5), even in association with other materials, as for instance, the use of hardwood rubbing strakes and fendering on concrete piers. Associated with marine work are problems of impact or abrasion due to movement of vessels, infestation by marine borers, fungal attack by fungi akin to the wet rot or cellar fungus types, erosion by chemical action of salt or by wave action, and dimensional movement caused by differentials in temperature and moisture content.

Figure 10.5 *Timber groynes for control of littoral drift*

The requirements for timber to be used in marine work are high density, close grain structure and natural durability and wear resistance. Generally, it is only in the hardwood classification that all these qualities are found and, traditionally, oak and greenheart have been used for marine work in the UK because they combined the required qualities with availability and economy. Balau, jarrah, ekki, iroko and opepe are other timbers which are suitable for use, although with all of these supply may be a problem. Where available timber does not meet natural durability requirements, heavy impregnation with suitable preservatives by pressure treatment methods can provide an acceptable alternative.

Heavy construction work.

Precast and/or *in situ* concrete piles and steel sheet piling have all served to reduce the amount of timber piling used in the construction industry. Timber is, however, still used for heavy constructional purposes where availability and cost of materials are favourable. Among other uses under this heading are pylons, gantries, bridges, shoring and abutments. In heavy constructional work there are many types of problems, complex and interrelated, varying often with the nature of the task to be performed and the location of the work. In the main, problems are of impact due to deliberate or accidental loading, infestation by airborne insects, chemical attack, fungal attack by fungi of the wet-rot type and dimensional movement due to temperature and moisture variability.

The properties required of timbers in many instances are the same as for marine work in that high density, closeness of grain and resistance to impact are all important, with the addition of resistance to acidity, alkalinity or other chemical nature of the soil for timber in ground contact. Cost and availability in the large sizes required for the work may also be limiting factors on the choice of timber. The timbers mentioned for marine work are again frequently used in heavy construction, namely greenheart, jarrah, opepe, iroko and oak, while keruing, larch, Douglas fir and pitch pine, are also suitable. If necessary, depending on the situation the latter timbers, which are classed only as moderately durable, can be given appropriate preservative treatment.

Medium/light constructional work

In recent years there has been little or no diminution in the range of uses for timber in this type of constructional work. Indeed, with the development of off-site pre-fabrication and rationalisation of traditional building techniques, for example in timber-framed housing (see figure 10.6), much more use has been made of the availability and ease of working and handling of the cheaper types of timber. Roof trusses, partitions, screens, floors and wall panels are often factory produced in large numbers, with reductions in man and machine fabrication time as well as site erection time, all of which lead to economic advantage.

Figure 10.6 *Timber-framed house construction (Photograph courtesy of Mr Mark Inglis)*

The requirements in this class of work are chiefly resistance to insect and fungal attack in position, together with a minimum of dimensional change due to the temperature and humidity variations likely to be encountered within the building. Again, individual significance varies with the location and type of building. Natural qualities of resistance to deterioration can be augmented by application of preservative. Because the location in this type of work will almost certainly be inside a building, it may be necessary to reinforce any natural resistance to fire or rate of flame spread by impregnation with a fire-retardant chemical.

Also, for this class of work, suitability is very much related to cost as well as the requirements described, owing to the very great amount of timber required for the multiplicity of uses in this classification. The timbers most frequently used are of the softwood type, principally whitewood, Douglas fir, Western hemlock and redwood, all of which are readily available in suitable sizes and quality at costs which are cheap by comparison with other softwoods and certainly most, if not all, of the hardwoods.

Falsework carpentry

This term is considered to include all uses of timber for the purpose of facilitating other types of construction, such as shuttering for *in situ* or precast concrete work, support formwork for brick or stone arch or shell forms, or jigs for glued-laminated timber beam or shell forms.

Although steel wall-forms and plastic pan-shutters are available, by far the bulk of shuttering is still of timber with plywood sheathing. This is because of ease of working, cost and availability, and ignores consideration of misuse by contractors as a result of poor fixings and the excessively green state of the timber in many instances. Although system scaffolding and adjustable steel props have taken over most of the main role in falsework (temporary support to timber shuttering), timber is still used in this area. In falsework carpentry, the problems are often specific to the location and the site environment but, in the main, are those mentioned in previous sections, namely, undue or unexpected loadings, dimensional movement, fixing and demountability.

The principal requirements for timber for this work are dimensional stability, ease of working, weight for handling and transportation, and resistance to impact and abrasion. All of these should combine to allow maximum re-use of moulds and formwork. Decisions on the most suitable timbers are very much a matter of availability and cost, especially related to the repeated use aspect. Cost dictates the use of softwoods, and the most common are whitewood, spruce and Western hemlock.

Finishing joinery

This category includes all of the non-structural timber used in buildings, including windows, doors, stairs, floorboards, facings, skirtings and panelling. Practice varies from country to country and also depends on the class of building and cost limitations. Virtually any softwood or hardwood timber can be used for elements within this class of work, including many of the more exotic, more expensive and, occasionally, less reliable timbers. The requirements for timbers for joinery uses are ease of working and finishing, good grain pattern and appearance when clear-finished, dimensional stability in conditions of variability of temperature and humidity, both internal and external, resistance to infestation and fungal attack (generally by applied treatment), availability and cost. The most important is often appearance.

Since most of the softwood timbers used have poor natural durability, inadequacy of finishing paint or varnish can be a problem in timber elements exposed to driving wind and rain or to excessive variations in temperature or humidity. Poor factory priming treatment of timber elements such as windows, doors or screens can produce trouble in the event of inadequate site storage or unduly long exposure to wind and rain before final paintwork is applied. Partly as a result of problems with premature failure, increasing use is being made of hardwoods for external joinery in anticipation of improved durability. The range of timbers available for use in finishing joinery is very wide, and covers a range of costs while still giving acceptability of performance in various building situations. Availability of suitable sizes is also rarely a problem. The selection of suitable materials may thus be dictated by project cost limitations, or, if appearance is an overriding consideration, it is simply a case of having to pay for what is required.

11

Processed Timber Products

While solid timber is undoubtedly a successful construction material, it is possible to point to a number of drawbacks. These include its inherent anisotropy, the many natural defects from which it suffers and which reduce its strength, the limited sizes (in section and length) which are commercially available, and the difficulty of drying large sections even when they are available. A range of processed products is available, which can overcome these defects to some extent; they are typically various sheet materials and glued-laminated structural members.

11.1 Sheet materials

The range of sheet, board or panel products includes veneer and core plywoods, chipboard and flakeboards, and a variety of fibreboards. The following sections give a brief outline of the manufacture and uses of these products; a more complete outline is given in TRADA (1992) while a detailed discussion of all stages of manufacture is presented in Walker (1993). The concept behind these products is that by rearranging the wood material in various ways, the effects of defects and anisotropy can be reduced or eliminated. The extent to which this is achieved may be gauged from table 11.1, which presents experimental bending test data for 75 mm wide sections cut from a range of panels, all approximately 25 mm thick, and tested under single point loading on a span of 450 mm. The data presented is not definitive for the different materials but should be taken as purely indicative of the relative strengths and degree of in-plane anisotropy for the different panels. A useful comparison can be made with the solid timber data from Chapter 10, table 10.1, which was obtained from the same size of specimens tested in the same way.

Veneer plywood

This is the correct name for what is commonly referred to simply as plywood. It is defined in BS 6566: Part 2 as a material in which all the plies are made of veneers (thin sheets of wood) oriented with their plane parallel to the surface of the panel.

150

TABLE 11.1
Strength in bending of various sheet materials

Material	Strength in bending (N mm^{-2})	
	Parallel*	Perpendicular
Plywood (9 ply)	26	50
Plywood (11 ply)	49	43
Chipboard	8	8
OSB	26	14
MDF	45	44

* This relates to the orientation of the face grain for the plywoods and OSB in relation to the span at test. For the chipboard and MDF it relates to the direction of cutting specimens, parallel or perpendicular to the long direction of the board.

Normally the direction of the veneer grain in a ply is at right angles to the direction in the immediately adjacent plies and there are an odd number of plies balanced in direction and thickness on either side of the core, see figure 11.1, although special panels can be made where this is not the case. The basic principle in veneer plywood is that by crossing the stronger and more stable longitudinal direction of wood cells in alternate layers, a panel is produced which has more or less equal properties in the plane of the panel. As can be seen from table 11.1, the extent to which this in-plane isotropy is realised depends on the make-up of the panel, particularly the number of plies and their thickness, the two plywoods tested for table 11.1 being the two shown on the right in figure 11.1.

Logs for plywood manufacture are first conditioned by soaking in water or steaming, which softens the timber, making it easier to cut without checking as it comes off the knife blade. The conditioned log is then centred on a rotary peeler (lathe) and rotated at quite high speed against the knife. To prevent the veneer breaking up as it is cut, a pressure bar is forced against the log just ahead of the knife blade. The constant gap between the pressure bar and the knife also helps to produce an even thickness of veneer. As it comes off the peeler, the continuous sheet of veneer is cut to size and then dried, usually to a moisture content below 10 per cent. Any unacceptable defects in the pieces can be cut out and patched, and narrow strips of veneer can be edge bonded to make up full sheets, before glue is applied and the plies assembled as required. The glues used depend on the intended use of the plywood, urea formaldehyde being suitable for interior use, phenol formaldehyde being used in exterior grades of plywood, see BS 1203. An initial cold press may be used to help spread the glue uniformly on each mating surface before the assembly is put into a hot press, where the glue cures. The cured panels are then cooled and trimmed to finished size.

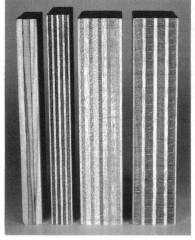

Figure 11.1 *Veneer plywood*

Veneer plywood is generally graded in terms of appearance, durability and bond (i.e. glue line) performance, see BS 6566. If necessary the durability of plywood may be improved by a suitable preservative treatment applied either to the veneers or the finished board. Veneer plywood can be produced from softwood or hardwood, the most common species being Douglas fir, spruce, pine, birch, utile, makore and sapele. Not all the plies are necessarily the same timber, for example birch faced ply may contain inner plies of softwood species. Veneer plywood is used extensively in construction in applications which include, for example, formwork sheathing (Pallet, 1994), plyweb and box beams, stressed skin panels in timber frame buildings and joint gussets in portal frames (Mettem, 1986), and a variety of joinery operations. Structural uses of plywood are covered by BS 5268: Part 2, which gives permissible stresses for design for a wide range of products, but assumes that all are manufactured using exterior-quality adhesives.

Core plywood

More commonly referred to as blockboard or laminboard, this is manufactured from strips of solid timber, normally edge-glued together to form a solid slab of material which is then surfaced with one or two cross-banded veneers on each face. The distinction between block- and laminboard is based on the width of the core strips, laminboard having strips less than 7 mm wide while in blockboard the core strips can be up to 25 mm wide. The difference is readily apparent from figure 11.2. The core strips can be cut from small roundwood logs, edge boards from solid timber conversion or the peeler cores remaining from veneer production. Generally two surface veneers are applied so that the face grain of the panel runs in the same direction as the grain in the core strips. Gluing, pressing and heat curing are similar to those processes used in the production of veneer plywood, the most commonly used glue being urea formaldehyde. These materials are normally used in interior applications and have only limited applications in construction, such as panelling, partitions and door blanks. Where they are intended to be used with a clear varnish finish, they may use matched sliced hardwood decorative veneers for the face, see Walker (1993).

Figure 11.2 *Core plywood: laminboard (left) and blockboard (right)*

Particleboards

This general description includes the material commonly referred to as wood chipboard, as well as the more recently developed oriented strand board (OSB) and cement bonded particle board, all of which are manufactured in accordance with BS 5669. The basis of these materials is that by breaking the wood down into relatively small pieces, and reassembling these pieces with a binder of some kind, the effects of the original grain structure and natural defects are removed and a more uniform product can be obtained.

Chipboard

The raw material for wood chipboard may be solid wood, such as sawmill offcuts, forest thinnings or logging waste, or wood residues such as planer shavings or sawdust. Softwood or hardwood can be used but softwoods are normally preferred, and roundwood is usually required debarked but this is not essential. Solid wood is fed into a chipper which cuts the required sizes and shape of particle, long and thin particles being preferred. Shavings and similar waste are ground to size. The chips are then dried and screened to separate out oversize chips and fine dust, and sometimes to classify the chips as face or core, face chips being finer. The chips are then mixed with a carefully controlled small quantity of resin, usually urea formaldehyde, and then formed into a mat on a flat plate, the chips being oriented totally at random. In some cases the mat may be layered with finer chips at the top and bottom to give a smoother surface to the finished board. The mat is then pressed to the required thickness in a high-temperature press which brings the particles into intimate contact and cures the glue. Manufacture is completed by trimming to size and sanding.

As is apparent from the data in table 11.1, chipboard is an essentially isotropic material in the plane of the board. BS 5669 describes six categories of chipboard and gives limits for such properties as strength, swelling on water contact, thermal conductivity and extractable formaldehyde content. The first two grades, C1 (the type used for the tests in table 11.1) and C1A, are furniture boards and are not used in construction except for internal fittings in buildings. Grades C2, C3 and C4 are generally used in construction applications such as flooring, sheathing of wall panels, flat roof construction or sarking to pitched roofs. The latter two grades are more moisture resistant than C2 and are more widely used. Grade C5 is the only grade of chipboard considered suitable for structural applications by BS 5268 and it can be used for similar applications as plywood in box beams, I-beams and stressed skin panels. Recommendations on the appropriate grades to use in a variety of applications are contained in BS 5669: Part 5.

Oriented strand board

This is a development of waferboard which is made by cutting small roundwood logs or plywood peeler cores into large flakes which are less than 1 mm in thickness. The flakes are cut so that the plane of the flakes is parallel to the grain of the timber. In waferboard the flakes were left as cut and the board was assembled with the plane of the flakes parallel to the faces of the board but totally random throughout its thickness, giving essentially similar properties in any direction in the plane of the board. In oriented strand board (OSB) the flakes are trimmed into rectangular pieces at least twice as long as they are broad, their length being in the order of 75 mm or more (Walker, 1993). The strands are oriented in the plane of the board such that in the centre layer of the board the strands lie across the width of the board, while in the outer surface layers the strands are arranged with their length generally parallel to the length of the board. This gives a board which has similar in-plane anisotropy as plywood, see table

Figure 11.3 *(left to right) Oriented strand board (OSB), wood chipboard and medium density fibreboard (MDF)*

11.1. The difference in size of the particles in chipboard and OSB is readily apparent from figure 11.3 which shows samples of each.

Aside from flake cutting, the rest of the manufacturing process is similar to that for chipboard: the cut strands are dried, spray coated with droplets of wax and powdered phenol formaldehyde resin (other resins can be used) which sticks to the wax droplets, laid up into a mat and pressed at high temperature to compress the strands and cure the resin. Two grades are recognised in BS 5669, F1 and F2, the latter being a moisture resistant grade. Like chipboard, OSB is susceptible to swelling on wetting. OSB can be used in most of the applications for which plywood is used and it is claimed to be taking over from plywood in some areas, such as wall sheathing in timber-frame construction (Timber Trades Journal, 1994), see figure 10.6, except in highly stressed zones.

Cement bonded particleboards

These are quite different from other boards, not simply because they are cement bonded rather than bonded with synthetic resins, but also because the wood content of the finished board is quite low. The density of this type of board is much higher than other boards and they have a high resistance to fire. The manufacturing process involves reducing the wood to small particles, screening these for size and then blending with cement, water and any required chemical admixtures. The wet mixture is then spread to thickness and held in a light press until the cement sets, after which the boards are unclamped and left to cure. Two types of board are listed in BS 5669, type T1 which is manufactured with magnesite cement, and type T2 which uses Portland cement. The two types have similar strength requirements but are distinguished by their reaction to moisture, type T2 is resistant and can be used internally or externally, while type T1 is not and is intended only for use internally in dry conditions. The range of applications for cement bonded particle board given in BS 5669 is similar to that for wood chipboard except as noted above.

Fibreboards

In these products the dismantling of the wood tissue is taken to its ultimate extreme by reducing it to the basic fibre, thereby eliminating all trace of structure, defects and anisotropy. There are two basic processes, wet and dry, both of which start off by reducing the raw material, which can be thinnings, peeler cores or other timber processing waste, to particles and then grinding the particles to fibres.

In the wet process the fibres are then added to water to form a slurry which is fed under thicknessing rollers on to a wire mesh conveyer where the water drains out again, leaving a felted sheet. For hardboards this sheet is then pressed on mesh plates, to allow the remaining water to escape, and heated to promote rebonding of the fibres by their residual lignin content. For softboard, or insulating board, the wet sheet is simply fed through drying ovens to remove the moisture and promote the fibre bonding process.

In the dry process the fibres are put into a dryer, where they are mixed with urea formaldehyde or other synthetic resin, before being formed into a mat on solid plates. After an initial pressing the mat is cut to size and then subjected to heat and pressure to compress the fibres and cure the resin. The material produced by this process is called medium density fibreboard (MDF) which has an exceptionally fine texture throughout, with a very smooth surface finish, see figure 11.3. A moisture resistant grade of MDF is made in addition to the standard product. As can be seen from table 11.1, MDF is also strong and effectively totally isotropic in the plane of the board.

Fibreboards are described in BS 1142 and generally are not considered for structural uses apart from one grade of tempered hardboard which is included in BS 5268 for use in box beams and stressed skin panels for dry exposure conditions. Most uses in construction relate to internal fitting and finishing and as insulation. MDF in particular is increasingly being marketed as a replacement for solid timber mouldings in finishing joinery since its fine texture makes it readily machinable to a range of profiles, its smooth surface gives an excellent substrate for painting and it is more stable in changing environments.

11.2 Glued-laminated sections

Unlike panel products in which the wood structure is taken apart and reassembled to improve properties in various ways, glued-laminated (gluelam) sections are assembled from unaltered solid timber sections which are glued together to form a single structural element. Generally sections are horizontally laminated, i.e. the glue lines are parallel to the neutral axis of the element, but vertical lamination is possible and the finished sections may be straight or curved.

The manufacturing process starts with timber grading to the rules given in BS 4978 for softwoods and BS 5756 for tropical hardwoods. The majority of gluelam members used in the UK are produced with whitewood, but laminated hardwood sections are produced and may be specified for their better natural durability. Three laminating grades are described LA, LB and LC; the most commonly used grade is LB, with grade LC sometimes being substituted for the less highly stressed inner laminations. The timber sections are kiln dried to a moisture content of about 12 per cent, after which they are finger jointed to the required lengths. This involves machining interlocking V-cuts across the width of the end section of each piece, applying glue and then forcing them together under longitudinal pressure while the glue cures. Other jointing methods are used but the finger joint is generally preferred (Walker, 1993). Advantage may be taken of the ability of finger jointing to give joint strength equal to timber strength to upgrade timber, by cutting out defects which bring a section into a lower grade, and rejoining the pieces.

Once any joints have cured, the sections are planed to ensure a uniform thickness throughout their length, which is important to ensure the thinnest possible glue lines. The thickness of laminations may vary from about 10 mm to 50 mm, thinner laminations enabling tighter bends to be produced and giving greater strength but at a cost premium. The glue used is mainly phenol resorcinol formaldehyde in a gap filling formulation to BS EN 301, the gap filling property being required since even planed timber surfaces are not completely flat. Once glued the individual laminations are laid up into the required section and clamped into shape while the glue is heat cured. The completed section is then planed to bring it to size and remove any glue which has been extruded under pressure. If the section is intended to be used in a potentially hazardous situation, preservative treatment would be applied at this stage, rather than treating the individual laminations, as preservatives do not necessarily penetrate deeply into a section and the skin of treated timber could be removed in the final planing operation.

One of the advantages of gluelam sections is the flexibility of shape and size which can be achieved, see figure 11.4, by joining together small sections into straight or curved beams, arches or frames, in section sizes which would be virtually impossible to saw from solid timber and enabling clear spans in excess of 50 m. Another advantage is the appearance of the elements – clear finished gluelam is just as warm and attractive as the timber from which it is produced. Gluelam is just as predictable in fire as solid timber and, given the normally very large section size, charring may have an insignificant effect on the overall section properties for short periods of burning. Additional protection in fire may be obtained by surface treatments, such as clear intumescent varnishes.

Figure 11.4 *Curved gluelam sections (Photograph courtesy of Tesco Stores Ltd)*

Gluelam elements are accepted in BS 5268 as being generally stronger and stiffer than solid timber. SS grade stresses are increased by factors ranging from 1.04 to 2.00 to obtain design stresses for LB grade elements depending on the number of laminations present and the strength property being considered. Even gluelam composed entirely of the lowest grade LC timber is upgraded to the equivalent of solid timber SS grade once there are at least 15 or more laminations. Mixed grade elements in which the inner laminations are of LC grade with LB for the outer laminations are considered almost equal to single grade LB members. This highlights the final significant advantage of gluelam: the possibility of matching the grade of timber to the stress regime within the depth of a section to make the most economical use of available material.

References

Attenborough, D. (1979) *Life on Earth*, Collins/British Broadcasting Corporation, London.

Baird, J.A. (1990) *Timber Specifiers' Guide*, BSP Professional Books, Oxford.

Bravery, A.F. (1992) *Wood Preservation in Europe,* Building Research Establishment, Watford.

Bravery, A.F., Berry, R.W., Carey, J.K. and Cooper, D.E. (1992) *Recognising Wood Rot and Insect Damage in Buildings*, Building Research Establishment, Watford.

British Wood Preserving Association (1986) *BWPA Manual*, BWPA, Stratford.

BS 144 Wood preservation using coal tar creosotes; Part 1: 1990 Specification for preservative; Part 2: 1990 Methods for timber treatment.

BS 373: 1957 (1986) Methods of testing small clear specimens of timber.

BS 476 Fire tests on building materials and structures; Part 7: 1987 Method for classification of the surface spread of flame of products.

BS 1142: 1989 Specification for fibre building boards.

BS 1203: 1979 (1991) Specification for synthetic resin adhesives (phenolic and aminoplastic) for plywood.

BS 1282: 1975 Guide to the choice, use and application of wood preservatives.

BS 4072 Wood preservation by means of copper/chromium/arsenic compositions; Part 1: 1987 Specifications for preservatives; Part 2: 1987 Wood preservation by means of copper/chromium/arsenic compositions.

BS 4169: 1988 Specification for manufacture of glued-laminated timber structural members.

BS 4978: 1988 Specifications for softwood grades for structural use.

BS 5268 Structural use of timber; Part 2: 1991 Code of practice for permissible stress design, materials and workmanship; Part 4: 1978 (1990) section 4.1. Recommendations for calculating fire resistance of timber members; Part 5: 1989 Code of practice for the preservative treatment of structural timber.

BS 5589: 1989 Code of practice for preservation of timber.

BS 5669 Particleboards; Parts 1–5: 1989–1993.

BS 5707 Solutions of wood preservative in organic solvent; Parts 1–3 1979–1980 (1986–1990).

BS 5756: 1980 (1985) Specifications for tropical hardwoods graded for structural use.

BS 6566: 1985 (1991) Plywood (in 8 parts).

BS DD ENV 1995-1-1: 1994, EuroCode 5 Design of timber structures.

BS EN 301: 1992 Adhesives, phenolic and aminoplastic, for load-bearing timber structures: classification and performance requirements.

BS EN 302: 1992 Adhesives for load-bearing timber structures: test methods (in 4 parts).

BS EN 335: 1992 Hazard classes of wood and wood-based products against biological attack (in 2 parts).

Building Research Establishment (1969) *The Movement of Timbers,* Technical Note No. 38, BRE, Princes Risborough Laboratory.

Building Research Establishment (1993) *Wood Preservatives: Application Methods,* Digest 378, HMSO, London.

Control of Pesticides Regulations 1986, HMSO, London.

Control of Substances Hazardous to Health (COSHH) Regulations 1988, HMSO, London.

Desch, H.E. and Dinwoodie, J.M. (1996) *Timber – Structure, Properties, Conversion and Use*, 7th edition, Macmillan, Basingstoke.

Dinwoodie, J.M. (1994) Timber, in *Construction Materials, their Nature and Behaviour*, ed. J.M. Illston, Spon, London.

Environmental Protection Act 1990, HMSO, London.

Everett, A. and Barritt, C.M.H. (1994) *Materials*, Longman Scientific & Technical, Harlow.

Gordon, J.E. (1988) *The Science of Structures and Materials*, Scientific American Library, New York.

Harding, T. (1988) *British Softwoods: Properties and Uses*, HMSO, London.

Hart, C. (1973) *British Trees in Colour*, Michael Joseph, London.

Health and Safety Executive, *Pesticides 1995, Pesticides Approved under The Control of Pesticides Regulations 1986* (updated annually), HMSO, London.

Hibberd, B.G. (Ed.) (1991) *Forestry Practice*, Forestry Commission Handbook 6, HMSO, London.

Lavers, G.M. (1983) *The Strength Properties of Timber*, Building Research Establishment, HMSO, London.

Madsen, B. (1992) *Structural Behaviour of Timber*, Timber Engineering Ltd, Vancouver.

Mettem, C.J. (1986) *Structural Timber Design and Technology*, Longman Scientific and Technical, Harlow.

Mettem, C.J. and Richens, A.D. (1991) *Hardwoods in Construction*, TBL 62, Timber Research and Development Association, High Wycombe.

Neville, A.M. (1986) *Properties of Concrete*, Longman Scientific & Technical, Harlow.

New Civil Engineer (1995) Forthright defence, *NCE*, 23 February 1995, p. 12.

Pallet, P.F. (1994) Wood-based panel products in formwork, *Concrete*, Vol. 28, No. 2, pp. 41–46.

Richardson, B.A. (1993) *Wood Preservation*, Spon, London.

Ridout, B.V. (1992) *An Introduction to Timber Decay and its Treatment*, Scientific and Educational Services, London.

Score, A.J. and Palfreyman, J.W. (1994) Biological control of the dry rot fungus *Serpula lacrymans* by *Trichoderma* Species: the effects of complex and synthetic media on interaction and hyphal extension rates, *International Biodeterioration & Biodegradation*, Vol. 33, pp. 115–128.

Timber Research and Development Association (1985) *Wood – Decorative and Practical*, Wood Information Sheet 2/3–6, TRADA, High Wycombe.

Timber Research and Development Association (1987) *Wood Preservation – A General Background*, Wood Information Sheet 2/3–30, TRADA, High Wycombe.

Timber Research and Development Association (1991a) *Timbers – their Properties and Uses*, Wood Information Sheet 2/3–10, TRADA, High Wycombe.

Timber Research and Development Association (1991b) *Visual Stress Grading*, TBL35, TRADA, High Wycombe.

Timber Research and Development Association (1992) *Introduction to Wood-based Panel Products*, Wood Information Sheet 2/3–23, TRADA, High Wycombe.

Timber Research and Development Association (1993) *Flame Retardant Treatments for Timber*, Wood Information Sheet 2/3–3, TRADA, High Wycombe.

Timber Trades Journal (1994) Emission control, *TTJ*, 30 July 1994, pp. 12–14.

Times Higher Education Supplement (1994) Portable sawmills to save forests, *THES* (London), August 1994, p. 4.

Walker, J.C.F. (1993) *Primary Wood Processing: Principles and Practice*, Chapman and Hall, London.

Wildlife and Countryside Act 1981, HMSO, London.

Further reading

W.H. Brown, *The Conversion and Seasoning of Wood*, Stobart & Son, London, 1988.

J.M. Dinwoodie, *Wood: Nature's Cellular, Polymeric Fibre-Composite*, Institute of Metals, London, 1989.

Various publications from TRADA and BRE.

W.H. Wilcox, E.E. Botsai and H. Kubler, *Wood as a Building Material*, Wiley, New York, 1991.

III CONCRETE

Introduction

Concrete is a man-made composite the major constituent of which is natural aggregate, such as gravel and sand or crushed rock. Alternatively artificial aggregates, for example, blast-furnace slag, expanded clay, broken brick and steel shot, may be used where appropriate. The other principal constituent of concrete is the binding medium used to bind the aggregate particles together to form a hard composite material. The most commonly used binding medium is the product formed by a chemical reaction between cement and water. Other binding mediums are used on a much smaller scale for special concretes in which the cement and water of normal concretes are replaced either wholly or in part by epoxide or polyester resins. These *polymer concretes* known as resin-based or resin-additive concretes respectively are costly and generally not suitable for use where fire-resistant properties are required but they are useful for repair work and other special applications. Resin-based concretes have been used, for example, for precast chemical-resistant pipes and lightweight drainage channels. This Part deals only with *normal concretes* in which cement and water form the binding medium.

In its hardened state, concrete is a rock-like material with a high compressive strength. By virtue of the ease with which fresh concrete in its plastic state may be moulded into virtually any shape, it may be used to advantage architecturally or solely for decorative purposes. Special surface finishes, for example, exposed aggregate, can also be used to great effect.

Normal concrete has a comparatively low tensile strength and for structural applications it is normal practice either to incorporate steel bars to resist any tensile forces (reinforced concrete) or to apply compressive forces to the concrete to counteract these tensile forces (prestressed concrete). Concrete is also used in conjunction with other materials, for example, it may form the compression flange of a box section the remainder of which is steel (composite construction). Concrete is used structurally in buildings for foundations, columns, beams and slabs, in shell structures, bridges, sewage-treatment works, railway sleepers, roads, cooling towers, dams, chimneys, harbours, off-shore structures, coastal protection works and so on. It is used also for a wide range of precast concrete products which includes concrete blocks, cladding panels, pipes and lamp standards.

The impact strength, as well as the tensile strength, of normal concretes is low and this can be improved by the introduction of randomly orientated fibres into the concrete mix. Steel, polypropylene, asbestos and glass fibres have all been used with some success in precast products, for example, pipes, building panels and piles. Steel fibres also increase the flexural strength, or modulus of rupture, of concrete and this particular type of *fibre-reinforced concrete* has been used in ground paving slabs for roads where flexural and impact strength are both important. Fibre-reinforced concretes are however essentially special-purpose concretes and for most purposes the normal concretes described in this book are used.

In addition to its potential from aesthetic considerations, concrete requires little maintenance and has good fire resistance. Concrete has other properties which may on occasions be considered less desirable, for example, the time-dependent deformations associated with drying shrinkage and other related phenomena. However, if the effects of environmental conditions, creep, shrinkage and loading on the dimensional changes of concrete structures and structural elements are fully appreciated, and catered for at the design stage, no subsequent difficulties in this respect should arise.

A true appreciation of the relevant properties of any material is necessary if a satisfactory end product is to be obtained and concrete, in this respect, is no different from other materials.

12

Constituent Materials

Concrete is composed mainly of three materials, namely, cement, water and aggregate, and an additional material, known as an admixture, is sometimes added to modify certain of its properties. Cement is the chemically active constituent but its reactivity is only brought into effect on mixing with water. The aggregate plays no part in chemical reactions but its usefulness arises because it is an economical filler material with good resistance to volume changes which take place within the concrete after mixing, and it improves the durability of the concrete.

A typical structure of hardened concrete and the proportions of the constituent materials encountered in most concrete mixes are shown in figure 12.1. In a properly proportioned and compacted concrete, the voids are usually less than 2 per cent. The properties of concrete in its fresh and hardened state can show large variation depending on the type, quality and proportions of the constituents, and from the discussion to follow students should endeavour to appreciate the significance of those properties of the constituent materials which affect concrete behaviour.

12.1 Cement

The different cements used for making concrete are finely ground powders and all have the important property that when mixed with water a chemical reaction (hydration) takes place which, in time, produces a very hard and strong binding medium for the aggregate particles. In the early stages of hydration, while in its plastic stage, cement mortar gives to the fresh concrete its cohesive properties.

The different types of cement and the related British Standards, in which certain physical and chemical requirements are specified, are given in figure 12.2; methods of testing cement for various properties are described in BS 4550. Of these, Portland cement is the most widely used, the others being used where concretes with special properties are required.

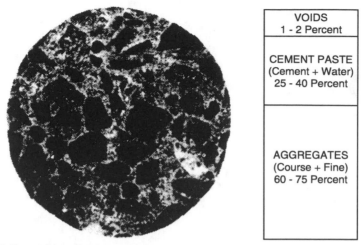

VOIDS 1 - 2 Percent
CEMENT PASTE (Cement + Water) 25 - 40 Percent
AGGREGATES (Course + Fine) 60 - 75 Percent

Figure 12.1 *Composition of concrete*

Portland cement

Portland cement was developed in 1824 and derives its name from Portland lime-stone in Dorset because of its close resemblance to this rock after hydration has taken place. The basic raw materials used in the manufacture of Portland cements are calcium carbonate, found in calcareous rocks such as limestone or chalk, and silica, alumina and iron oxide found in argillaceous rocks such as clay or shale. Marl, which is a mixture of calcareous and argillaceous materials, can also be used.

Manufacture

Cement is prepared by first intimately grinding and mixing the raw constituents in certain proportions, burning this mixture at a very high temperature to produce clinker, and then grinding it into powder form, figure 12.3. Since the clinker is formed by diffusion between the solid particles, intimate mixing of the ingredients is essential if a uniform cement is to be produced. This mixing may be in a dry or wet state depending on the hardness of the available rock.

The wet process is used, in general for the softer materials such as chalk or clay. Water is added to the proportioned mixture of crushed chalk and clay to produce a slurry which is eventually led off to a kiln. This is a steel cylinder, with a refractory lining, which is slightly inclined to the horizontal and rotates continuously about its own axis. It is usually fired by pulverised coal, although gas or oil may also be used. It may be as large as 3.5 m in diameter and 150 m long and handle up to 700 t of cement in a day. The slurry is fed in at the upper end of the kiln and the clinker is discharged at the lower end where fuel is injected. With its temperature increasing progressively, the slurry undergoes a number of changes as it travels down the kiln. At 100°C the water is driven off, at about 850°C carbon dioxide is given off and at about 1400°C incipient fusion takes place in the firing zone where calcium silicates and calcium aluminates are formed in the resulting

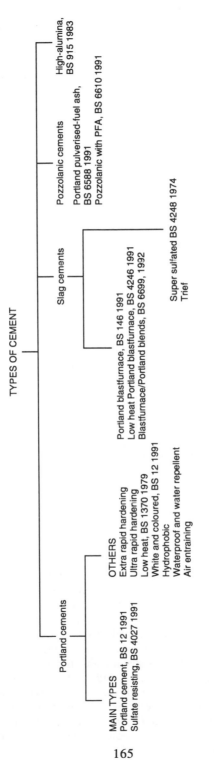

TYPES OF CEMENT

Portland cements

MAIN TYPES
Portland cement, BS 12 1991
Sulfate resisting, BS 4027 1991

OTHERS
Extra rapid hardening
Ultra rapid hardening
Low heat, BS 1370 1979
White and coloured, BS 12 1991
Hydrophobic
Waterproof and water repellent
Air entraining

Slag cements

Portland blastfurnace, BS 146 1991
Low heat Portland blastfurnace, BS 4246 1991
Blastfurnace/Portland blends, BS 6699, 1992

Super sulfated BS 4248 1974
Trief

Pozzolanic cements

Portland pulverised-fuel ash,
BS 6588 1991
Pozzolanic with PFA, BS 6610 1991

High-alumina,
BS 915 1983

Figure 12.2 *Different types of cement used for making concrete*

165

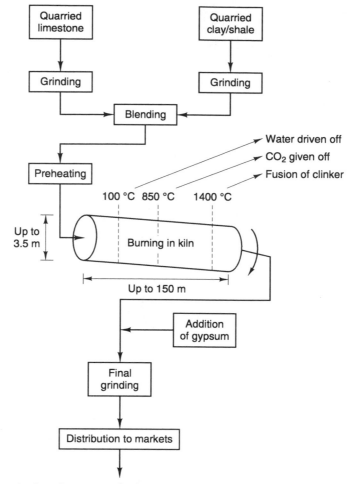

Figure 12.3 *Schematic view of cement production*

clinker. The clinker is allowed to cool and then ground, with 1 to 5 per cent gypsum, to the required fineness. Different types of Portland cement are obtained by varying the proportions of the raw materials, the temperature of burning and the fineness of grinding, and in some cases by intergrinding the clinker with other recognised materials such as pulverised-fuel ash (PFA), ground-granulated blast-furnace slag (GGBS), and condensed silica fume (CSF) which are allowed in quantities of up to 5 per cent. Gypsum is added to control the setting of the cement, which would otherwise set much too quickly for general use. Certain additives may also be introduced for producing special cements, for example, calcium chloride is added in the manufacture of extra-rapid-hardening cement.

The dry or semi-dry process is used for the harder rocks such as limestone and shale. The constituent materials are crushed into powder form and, with a minimum amount of water, passed into an inclined rotating nodulising pan where nodules are formed. These are known as *raw meal*. This is fed into a kiln and thereafter the manufacturing process is similar to the wet process although a much shorter length of kiln is used. It should be noted that the dry and semi-dry processes are more energy efficient than the wet process.

The grinding of the clinker produces a cement powder which is still hot and this cement is usually allowed to cool before it leaves the cement works.

Basic characteristics of Portland cements

Differences in the behaviour of the various Portland cements are determined by their chemical composition and fineness. The effect of these on the physical properties of cement mortars and concrete are considered here.

Chemical composition

As a result of the chemical changes which take place within the kiln several compounds are formed in the resulting cement although only four (see table 12.1) are generally considered to be important. A direct determination of the actual proportion of these principal compounds is a very tedious process and it is more usual to calculate these from the proportions of their oxide constituents, which can be determined more easily. A typical calculation using Bogue's method is shown in table 12.2. The limitations on chemical composition specified in British Standards for the various main Portland cements are summarised in table 12.3.

The two silicates, C_3S and C_2S, which are the most stable of these compounds, together form 70 to 80 per cent of the constituents in the cement and contribute most to the physical properties of concrete. When cement comes into contact with water, C_3S begins to hydrate rapidly, generating a considerable amount of heat and making a significant contribution to the development of the early strength, particularly during the first 14 days. In contrast C_2S, which hydrates slowly and is mainly responsible for the development in strength after about 7 days, may be active for a considerable period of time. It is generally believed that cements rich in C_2S result in a greater resistance to chemical attack and a smaller drying shrinkage than do other Portland cements. It may be noted from table 12.2 that the C_3S and C_2S contents are interdependent. The hydration of C_3A is extremely exothermic and takes place very quickly, producing little increase in strength after about 24 hours. Of the four principal compounds tricalcium aluminate, C_3A, is the least stable and cements containing more than 10 per cent of this compound produce concretes which are particularly susceptible to sulfate attack. Tetracalcium aluminoferrite,

TABLE 12.1
Main chemical compounds of Portland cements

Name of compounds	Chemical composition	Usual abbreviation
Tricalcium silicate	$3CaO.SiO_2$	C_3S
Dicalcium silicate	$2CaO.SiO_2$	C_2S
Tricalcium aluminate	$3CaO.Al_2O_3$	C_3A
Tetracalcium aluminoferrite	$4CaO.Al_2O_3.Fe_2O_3$	C_4AF

TABLE 12.2

A typical chemical composition of ordinary Portland cement

Oxide composition (per cent)		Calculation of percentage proportion of main compounds in cement	
Lime, CaO	64.73	C_3S	$= 4.07 (CaO - $ free $CaO) - (7.60 \times SiO_2$ $+ 6.72 \times Al_2O_3 + 1.38 \times Fe_2O_3 + 2.85 \times SO_3)$
Silica, SiO_2	21.20		
Alumina, Al_2O_3	5.22		$= 50.7$
Iron oxide, Fe_2O_3	3.08		
		C_2S	$= 2.87 \times SiO_2 - 0.754 \times C_3S$
Magnesia, MgO	1.04		$= 22.5$
Sulphur trioxide, SO_3	2.01		
Soda, Na_2O	0.19		
		C_3A	$= 2.65 \times Al_2O_3 - 1.69 \times Fe_2O_3$
Potash, K_2O	0.42		$= 8.6$
Loss on ignition, LOI	1.45		
Insoluble residue, IR	0.66	C_4AF	$= 3.04 \times Fe_2O_3$
	100.00		$= 9.4$
Free lime, CaO	1.60		

TABLE 12.3

British Standard requirements for the chemical composition of the principal Portland cements

Portland cement	Chemical constituent						
	C_3A	IR	MgO	SO_3	Cl	LOI	Additives
				not exceeding, %			
BS 12 Classes 32.5–62.5 (N and R)	N/A	1.5 (5.0)[1]	5.0[2]	3.5	0.1	3.0 (5.0)[3]	1.0[4]
Sulfate resisting Classes 32.5–52.5 (N and R)	3.5	1.5	5.0[2]	2.5	0.1	3.0	1.0[4]

[1] If minor constituents are included.
[2] In clinker.
[3] Where calcareous minor constituents are added.
[4] Must be stated on packaging/delivery note.

C_4AF, is of less importance than the other three compounds when considering the properties of hardened cement mortars or concrete.

From the foregoing, certain conclusions may be drawn concerning the nature of various cements. The increased rate of strength development of rapid-hardening Portland cement arises from its generally high C_3S content and also from its increased fineness which, by increasing the specific surface of the cement, increases the rate at which hydration can occur. The low rate of strength development of low-heat Portland cement is due to its relatively high C_2S content and low C_3A and C_3S contents. An exceptionally low C_3A content contributes to the increased resistance to sulfate attack of sulfate-resisting cement. It should be noted that while there can be large differences in the early strength of concretes made with different Portland cements, their final strengths will generally be very much the same (see Chapter 14).

Fineness

The reaction between the water and cement starts on the surface of the cement particles and in consequence the greater the surface area of a given volume of cement the greater the hydration. It follows that for a given composition, a fine cement will develop strength and generate heat more quickly than a coarse cement. It will, of course, also cost more to manufacture as the clinker must be more finely ground. Fine cements, in general, improve the cohesiveness of fresh concrete and can be effective in reducing the risk of bleeding (see Chapter 13), but they increase the tendency for shrinkage cracking.

Several methods are available for measuring the fineness of cement, for example BS 4550: Part 3 prescribes a permeability method which is a measure of the resistance of a layer of cement to the passage of air. The measured fineness is an overall value known as specific surface and is expressed in square metres per kilogram ($m^2 \, kg^{-1}$). Cements manufactured in the UK have a typical fineness range of $325-385 \, m^2 \, kg^{-1}$.

Hydration

The chemical combination of cement and water, known as hydration, produces a very hard and strong binding medium for the aggregate particles in concrete and is accompanied by the liberation of heat, normally expressed as joules per gram. The rate of hydration depends on the relative properties of silicate and aluminate compounds, the cement fineness and the ambient conditions (particularly temperature and moisture). The time taken by the main constituents of cement to attain 80 per cent hydration is given in table 12.4. Factors affecting the rate of hydration have a similar effect on the liberation of heat. It can be seen from table 12.5 that the heat associated with the hydration of each of the principal compounds of cement is very different and in consequence cements having different compositions also have different heat characteristics (see figure 12.4).

TABLE 12.4
Time taken to achieve 80 per cent hydration of the main compounds of Portland cement, based on Goetz (1969)

Chemical compounds	Time (days)
C_3S	10
C_2S	100
C_3A	6
C_4AF	50

Concrete is a poor conductor of heat and the heat generated during hydration can have undesirable effects on the properties of the hardened concrete as a result of microcracking of the binding medium. The possible advantages associated with the increased rate of hydration may in these circumstances be outweighed by the loss in durability of the concrete resulting from the microcracking. Other factors which affect the temperature of the concrete are the size of the structure, the ambient conditions, the type of formwork and the rate at which concrete is placed. It should be noted that it is the rate at which heat is generated and not the total lib-

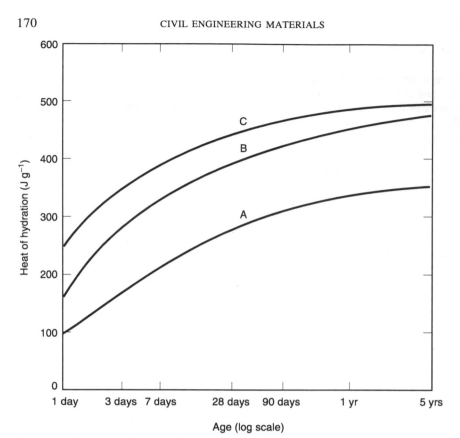

Figure 12.4 *Typical results for the heat evolution at 20°C of different Portland cements: (A) low heat, (B) ordinary and (C) rapid hardening, based on Lea (1970)*

TABLE 12.5
Heat of hydration of the main chemical compounds of Portland cement, based on Goetz (1969)

Chemical compounds	Heat of hydration	
	cal g^{-1}	*J g*$^{-1}$
C$_3$S	120	502
C$_2$S	62	251
C$_3$A	207	837
C$_4$AF	100	419

erated heat which in practice affects the rise in temperature. The heat characteristics must be considered when determining the suitability of a cement for a given job.

Setting and hardening

Setting and hardening of the cement paste are the main physical processes associated with hydration of the cement. Hydration results in the formation of a gel around each of the cement particles and in time these layers of gel grow to the extent that they come into contact with each other. At this stage the cement paste begins to lose its fluidity. The beginning of a noticeable stiffening in the cement paste is known as the *initial* set. Further stiffening occurs as the volume of gel increases and the stage at which this is complete and the final hardening process, responsible for its strength, commences is known as the *final* set. The time from the addition of the water to the initial and final set are known as the setting times (BS 4550: Part 3) and the specific requirements in this respect for the different cements are given in the appropriate British Standards. The setting times for some of the more important Portland cements are given in table 12.6. In practice, when mixes have a higher water content than that used in the standard tests, the cement paste takes a correspondingly longer time to set. Setting time is affected by cement composition and fineness, and also, through its influence on the rate of hydration by the ambient temperature.

TABLE 12.6
Typical initial and final setting times for the main Portland cements

Portland cement	Typical results		British Standard requirements	
	Initial setting time	Final setting time	Initial setting time not less than	Final setting time not more than
	(min)	(min)	(min)	(h)
32.5N, 32.5R, 42.5N, 42.5R	162	214	60	N/A
52.5N, 62.5N	97	149	45	N/A

Two further phenomena are a *flash set* and a *false set*. The former takes place in cement with insufficient gypsum to control the rapid reaction of C_3A with water. This reaction generates a considerable amount of heat and causes the cement to stiffen within a few minutes after mixing. This can only be overcome by adding more water and reagitating the mix. The addition of water results in a reduction in strength. A false set also produces a rapid stiffening of the paste but is not accompanied by excessive heat. In this case remixing the paste without further addition of water causes it to regain its plasticity and its subsequent setting and hardening characteristics are quite normal. False set is thought to be the result of intergrinding gypsum with very hot clinker in the final stages of the manufacture of cement.

Strength

The strength of hardened cement is generally its most important property. The British Standard strength requirements for Portland cements, obtained from mortar or concrete tests carried out in accordance with BS 4550: Part 3, are summarised in table 12.7. It should be understood that cement paste alone is not used for this test because of the unacceptably large variations of strength thus obtained. Standard aggregates are used for making prescribed mortar or concrete test mixes to eliminate aggregate effects from the measured strength of the cement.

TABLE 12.7
British Standard requirements for strength of the principal Portland cements

| Portland cement class | Minimum compressive strength ($N\,mm^{-2}$) | | | |
| | Mortar prisms | | | |
	3 days	7 days	28 days	
32.5N	—	≥16	≥32.5	≤52.5
32.5R	≥10	—	≥32.5	≤52.5
42.5N	≥10	—	≥42.5	≤62.5
42.5R	≥20	—	≥42.5	≤62.5
52.5N	≥20	—	≥52.5	≤72.5*
62.5N*	≥20	—	≥62.5	

*Not applicable for sulfate-resisting cements.

Soundness

An excessive change in volume, particularly expansion, of a cement paste after setting indicates that the cement is unsound and not suitable for the manufacture of concrete. In general, the effects of using unsound cement may not be apparent for some considerable period of time, but usually manifest themselves in cracking and disintegration of the surface of the concrete. One of the methods for testing the soundness of cement is that developed by Le Châtelier as described in BS 4550: Part 3. The British Standard limitations specified for various Portland cements require that the measured expansion in this test be not more than 10 mm.

Types of cement

The different types of cement are shown in figure 12.2 and their main properties are summarised in table 12.8. A brief description of typical properties of each type of cement is given here. For more detailed information the reader is referred to Neville (1986), Orchard (1979) and Harrison and Spooner (1986).

TABLE 12.8
Main properties of different cements

Type of cement	Rate of strength development	Rate of heat evolution	Drying shrinkage	Resistance to sulfate attack
Portland cements				
All BS 12 classes with N suffix	medium	medium	medium	low
All BS 12 classes with R suffix	high	high	medium	low
Sulfate resisting	low to medium	low to medium	medium	high
Extra rapid hardening	high to very high	high to very high	medium to high	low
Ultra rapid hardening	high to very high	high to very high	high	low
Low heat	low	low	medium	medium to high
White and coloured	medium	medium	medium	low
Hydrophobic	medium	medium	medium	low
Waterproof and water repellent	medium	medium	medium	low
Air entraining	medium	medium	medium	low
Slag cements				
Portland blastfurnace	low to medium	low to medium	above medium	medium
Low heat Portland blastfurnace	low	low	above medium	medium to high
Supersulfated	medium	very low	medium	high to very high
Pozzolanic cements				
Portland pulverised-fuel ash	low to medium	low to medium	medium	high
Pozzolanic with PFA	low	low	medium	high
High alumina cement*	very high	very high	medium	very high

* Subject to *conversion* in most environments.

Portland cements

All BS 12 cement classes with the suffix N have a medium rate of hardening, making them suitable for most concrete work. They have, however, a low resistance to chemical attack. All BS 12 cement classes with the suffix R are similar to cement classes with the suffix N but produce a much higher early strength. The increased rate of hydration is accompanied by a high rate of heat development which makes them unsuitable for large masses of concrete, although they may be used to advantage in cold weather. *Low-heat Portland cement* has a limited use but is suitable for very large structures, such as concrete dams, where the use of ordinary cement would result in unacceptably large temperature gradients within the concrete. Its slow rate of hydration is accompanied by a much slower rate of increase in strength than for ordinary Portland cement although its final strength is very similar. Its resistance to chemical attack is greater than that of ordinary Portland cement. *Sulfate-resisting Portland cement,* except for its high resistance to sulfate attack, has principal properties similar to those of ordinary Portland cement. Calcium chloride should not be used with this cement as it reduces its resistance to sulfate attack. *Extra-rapid-hardening Portland cement* is used when very high early strength is required or for concreting in cold conditions. Because of its rapid setting and hardening properties the concrete should be placed and compacted within about 30 minutes of mixing. Since the cement contains approximately 2 per cent calcium chloride, dry storage is essential. Its use in reinforced or prestressed concrete is not recommended (BS 8110: Part 1). *Ultra-high early-strength Portland cement*, apart from its much greater fineness and larger gypsum content, is similar in composition to ordinary Portland cement. Although the early development in strength is considerably higher than with rapid-hardening cement there is little increase after 28 days. It is suitable for reinforced and prestressed concrete work. *White and coloured Portland cements* are similar in basic properties to ordinary Portland cement. White cement requires special manufacturing methods, using raw materials containing less than 1 per cent iron oxide. Coloured cements are produced by intergrinding a chemically inert pigment with ordinary clinker. Because of its inert characteristics, the presence of a pigment slightly reduces the concrete strength. These cements are used for architectural purposes. *Hydrophobic Portland cement,* owing to the presence of a water-repellent film around its grain, can be stored under unfavourable conditions of humidity for a long period of time without any significant deterioration. The protective coating is broken off during mixing and normal hydration then takes place. *Waterproof and water-repellent Portland cements* produce a more impermeable fully compacted concrete than BS 12 Portland cements. *Air-entraining Portland cement* produces concrete with a greater resistance to frost attack. The cement is produced by intergrinding an air-entraining agent with ordinary clinker during manufacture. However in practice it is more advantageous to add an air-entraining agent during mixing since its quantity can be varied to meet particular requirements.

Artificial pozzolans

Artificial pozzolans, such as ground-granulated blastfurnace slag (GGBS), pulverised-fuel ash (PFA) and condensed silica fume (CSF) are now being used extensively in concrete construction. Comparison of the chemical and physical properties of these materials with Portland cement is given in tables 12.9 and 12.10.

Ground-granulated blastfurnace slag

Blastfurnace slag is formed as a by-product of the manufacture of iron in the blast-furnace. The slag results from fusion of the lime arriving from the limestone added to the furnace with the siliceous and aluminous residues from the iron ore and from the coke used for its reduction.

The quality of the iron produced in the blastfurnace is related to the chemistry of the fluxing materials and the slag. In order to manufacture iron of consistent quality, the composition of the slag is monitored by frequent chemical analyses so that suitable modifications can be made to the composition of the raw materials and to the operating conditions of the blastfurnace.

In order to produce ground-granulated blastfurnace slag (GGBS), the molten slag from the blastfurnace is rapidly quenched with water so that a high proportion of the slag solidifies as a glassy granulated product with a consistent degree of vitrification and chemical composition. The rapidly cooled product is de-watered and dried; it is then ground in conventional cement clinker grinding mills to cement fineness. No additional materials are added and the uniformity of the product is monitored by measuring, using microscopy, the proportion of glassy particles which are present.

Different types of slag cement can be produced by intergrinding varying portions of granulated blastfurnace slag with activators such as ordinary Portland cement and gypsum. The chemical composition of the granulated slag is similar to that of Portland cement but the proportions are different. One general requirement is that the slag used must have a high lime content. Of the slag cements listed in figure 12.2 only Portland blastfurnace cement has been used to any great extent.

Portland blastfurnace cement is produced by mixing up to 65 per cent granulated blastfurnace slag with ordinary Portland cement. The basic characteristics are similar to those of ordinary Portland cement although the rate of hydration is lower. It is particularly suited for structures involving large masses of concrete. Its resistance to chemical attack, particularly seawater, is somewhat better than that of ordinary Portland cement. *Low-heat Portland blastfurnace* cement, except for its greater slag content, is similar to Portland blastfurnace cement and can therefore be effectively employed where a control in the rise of temperature is the main requirement. It has a greater resistance to chemical attack than Portland blastfurnace and ordinary Portland cements. *Supersulfated cement* contains up to 85 per cent slag, 10 to 15 per cent gypsum and a small percentage of ordinary Portland cement. Its resistance to chemical attack is similar to that of sulfate-resisting Portland cement. Because of its low heat properties it can also be used for mass concrete work. In order to prevent the friable and dusty surface appearance often associated with this cement, careful initial curing is required.

Pulverised-fuel ash (PFA)

In the UK the fine ashes which are precipitated from the exhaust gases produced by the combustion of pulverised bituminous coal used for electricity generation are known as pulverised-fuel ash (PFA). The coal consists of a mixture of carbonaceous matter and various minerals, e.g. shales, clays, sulphides and carbonates. At the high temperatures in the boiler furnace, the minerals undergo various physical and chemical changes which are dependent not only on the temperature but also on the source of the coal and the length of time it is maintained at a high temperature. All of the minerals are converted into oxides and many of them are fused into tiny glass spheres of complex silicates. The overall chemical composition of an ash from any one source is fairly constant with time. The methods of extraction of PFA from the flue gases vary with the design of the plant. In some cases cyclones are used to remove most of the coarser particles before the electrostatic precipitation of the finer material. In other systems only the latter type of precipitation is used. Normally the quantity of ash extracted decreases as the flue gases pass through the various extraction stages; at the same time the fineness of the ash increases.

Although from a particular coal source, the overall chemical composition and the glass content of an ash are likely to be fairly constant with time, there can be, because of changes in power station operating conditions, changes in fineness and carbon content even for ashes taken from one part of the extraction system. These parameters are of particular importance in relation to the use of PFA in concrete. It is therefore necessary for selection, classification and blending to use procedures which ensure that material complying with British Standards can be supplied on a regular basis.

Pozzolanic cements are made by grinding together a pozzolanic material with Portland cement clinker. Pozzolanic materials contain alumino-silicate glass, which reacts with the lime (calcium hydroxide) released by the hydration of the cement and forms cementitious materials. Pozzolanas occur naturally, for example, volcanic ash, or are obtained in other forms such as pulverised-fuel ash (PFA), which is now increasingly used in the manufacture of cements. In the UK its use is permitted in the manufacture of *Portland pulverised-fuel ash cement* (BS 6588) with between 15 to 35 per cent PFA content and *pozzolanic cement with pulverised-fuel ash as pozzolana* (BS 6610) with 35 to 50 per cent PFA content. The use of these cements reduces the water demand of a concrete mix, when compared with that using the corresponding Portland cement. The rate of gain in strength and liberation of heat are lower than for Portland cement and this can be useful for mass concrete work; the use of pozzolanic cements (BS 6610) is particularly beneficial in this respect. However, an equal 28 day strength can generally be achieved by increasing the cementitious content of a Portland pulverised-fuel ash cement concrete by approximately 10 per cent compared with that of a Portland cement mix. Like sulfate-resisting Portland cement, the pozzolanic cements have high resistance to chemical attack.

Condensed silica fume (CSF)

Condensed silica fume is a by-product of the manufacture of silicon and ferro-silicon using processes which involve the high-temperature reduction of quartz, silicon dioxide. During the manufacturing process, some silicon monoxide vapour leaves the high-temperature reducing parts of the furnace where it was formed; in the upper cooler parts of the furnace, the silicon monoxide is converted into micro-spheres of amorphous silica. After removing the coarser particles using a cyclone, the finely divided silica microspheres, condensed silica fume, are collected by baghouse filters.

Chemical composition

Typical chemical composition (expressed as oxides) for GGBS, PFA and CSF are given in table 12.9 which shows, particularly with respect to the proportions of CaO and SiO_2 present, the dissimilarity between the materials. Although all three, when blended with Portland cement, behave as cementitious materials and give rise to hydrated calcium silicates and aluminates, their reactions are different. PFA and CSF contain very little CaO and produce hydrated cementitious products by undergoing a pozzolanic reaction with the calcium hydroxide liberated by the Portland cement. On the other hand, GGBS can contribute a considerable amount of CaO to the reaction processes.

The percentage of SiO_2 in all three materials is greater than that of the Portland cement with which they are blended. It therefore follows that the hydrated calcium silicates formed after reaction have lower $CaO:SiO_2$ ratios than those formed from Portland cement alone.

TABLE 12.9

Typical chemical compositions of cementitious materials

	Cementitious material (% by weight)			
	Portland cement	GGBS	PFA	CSF
SiO_2	20	37	48	92
Al_2O_3	5	11	26	0.7
Fe_2O_3	3	0.3	10	1.2
CaO	65	40	3	0.2
MgO	1.1	7	2	0.2
SO_3	2.4	0.3	0.7	—
S^{2-}	—	1.0	—	—
Na_2O	0.2	0.4	1.0	1.2
K_2O	0.9	0.7	3.0	1.9
Other oxides	1.4	2.3	1.3	2.6
LOI	1	—	5	—

GGBS differs from PFA and CSF in that it contains a proportion of sulphide, mainly as calcium sulphide. This does not adversely affect its use provided that the proportion is kept below 2 per cent, as recommended by British Standards (BS 6699).

All of the cementitious materials contain alkali metal ions. Unlike Portland cement where most of the alkali metal ions are present as water-soluble sulfates, the alkali metal ions in GGBS, PFA and silica fume are concentrated in the glassy structure of their particles, and their soluble alkali metal ion content is low. As the glass particles react they liberate alkali metal ions, but these do not affect the alkalinity of the concrete in the same way as those derived from Portland cement. Because of the lower $CaO:SiO_2$ ratio in the hydrated calcium silicates, a greater proportion of alkali metal ions is removed from solution.

PFA contains some unburnt carbonaceous matter, normally taken as equivalent to its loss on ignition, LOI. It is important that this is maintained below that recommended in British Standards (max. of 7.0 per cent) both for aesthetic and technical reasons. High carbon ashes may not show satisfactory water reduction and, since they absorb admixtures, they may reduce the effectiveness of air-entraining agents.

Physical characteristics

Typical values for fineness, bulk density and density are given in table 12.10. The following should be noted:

1. Portland cement, GGBS and PFA (suitable for use in concrete) have similar finenesses. However the spherical glass particles in silica fume are very much finer.
2. The relative densities show that when GGBS, PFA and SCF are blended with Portland cement there is, on a mass basis, an increase in the volume of cementitious powder compared with that of the same total mass of Portland cement. This is particularly significant with Portland cement/PFA blends where 70:30 proportions are common and has important implications on concrete workability.

TABLE 12.10
Typical physical characteristics

Physical characteristic	Cementitious material			
	Portland cement	GGBS	PFA	Silica fume
Fineness ($m^2 kg^{-1}$)	340	350	380	15 000
Bulk density ($kg m^{-3}$)	1400	1200	900	240
Density ($kg m^{-3}$)	3150	2900	2300	2200

High-alumina cement

This cement, which is manufactured by melting a mixture of limestone or chalk and bauxite (aluminium ore) at about 1450°C and then grinding the cold mass, has different composition and properties from those of Portland cements. The high proportion of aluminate, about 40 per cent, brings about an exceptionally high early strength and consequently it often becomes necessary to keep concrete, in which this cement is used, continuously wet for at least 24 hours to avoid damage from the associated heat of hydration.

Structural concrete made with high-alumina cement (commonly known as HAC) presents a serious problem however, as it can suffer a substantial reduction in its strength in most normal environments, owing to the conversion of the hydrated cement to a more porous form. Thus, the use of high-alumina cement in structural concrete is not permitted in BS 8110.

The conversion of high alumina cement is particularly sensitive to temperature, the water–cement ratio and the richness of the mix at a given water–cement ratio, increasing as all these factors increase. In certain circumstances, the residual strength of concrete can be considerably lower than its one day strength. Water–cement ratios higher than 0.4 should generally be avoided.

The exceptionally high early strength property of high-alumina cement makes it well suited for repair work of limited life and for temporary works. It also has a wide application in refractory concrete and its resistance to chemical attack, particularly sulfate attack, is greater than that of Portland cements, although in its converted, more porous, state it is very susceptible to alkali and sulfate attack.

12.2 Aggregate

Aggregate is much cheaper than cement and maximum economy is obtained by using as much aggregate as possible in concrete. Its use also considerably improves both the volume stability and the durability of the resulting concrete. The commonly held view that aggregate is a completely inert filler in concrete is not true, its physical characteristics and in some cases its chemical composition affecting to a varying degree the properties of concrete in both its plastic and hardened states.

Basic characteristics of aggregate

The criterion for a good aggregate is that it should produce the desired properties in both the fresh and hardened concrete. In testing aggregate it is important that a truly representative sample is used. The procedure for obtaining such a test sample is described in BS 812: Part 102.

Physical properties

The properties of the aggregate known to have a significant effect on concrete behaviour are its strength, deformation, durability, toughness, hardness, volume change, porosity, relative density and chemical reactivity.

The *strength* of an aggregate limits the attainable strength of concrete only when its compressive strength is less than or of the same order as the design strength of concrete. In practice the majority of rock aggregates used are usually considerably stronger than concrete. While the strength of concrete does not normally exceed

80 N mm^{-2} and is generally between 30 and 50 N mm^{-2}, the strength of the aggregates commonly used is in the range 70 to 350 N mm^{-2}. In general, igneous rocks are very much stronger than sedimentary and metamorphic rocks. Because of the irregular size and shape of aggregate particles a direct measurement of their strength properties is not possible. These are normally assessed from compressive strength tests on cylindrical specimens taken from the parent rock and from crushing value tests on the bulk aggregate. For weaker materials, that is, those with crushing values greater than 30, the crushing value may be unreliable and the load required to produce 10 per cent fines in the crushing test should be used (BS 812). It should be noted that the strength test on rock specimens was deleted from BS 812: Part 3 in 1975 and is consequently less used than the tests on bulk aggregate. The results of these tests for the strength properties of aggregates are only a guide to aggregate quality however, which may also be assessed from the intensity of aggregate fracturing in ruptured concrete specimens.

The *deformation* characteristics of an aggregate are seldom considered in assessing its suitability for concrete work, although they can easily be determined from compression tests on specimens from the parent rock. In general, the modulus of elasticity of concrete increases with increasing aggregate modulus. The deformation characteristics of the aggregate also play an important part in the creep and shrinkage properties of concrete as the restraint afforded by the aggregate to the creep and shrinkage of the cement paste depends on their relative moduli of elasticity.

A commonly used definition for aggregate *toughness* is its resistance to failure by impact and this is normally determined from the aggregate impact test (BS 812: Part 112). The aggregate impact value (AIV) has a close linear correlation with the aggregate crushing value (ACV) and can therefore be employed as the test for assessing aggregate strength (Spence *et al.*, 1974). The AIV test has the advantages over the ACV test of simplicity and cheapness in operation. It does not require the more elaborate facilities of a testing laboratory, the equipment is portable and the small sample required makes it a particularly useful test for quality control purposes. *Hardness* is the resistance of an aggregate to wear and is normally determined by an abrasion test (BS 812: Part 113). Toughness and hardness properties of an aggregate are particularly important for concrete used in road pavements.

Volume changes due to moisture movements in aggregates derived from sandstones, greywackes and some basalts may result in considerable shrinkage of the concrete (BRE, 1968; Dhir *et al.*, 1978). If the concrete is restrained this produces internal tensile stresses, possible tensile cracking and subsequent deterioration of the concrete. If the coefficient of thermal expansion of an aggregate differs considerably from that of the cement paste this too may adversely affect the concrete performance.

Aggregate *porosity* is an important property since it affects the behaviour of both freshly mixed and hardened concrete through its effect on the strength, water absorption and permeability of the aggregate. An aggregate with high porosity will tend to produce a less durable concrete, particularly when subjected to freezing and thawing, than an aggregate with low porosity. Direct measurement of porosity is difficult and in practice a related property, namely, water absorption, is measured. The water absorption is defined as the weight of water absorbed by a dry aggregate in reaching a saturated surface-dry state and is expressed as a percentage of the weight of the dry aggregate. In general, sedimentary rock materials have the highest absorption values. A gravel aggregate with a similar petrology to that of a crushed-rock aggregate will absorb more water because of its greater weathering. It should be noted that, for a given rock type, absorption can vary depending on the way in which it is measured and also on the size of the aggregate particles.

The total moisture content, that is, the absorbed moisture plus the free or surface water, of aggregates used for making concrete varies considerably. The water added at the mixer must be adjusted to take account of this if the free water content is to be kept constant and the required workability and strength of concrete maintained. Concrete mix proportions are normally based on the weight of the aggregates in their saturated surface-dry condition and any change in their moisture content must be reflected in adjustments to the weights of the aggregates used in the mix. Several methods for the determination of moisture content and absorption are described in BS 812: Part 2.

The *relative density* of a material is the ratio of its unit weight to that of water. Since aggregates incorporate pores the value of relative density varies depending on the extent to which the pores contain (absorbed) water when the value is determined (BS 812: Part 2). For the purposes of mix design the relative density on a 'saturated and surface-dry' basis is used. This is given by $A/(A - B)$ where A is the weight of the saturated surface-dry sample in air and B the weight of the saturated sample in water. The relative density of most natural aggregates falls within the range 2.5–3.0. For artificial aggregates the relative density varies over a much wider range. Aggregate is the major constituent of concrete and as such its relative density is an important factor affecting the density of the resulting concrete.

Shape and surface texture

Aggregate shape and surface texture can affect the properties of concrete in both its plastic and hardened states. These external characteristics may be assessed by observation of the aggregate particles and classification of their particle shape and texture in accordance with tables 12.11 and 12.12 from BS 812: Part 1. The classification is somewhat subjective, however, and the particle shape may also be assessed by direct measurement of the aggregate particles to determine the flakiness, elongation and angularity.

The angularity is expressed in terms of the angularity number. This is the difference between the solid volume of rounded aggregate particles after compaction in a standard cylinder, expressed as a percentage of the volume of the cylinder, and the solid volume of the particular aggregate being investigated when compacted in a similar manner. It is a measure of the increased voids in the compacted non-rounded aggregate, expressed as a percentage of the total volume. The angularity number ranges from zero for a perfectly rounded aggregate to about 12. Hughes (1966) proposed an alternative method for determining shape based on an *angularity* factor. This is defined as the ratio of the solid volume of loose single-size spheres to the solid volume of a single-size aggregate of similar nominal dimensions placed under identical conditions. The method has been shown to be suitable for both fine and coarse aggregates.

TABLE 12.11

Shape of aggregates

Classification	Description
Rounded	Fully water-worn or completely shaped by attrition
Irregular	Naturally irregular, or partly shaped by attrition and having rounded edges
Angular	Possessing well-defined edges formed at the intersection of roughly planar faces
Flaky	Material of which the thickness is small relative to the other two dimensions
Elongated	Material, usually angular, in which the length is considerably larger than the other two dimensions
Flaky and elongated	Material, having the length considerably larger than the width, and the width considerably larger than the thickness

TABLE 12.12

Surface texture of aggregates

Surface texture	Characteristics
Glassy	Conchoidal fracture
Smooth	Water-worn, or smooth due to fracture of laminated or fine-grained rock
Granular	Fracture showing more or less uniform rounded grains
Rough	Rough fracture of fine or medium-grained rock containing no easily visible crystalline constituents
Crystalline	Containing easily visible crystalline constituents
Honeycombed	With visible pores and cavities

Grading

The grading of an aggregate defines the proportions of particles of different size in the aggregate. The size of the aggregate particles normally used in concrete varies from 37.5 to 0.15 mm. BS 882 places aggregates in three main categories: *fine aggregate* or *sand* containing particles the majority of which are smaller than 5.00 mm, *coarse aggregate* containing particles the majority of which are larger than 5.00 mm, and *all-in aggregate* comprising both fine and coarse aggregate.

The grading of an aggregate can have a considerable effect on the workability and stability of a concrete mix (see Chapter 13) and is a most important factor in concrete mix design. In the UK, the grading of natural aggregates is generally required to be within the limits specified in BS 882. The fine aggregates are generally required to conform to any one of three standards grading limits, all of which are suitable for concrete provided suitable mix proportions are used.

A sieve analysis is used for determining the grading of an aggregate (BS 812: Part 103). The aggregate sample must be air dried and the weight of the material retained on each sieve should not exceed the specified maximum values as overloading will give erroneous results. Sieving may be performed by hand or machine and the results of a typical analysis are shown in table 12.13 from which it will also be seen that successive sieve sizes decrease by a factor of about 2. A conve-

TABLE 12.13

A typical example of calculation for aggregate grading

BS 410 sieve size (mm)	Weight of aggregate retained (g)	Percentage retained	Cumulative percentage retained	Cumulative percentage passing
5.00	0	0	0	100
2.36	32	16	16	84
1.18	40	20	36	64
0.60	42	21	57	43
0.30	46	23	80	20
0.15	32	16	96	4
Pan	8	4	100	
Total	200			

nient visual assessment of the particle size distribution can be obtained from a grading chart (figure 12.5). Curve A represents a continuously graded aggregate and curve B, in which two of the sizes are missing, represents a gap-graded aggregate.

In practice, fine and coarse aggregates are batched separately, their proportions being governed largely by their respective gradings. One method for determining the required aggregate contents (see Chapter 17) uses the grading modulus and the equivalent mean diameter of both fine and coarse aggregate.

The *grading modulus* of an aggregate is the mean specific surface of spheres which pass through the same sieve size as the actual aggregate particles and whose size distribution, or grading, corresponds to that of those aggregate particles.

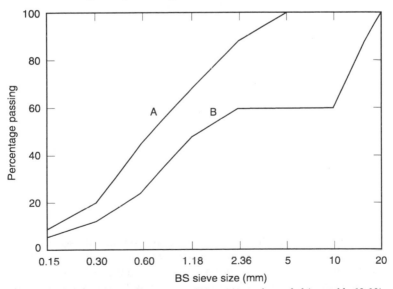

Figure 12.5 *Typical aggregate grading curves: (A) continuously graded (see table 12.13) and (B) gap-graded aggregates*

For spheres with a constant diameter D, the grading modulus G and the specific surface of a sphere are identical, that is, $G = (\pi D^2) / (\pi D^3 /6) = 6/D$.

Consider now the grading modulus of aggregate particles lying wholly between successive sieve sizes D_1 and D_2 corresponding to the smallest and largest diameters of the associated spheres. If, between successive sieve sizes, the proportion by volume of particles of diameter D is assumed to be inversely proportional to D, then the grading modulus is given by

$$G = \frac{\int_{D_1}^{D_2} \left(\frac{6}{D} \times \frac{1}{D} \right) dD}{\int_{D_1}^{D_2} \frac{1}{D} \, dD} = \frac{6 \left(\frac{1}{D_1} - \frac{1}{D_2} \right)}{\log_e \left(\frac{D_2}{D_1} \right)}$$

The grading moduli for particles between different sieve sizes are given in Chapter 17, table 17.3.

The grading modulus of an aggregate whose particles cover a range of sieve sizes is given by

$$G_a \text{ or } G_b = \Sigma \, [G(\text{per cent retained on each sieve})/100]$$

where G_a and G_b are the grading moduli of the coarse and fine aggregate respectively.

The *equivalent mean diameter* of particles lying wholly between successive sieve sizes D_1 and D_2 is conveniently assumed (Hughes, 1960) to be given by $D = (D_1 + D_2) / 2$. The equivalent mean diameters for particles between different sieve sizes are given in Chapter 17, table 17.3, and values for aggregates whose particles cover a range of sieve sizes are obtained in the same way as the grading modulus (see Chapter 17).

Types of aggregate

In the previous sections discussion has been mainly confined to rock aggregates. Although other types of aggregate are used for making concrete, their contribution is very small in comparison with rock aggregates. The general classifications of aggregates and the related British Standards are shown in figure 12.6. For a more detailed discussion, the reader is referred to Taylor (1965), BRE (1969, 1970), Short and Kinniburgh (1978) and Smith and Collis (1993).

Heavyweight aggregate

Heavyweight aggregates provide an effective and economical use of concrete for radiation shielding, by giving the necessary protection against X-rays, gamma-rays and neutrons, and for weight coating of submerged pipelines. The effectiveness of heavyweight concrete, with a density from 4000 to 8500 kg m^{-3}, depends on the aggregate type, the dimensions and the degree of compaction. It is frequently

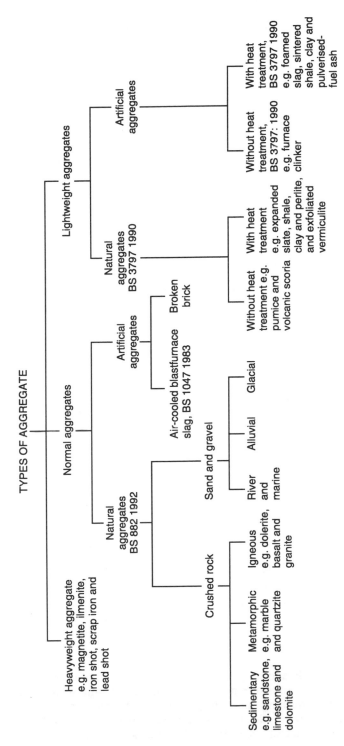

Figure 12.6 *A classification of aggregates used for making concrete*

185

difficult with heavyweight aggregates to obtain a mix which is both workable and not prone to segregation.

Normal aggregate

These aggregates are suitable for most purposes and produce concrete with a density in the range 2300–2500 kg m^{-3}. Rock aggregates are obtained by crushing quarried rock to the required particle size or by extracting the sand and gravel deposits formed by alluvial or glacial action. Some sands and gravels are also obtained by dredging from sea and river beds. Aggregates, in particular sands and gravels, should be washed to remove impurities such as clay and silt. In the case of river and marine aggregates the chloride content should generally be less than 1 per cent if these are to be used for structural concrete.

The properties of rock aggregates depend on their composition, grain size and texture. For example, granite has a low fire resistance because of the high coefficient of expansion of its quartz content. Sandstone aggregates generally produce concretes with a high drying shrinkage because of their high porosity. Crushed aggregates tend to be angular and gravels irregular or rounded in shape.

Air cooled blastfurnace slag aggregates produce concretes with similar strength to natural aggregates but with improved fire resistance. Broken-brick aggregates are also very fire resistant, but should not be used for normal concrete if its soluble sulphate content exceeds 1 per cent.

Lightweight aggregate

Lightweight aggregates find application in a wide variety of concrete products ranging from insulating screeds to reinforced or prestressed concrete although their greatest use has been in the manufacture of precast concrete blocks. Concretes made with lightweight aggregates have good fire resistance properties. The most commonly used artificial lightweight aggregates in the UK are sintered shale (Aglite), foamed slag, expanded clay (Leca), sintered pulverised-fuel ash (Lyntag and Taclite), pelletised expanded slag (Pellite and Lycrete) and sintered colliery shale (Sintag). They are highly porous and absorb considerably greater quantities of water than do normal aggregates. For this reason they should normally be batched by volume owing to the large variations that can occur in their moisture content. Their bulk density normally ranges from 350 to 850 kg m^{-3} for coarse aggregates and from 750 to 1100 kg m^{-3} for fine aggregates. The methods for testing lightweight aggregates are described in BS 3797: 1990.

12.3 Water

Water used in concrete, in addition to reacting with cement and thus causing it to set and harden, also facilitates mixing, placing and compacting of the fresh concrete. It is also used for washing the aggregates and for curing purposes. The effect of water content on the properties of fresh and hardened concrete is discussed in Chapters 13 and 14. In general, water fit for drinking, such as tap water, is acceptable for mixing concrete. The impurities that are likely to have an adverse effect when present in appreciable quantities include silt, clay, acids, alkalis and other salts, organic matter and sewage. The use of sea water does not appear to have any adverse effect on the strength and durability of Portland cement concrete but it is known to cause surface dampness, efflorescence and staining and should be avoided where concrete with a good appearance is required. Sea water also increases the risk of corrosion of steel and its use in reinforced concrete is not recommended. When the suitability of mixing water is in question, it is desirable to test for both the nature and extent of contamination as prescribed in BS 3148. The quality of the water may also be assessed by comparing the setting time and soundness of cement pastes made with water of known quality and the water whose quality is suspect.

The use of impure water for washing aggregates can adversely affect strength and durability if it deposits harmful substances on the surface of the particles. In general, the presence of impurities in the curing water does not have any harmful effects, although it may spoil the appearance of concrete. Water containing appreciable amounts of acid or organic materials should be avoided.

12.4 Admixtures

Admixtures are substances introduced into a batch of concrete, during or immediately before its mixing, in order to alter or improve the properties of the fresh or hardened concrete or both. Although certain finely divided solids, such as pozzolanas and slags, fall within the above broad definition of admixtures they are distinctly different from what is commonly regarded as the main stream of admixtures and therefore should be treated separately. It should also be noted that the materials used by the cement manufacturers to modify the properties of cement are normally described as additives.

In general, the changes brought about in the concrete by the use of admixtures are effected through the influence of the admixtures on hydration, liberation of heat, formation of pores and the development of the gel structure. Concrete admixtures should only be considered for use when the required modifications cannot be made by varying the composition and proportion of the basic constituent materials, or when the admixtures can produce the required effects more economically.

Since admixtures may also have detrimental effects, their suitability for a particular concrete should be carefully evaluated before use, based on a knowledge of their main active ingredients, on available performance data and on trial mixes. The specific effects of an admixture generally vary with the type of cement, mix composition, ambient conditions (particularly temperature) and its dosage. Since the quantity of admixture used is both small and critical the required dose must be carefully determined and administered. Where related British Standards exist (BS 5075) the admixtures should comply with their specifications such as those for acceptance test requirements given in table 12.14. It should be remembered that admixtures are not intended to replace good concreting practice and should not be used indiscriminately.

Types of admixture

Several hundred proprietary admixtures are available and since a great many usually contain several chemicals intended simultaneously to change several properties of concrete, they are not easy to classify. Moreover, as many of the individual constituents and their proportions are not widely known the selection of an admixture must frequently be based on the information provided by the suppliers. Different types of admixture are listed in table 12.15 and for a more detailed discussion on these admixtures the reader is referred to Rixom (1978), Concrete Society (1980) and Hewlett (1988).

Air-entraining agents

These are probably the most important group of admixtures. They improve the durability of concrete, in particular its resistance to the effects of frost and de-icing salts. The entrainment of air in the form of uniformly dispersed, very small and stable bubbles of predominantly between 0.25 and 1 mm diameter can be achieved by using foaming agents based on natural resins, animal or vegetable fats and synthetic detergents which promote the formation of air bubbles during mixing, or by using gas-producing chemicals such as zinc or aluminium powder which react with cement to produce gas bubbles. The first method is generally more effective and is the most widely used. The beneficial effects of entrained air are produced in two ways: first, by disrupting the continuity of capillary pores and thus reducing the permeability of concrete, and second, by reducing the internal stresses caused by the expansion of water on freezing.

Air-entraining agents also improve the workability and cohesiveness of fresh concrete and tend to reduce bleeding and segregation; this is particularly useful when aggregates with poor gradings are used. However, entrained air results in some reduction in concrete strength. Since improvements in workability can permit a reduction in the water content the loss in strength can be minimised. The amount of entrained air in concrete is dependent on the type of admixture and dosage used, as well as on the cement type, aggregate type and grading, mix proportions, ambient temperature, type of mixer and mixing time. Thus, it should only be used when adequate supervision is assured. The density of air-entrained concrete is reduced in direct proportion to the amount of air entrained. In the UK, air-entraining admixtures are required to comply with BS 5075: Part 2 (see table 12.14) and a method for determining the amount of air entrained in concrete is described in BS 1881: Part 106.

Accelerating agents

These can be divided into two groups, namely, setting accelerators and setting and hardening accelerators. The first of these are alkaline solutions which can considerably reduce the setting time and are particularly suitable for repair work involving water leakage. Because of their adverse effect on subsequent strength development these admixtures should not be used where the final concrete strength is an important consideration. Setting and hardening accelerators increase

TABLE 12.14

Admixture acceptance test requirements as specified in BS 5075*

Admixture	Compacting factor (CF) or slump with respect to control mix	Stiffening time from completion of mixing to reach a resistance to penetration of 0.5 N mm⁻²	3.5 N mm⁻²	Minimum compressive strength % control mix 1 day	7 days	28 days
Accelerating	(A) CF not more than 0.02 below	(A) More than 1 hour	(A) At least 1 hour less than control mix	(A) 125	(A) —	(A) 95
Retarding	(A) CF not more than 0.02 below	(A) At least 1 hour longer than control mix	—	(A) —	(A) 90	(A) 95
Normal water reducing	(A) CF at least 0.03 above (B) CF not more than 0.02 below	(B) Within ± 1 hour of control mix	(B) Within ± 1 hour of control mix	(A) — (B) —	(A) 90 (B) 110	(A) 90 (B) 110
Accelerating water reducing	(A) CF at least 0.03 above (B) CF not more than 0.02 below	(B) More than 1 hour	(B) At least 1 hour less than control mix	(A) 125 (B) 125	(A) — (B) —	(A) 90 (B) 110
Retarding water reducing	(A) CF at least 0.03 above (B) CF not more 0.02 below	(B) At least one hour longer than control mix	—	(A) — (B) —	(A) 90 (B) 110	(A) 90 (B) 110
Air-entraining **	—	Within ± 1 hour of control mix	Within ± 1 hour of control mix		—	70
Superplasticising ***	(B) Slump not more than 15 mm below	(B) Within ± 1 hour of control mix	(B) Within ± 1 hour of control mix	(A) — (B) 140	(A) 90 (B) 125	(A) 90 (B) 115
Retarding superplasticising†	(B) Slump not more than 15 mm below	(B) 1 to 4 hours longer	—	(A) — (B) —	(A) 90 (B) 125	(A) 90 (B) 115

*(A) Mix water content as control mix concrete. (B) Mix water content 92% of control mix concrete.

** Additional requirements: (i) Repeatability of air content (for three identical and consecutive test mixes, 4 to 6%), (ii) saturated density (at 3 days for two batches of test mix within 25 kg m⁻³ and at 28 days the test mix at least 50 kg m⁻³ less than control mix) and (iii) resistance to freezing and thawing (extension in length not more than 0.05% after 50 cycles).

*** Additional requirements: Mix A – (i) flow 510 to 620 mm, (ii) slump at 45 minutes not less than, and at 4 hours not more than that of the control mix at 10 to 15 minutes.

†Additional requirements: Mix A – (i) flow 510 to 620 mm, (ii) slump at 4 hours not less than that of the control mix at 10 to 15 minutes.

TABLE 12.15
Types of admixture

Type	Air entrainers	Accelerators		Retarders	Water reducers	Super-plasticisers	Bonding agents	Water repellers
		Setting	Setting and hardening					
Principal constituents	Foaming agents e.g. wood resins, synthetic detergent or gas generators, e.g. zinc powder	Highly alkaline solutions, e.g. aluminium chloride	Calcium chloride, calcium formate	Lignosulphonic or hydroxylated-carboxylic acids with cellulose or starch	Lignosulphonic or hydroxylated-carboxylic acids	Sulphonated melamine formaldehyde condensates or sulphonated naphthalene formaldehyde condensates or modified lignosulphonates	Polyvinyl acetate or styrene butadiene or acrylic	Metallic soaps or mineral and vegetable oil derivatives
Action with concrete	Formation of small stable discrete bubbles	Very rapid setting	Increased rate of setting and hardening	Delayed onset of setting and hardening	Increased workability or decreased water content	As for water-reducers but more effective	Increased bond strength	Decreased permeability
Main uses or effects	Increase resistance to frost action. Used in road and runway construction	Used for emergency repair work	High heat evolution. Used for concreting in winter conditions and early strength for rapid removal of formwork	Used for concreting in hot weather	Used to facilitate placing or to give higher strength and durability	As for water-reducers, but rapid placement and extremely high strength and durability	Patching and remedial work	Prevent absorption of rain water
Adverse effects	Some reduction in strength	Reduction in strength	Increased drying shrinkage. Decreased resistance to sulfate attack. Danger of metal corrosion with calcium chloride	Increased bleeding with some types. Increased drying shrinkage with others	As for retarders	Increased bleeding and segregation tendency. Increased shrinkage	Some reduction in compressive strength	

the ratc of both setting and early strength development. The most common admixture for this purpose is calcium chloride. Since its use may result in several adverse effects such as increased drying shrinkage, reduced resistance to sulfate attack and increased risk of corrosion of steel reinforcement, it should only be used with extreme caution and in accordance with any relevant specifications. Indeed, in line with many other countries, the use of calcium chloride in concrete containing steel reinforcement is no longer permitted in the UK (BS 8110). This has resulted in a new breed of chloride-free accelerating admixtures for use in reinforced concrete (McCurrich et al., 1979). These admixtures are based on calcium formate, sometimes blended with corrosion inhibitors such as soluble nitrates, benzoates and chromates. However, calcium chloride remains a most effective and economical material for use in plain concrete, particularly under winter conditions, for emergency repair work, or where early removal of formwork is required.

The accelerating admixtures are required to comply with BS 5075: Part 1 (see table 12.14) and their effect on concrete is dependent on dosage, cement type and mix proportions used and the ambient temperature of concrete. There are many accelerators which can achieve an increase in 1 day strength of up to 100 per cent over the corresponding normal mix, which is well in excess of the value specified in BS 5075: Part 1 (see table 12.14).

Retarders

Most admixtures in this group are based on lignosulphonic or hydroxylated-carboxylic acids and their salts with cellulose or starch. They are used mainly in hot countries where high temperatures can reduce the normal setting and hardening times. Other notable applications include situations where large concrete pour, sliding formwork, or ready-mixed concrete is used. The extended setting time prevents the formation of cold joints, allows time for steel-fixing and sometimes for overnight stopping of the formwork and compensates for the time lost in transit of ready-mixed concrete. A slightly reduced water content may be used when using these retarding agents, with a corresponding increase in final concrete strength. The lignin-based retarders result in some air-entrainment and tend to increase cohesiveness and reduce bleeding although drying shrinkage may be increased. The hydroxy-carboxylic retarders, however, tend to increase bleeding.

The use of retarders on their own is declining however, with retarding water-reducing admixtures gaining popularity. The effect of retarders is dependent on dosage, cement type and mix proportions used, as well as the time of addition and the ambient temperature. The effectiveness of retarding water-reducing admixtures is also influenced by aggregate type and grading. In the UK, both these admixtures are required to comply with BS 5075: Part 1 (see table 12.14).

Water reducers or plasticisers

These admixtures are also based on lignosulphonic and hydroxylated-carboxylic acids. Their effect is thought to be due to an increased dispersion of cement particles causing a reduction in the viscosity of the concrete. This effect can be used in three ways. First, to increase concrete workability for a given water–cement ratio and nominal strength; this allows easier placing and compaction of concrete. Second, to increase concrete strength without the addition of further cement owing to the reduced water requirement of a mix at a given workability: this allows the production of high strength concrete. Third, to reduce cement content of a mix at a given workability and strength by reducing its water content while maintaining the original water–cement ratio; this reduces the cost of a mix. The reduction in cement content will result in lower maximum temperatures and hence the risk of shrinkage cracking.

The effectiveness of a water-reducing admixture at a given dosage is dependent on cement type, aggregate type and grading, mix proportions, and ambient temperature. At a normal dose of admixture, water and cement reductions up to about 10 per cent can be achieved. The British specifications for this admixture are given in BS 5075: Part 1 (see table 12.14).

Superplasticisers

These are a relatively new category of water-reducing admixtures, and are most effective in dispersing cement particles and thus in increasing the concrete fluidity. The superplasticisers, which are mainly based on sulphonated melamine formaldehyde condensates or sulphonated naphthalene formaldehyde condensates, do not have the problem of retardation and excessive air entrainment associated with high rates of addition of normal plasticisers.

The superplasticisers are commonly used to produce flowing concrete, defined as having slump in excess of 200 mm or a flow value within 510 to 620 mm (BS 5075: Part 3), without having to change the original mix composition and without causing a strength reduction. The self-levelling property of flowing concrete means that it can be placed with little or no compaction and is therefore particularly suitable for heavily reinforced and inaccessible sections, or where rapid placement of concrete is required. To avoid bleeding, segregation and other adverse effects which tend to occur in high workability mixes it is important that the flowing concrete mix design, production and handling are all properly controlled. It should also be noted that the flowing characteristics of a mix are retained only for a short period of time, about 30 minutes after the addition of the superplasticiser, and for this reason superplasticisers should be added to the mix immediately prior to its placing. Although it is claimed that the properties of flowing concrete in its hardened state are comparable to those of the corresponding original mix, recent research has shown that this is not always so (Dhir and Yap, 1984a, 1984b) and the differences should be taken into account at the design stage of the structure.

An alternative use of superplasticisers is in the production of high strength concrete by reducing the mix water content, and hence water–cement ratio, while maintaining the original workability. A reduction in the water content up to 25 per cent is possible.

Bonding admixtures

These are organic polymer emulsions used to enhance the bonding properties of concrete, particularly for patching and remedial work. The bonding admixtures are known also to increase the abrasion resistance of concrete and its tensile strength, but some reduction in compressive strength also occurs.

Water-repelling agents

These are thc least-effective of all admixtures and are based on metallic soaps or vegetable or mineral oils. Their use gives a slight temporary reduction in concrete permeability.

Pigments

Colouring pigments, in powder form, are normally used in concrete for architectural purposes and the best effect is produced when they are interground with the cement clinker rather than added during mixing. Pigments used for this purpose are formulated from both natural and synthetic materials and in the UK should conform to BS 1014. Pigments do not normally affect the concrete properties, although those based on carbon may cause some loss of strength at early ages and can also reduce the effectiveness of air-entraining admixtures.

Pore fillers

These are chemically inactive finely ground materials such as bentonite, kaolin or rock flow. These admixtures are thought to improve workability, stability and impermeability of concrete, and are used with poorly graded aggregates, but their use may result in some reduction in concrete strength.

Pozzolanas

The most commonly used pozzolanas are pumicite and pulverised-fuel ash (PFA). Because of their reaction with lime, which is liberated during the hydration of Portland cement, these materials can improve the durability when added to concrete. Since they often retard the rate of setting and hardening and thus the rate of heat evolution, they can be useful in mass concrete work.

Pulverised-fuel ash (see section 12.1, Cement), the most common artificial pozzolana, can be divided into two generic types dependent on the parent coal source. Low-lime PFA, generally containing less than 10 per cent CaO, is derived from anthracite and bituminous coals, whereas high-lime PFA, generally containing more than 10 per cent CaO, is derived from sub-bituminous and lignitic coals. In the UK, only low-lime PFA is generally produced. The main difference between the two types of PFA is that the high-lime PFA is intrinsically hydraulic, although in other respects its activity is similar to the low-lime PFA. The use of PFA in concrete as a partial replacement of Portland cement is now recognised by the various British Standards (for example, BS 5328 and BS 8110) and other specifying authorities such as the Department of Transport (Specification for Highway Works, 1991). For use in concrete as a cementitious component in the UK, however, PFA is required to conform to BS 3892: Part 1 (see table 12.16). It should be noted that, while PFA conforming to BS 3892: Part 2 (table 12.16) may be used in concrete, it is not considered as a part of the cementitious component.

The main benefits obtained from the inclusion of PFA in fresh concrete are increased workability, decreased bleeding and reduced tendency to segregation. In the hardened concrete, PFA promotes a denser matrix structure and gives good long-term strength development. For detailed information on the mechanisms by which PFA affects the fresh and hardened concrete, the reader is referred to Dhir (1986), which also provides information concerning the characterisation of PFA and the mix proportioning, engineering properties and durability of concrete containing PFA. Briefly, a concrete mix incorporating PFA can be designed for a given workability and strength comparable with the corresponding mix in which Portland cement alone is used, provided that the variations in PFA characteristics (as in the case of Portland cement) are taken into account. A concrete mix designed in this manner can generally be expected to be similar to or better than the corresponding Portland cement concrete as a construction material.

The use of PFA in most structural grade concrete mixes can result in a saving in the Portland cement content of up to 100 kg m^{-3}, depending upon the quality of

TABLE 12.16

Specification for PFA for use in concrete

Characteristics	BS 3892 Part 1: 1993	BS 3892 Part 2: 1991		EN 450
		Grade A	Grade B	
Fineness, retention on 45 μm sieve (max.), %	12.5	30	60	40
Loss on ignition (max.), %	6.0	7.0	12.0	5.0
Moisture content (max.), %	0.5	0.5	0.5	N/A
Particle density (min.), kg m^{-3}	2000	N/A	N/A	N/A
Water requirement (max.), %	95	N/A	N/A	N/A
Strength factor (activity index*), min., %	80	N/A	N/A	75**
MgO (max.), %	N/A	4.0	4.0	N/A
SO$_3$ (max.), %	2.0	2.5	2.5	3.0
Calcium oxide (max.), %	10	N/A	N/A	1.0
Chloride (max.), %	0.1	N/A	N/A	0.1

* For EN 450.
** At 28 days.

PFA and Portland cement. Since pulverised-fuel ash is considerably cheaper than the cement, its use in concrete will result in direct economic savings, although indirect savings resulting from the conservation of energy (owing to reduced cement production) and environment (owing to PFA consumption rather than disposal) are of greater national interest.

13

Properties of Fresh Concrete

Fresh concrete is a mixture of water, cement, aggregate and admixtures, which may be used in certain cases to control rheology, rate of setting and hardening and durability. After mixing of these constituent materials to produce a uniform blend, operations such as transporting, placing, compacting and finishing of fresh concrete can all considerably affect the properties of hardened concrete. It is important that the constituent materials remain uniformly distributed within the concrete mass during the various stages of its handling and that full compaction is achieved. When either of these conditions is not satisfied the properties of the resulting hardened concrete, for example, strength and durability, are adversely affected.

The characteristics of fresh concrete which affect full compaction are its consistency, mobility and compactability. In concrete practice these are often collectively known as workability. The ability of concrete to maintain its uniformity is governed by its stability, which depends on its consistency and its cohesiveness. Since the methods employed for conveying, placing and consolidating a concrete mix, as well as the nature of the section to be cast, may vary from job to job it follows that the corresponding workability and stability requirements will also vary. The assessment of the suitability of a fresh concrete for a particular job will always to some extent remain a matter of personal judgement.

In spite of its importance, the behaviour of plastic concrete often tends to be overlooked. It is recommended that students should learn to appreciate the significance of the various characteristics of concrete in its plastic state and know how these may alter during operations involved in casting a concrete structure.

13.1 Workability

Workability of concrete has never been precisely defined. For practical purposes it generally implies the ease with which a concrete mix can be handled from the mixer to its finally compacted shape. The three main characteristics of the property are consistency, mobility and compactability. Consistency is a measure of wetness or fluidity. Mobility defines the ease with which a mix can flow into and completely fill the formwork or mould. Compactability is the ease with which a given mix can be fully compacted to remove all trapped air. In this context the required workability of a mix depends not only on the characteristics and relative proportions of the constituent materials but also on (1) the methods employed for conveyance and compaction, (2) the size, shape and surface roughness of form-work or moulds and (3) the quantity and spacing of reinforcement.

Another commonly accepted definition of workability is related to the amount of useful internal work necessary to produce full compaction. It should be appreciated that the necessary work again depends on the nature of the section being cast. Measurement of internal work presents many difficulties and several methods have been developed for this purpose but none gives an absolute measure of workability.

The tests commonly used for measuring workability do not measure the individual characteristics (consistency, mobility and compatability) of workability. However, they do provide useful and practical guidance on the workability of a mix. Workability affects the time and labour required for full compaction and so unworkable concrete directly increases cost. It is most important that a realistic assessment is made of the workability required for given site conditions before any decision is taken regarding suitable concrete mix proportions.

13.2 Measurement of workability

Four tests widely used for measuring workability are the slump, compacting factor, Vebe time and flow tests (figure 13.1). These are standard tests in the UK and are described in detail in BS 1881: Parts 102, 103, 104 and 105 respectively. Their use is recommended in BS 5328, which also requires the measured workability of concrete to be within certain limits of the required value as given in table 13.1, which also shows the suitability of each test for the range of concrete workabilities. It is important to note that there is no single relationship between the slump, compacting factor, Vebe time and flow results for different concretes. In the following sections the salient features of these tests together with their merits and limitations are discussed.

TABLE 13.1

Suitability and allowable tolerances of workability tests for concrete

Method	Workability range	Allowable tolerance
Slump	Medium–high	greater of $\pm$ 25 mm or 1/3 of required value
Compacting factor	Low–high	$\pm$ 0.03 for required value > 0.90
		$\pm$ 0.04 for required value > 0.80 and < 0.90
		$\pm$ 0.05 for required value < 0.80
Vebe time	Very low–low	greater of $\pm$ 3 s or 1/5 of required value
Flow	Very high	$\pm$ 50 mm about the required value

(a)

(b)

(c)   (d)

Figure 13.1 *Apparatus for workability measurement: (a) slump cone; (b) compacting factor; (c) Vebe consistometer; (d) flow test*

Slump test

This test was developed by Chapman in the United States in 1913. A 300 mm high concrete cone, prepared under standard conditions (BS 1881: Part 102) is allowed to subside and the slump or reduction in height of the cone is taken to be a measure of workability. The apparatus is inexpensive, portable and robust and is the simplest of all the methods employed for measuring workability. It is not surprising that, in spite of its several limitations, the slump test has retained its popularity.

The test primarily measures the consistency of plastic concrete and although it is difficult to see any significant relationship between slump and workability as defined previously, it is suitable for detecting changes in workability. For example, an increase in the water content or deficiency in the proportion of fine aggregate results in an increase in slump. Although the test is suitable for quality-control purposes it should be remembered that it is generally considered to be unsuitable for mix design since concretes requiring varying amounts of work for compaction can have similar numerical values of slump. The sensitivity and reliability of the test for detecting variation in mixes of different workabilities is largely dependent on its sensitivity to consistency. The test is not suitable for very dry or wet mixes. For very dry mixes, with zero or near-zero slump, moderate variations in workability do not result in measurable changes in slump. For wet mixes, complete collapse of the concrete produces unreliable values of slump.

The three types of slump usually observed are true slump, shear slump and collapse slump, as illustrated in figure 13.2. A true slump is observed with cohesive and rich mixes for which the slump is generally sensitive to variations in workability. A collapse slump is usually associated with very wet mixes and is generally indicative of poor quality concrete and most frequently results from segregation of its constituent materials. Shear slump occurs more often in leaner mixes than in rich ones and indicates a lack of cohesion which is generally associated with harsh mixes (low mortar content). Whenever a shear slump is obtained the test should be repeated and, if persistent, this fact should be recorded together with test results, because widely different values of slump can be obtained depending on whether the slump is of true or shear form.

The standard slump apparatus is only suitable for concretes in which the maximum aggregate size does not exceed 40 mm. It should be noted that the value of slump changes with time after mixing owing to normal hydration processes and evaporation of some of the free water, and it is desirable therefore that tests are

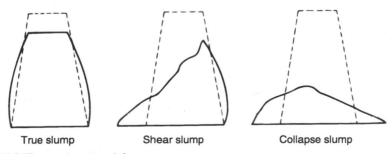

True slump   Shear slump   Collapse slump

Figure 13.2 *Three main types of slump*

performed within a fixed period of time. It is also advisable to delay testing for around ten minutes after the addition of water to allow for the absorption of water by dry aggregates.

Compacting factor test

This test, developed in the UK by Glanville *et al.* (1947), measures the degree of compaction imparted to concrete for a standard amount of work and thus offers a direct and reasonably reliable assessment of workability of concrete as previously defined. The apparatus is a relatively simple mechanical contrivance (figure 13.1) and is fully described in BS 1881: Part 103. The test requires measurement of the weights of partially and fully compacted concrete and the ratio of partially compacted weight to fully compacted weight, which is always less than 1, is known as the compacting factor. For the normal range of concretes the compacting factor lies between 0.80 and 0.92. The test is particularly useful for drier mixes for which the slump test is not satisfactory. The sensitivity of the compacting factor is reduced outside the normal range of workabilities. However, BS 1881: Part 103 only regards the test unsuitable for concrete having measured compacting factor below 0.70 or above 0.98.

It should also be appreciated that, strictly speaking, some of the basic assumptions of the test are not correct. The work done to overcome surface friction of the measuring cylinder probably varies with the characteristics of the mix. It has been shown by Cusens (1956) that for concretes with very low workability, the actual work required to obtain full compaction depends on the richness of a mix while the compacting factor remains sensibly unaffected. Thus it follows that the generally held belief that concretes with the same compacting factor require the same amount of work for full compaction cannot always be justified. One further point to note is that the procedure, for placing concrete in the measuring cylinder bears no resemblance to methods commonly employed on the site. As in the slump test, the measurement of compacting factor must be made within a certain specified period. The standard apparatus is suitable for concrete with a maximum aggregate size of up to 40 mm.

Vebe test

This test was developed in Sweden by Bährner (1940) (see figure 13.1). Although generally regarded as a test primarily used in research, its potential is now more widely acknowledged in the industry and the test is gradually being accepted. In this test (BS 1881: Part 104) the time taken to transform, by means of vibration, a standard cone of concrete to a compacted flat cylindrical mass is recorded in seconds, to the nearest 0.5 s. Unlike the two previous tests, the treatment of concrete in this test is comparable to the method of compacting concrete in practice. Moreover, the test is sensitive to change in consistency, mobility and compactability, and therefore a reasonable correlation between the test results and site assessment of workability can be expected.

The test is suitable for a wide range of mixes and, unlike the slump and compacting factor tests, it is sensitive to variations in workability of very dry and also air-entrained concretes. It is also more sensitive to variation in aggregate characteristics such as shape and surface texture. The reproducibility of results is good. As for other tests, its accuracy tends to decrease with increasing maximum size of aggregate; above 20 mm the tests results become somewhat unreliable. However, BS 1881: Part 104 permits its use for concrete having aggregate of maximum size up to 40 mm. For concretes requiring very little vibration for compaction the Vebe time is only about 3 s. Such results are likely to be less reliable than for larger Vebe times because of the difficulty in estimating the time of the end point (concrete in contact with the whole of the underside of the plastic disc). At the other end of the workability range, such as with dry mixes, the recorded Vebe times are likely to be in excess of their *true* workability since prolonged vibration is required to remove the entrapped air bubbles under the transparent disc. To overcome this difficulty an automatic device which records the vertical settlement of the disc with respect to time can be attached to the apparatus. This recording device can also assist in eliminating human error in judging the end point. The apparatus for the Vebe test is more expensive than that for the slump and compacting factor tests, requiring an electric power supply and greater experience in handling; all these factors make it more suitable for the precast concrete industry and ready-mixed concrete plants than for general site use.

Flow test

This test, a slightly modified form of the DIN 1048 (German Standard) test, is used to measure the ability of concrete to flow (BS 1881: Part 105). The apparatus consists essentially of a 700 mm square wooden table (covered by a flat steel plate) which is hinged along one side to permit a vertical movement of 40 mm, a slump cone (200 mm high, internal diameters 200 mm bottom and 130 mm top) and a wooden tamper (40 mm square base). The cone is placed at the centre of the table and filled with concrete in two equal layers, each layer being compacted by 10 light strokes of the tamper. The excess concrete is struck off the cone and the table top cleaned immediately; 30 seconds later the cone is removed gently. The free side of the table is then raised 40 mm (upper stop) and allowed to fall freely 15 times in about 45 to 75 seconds. The resulting spread of concrete is measured along its two diameters parallel to the sides of the table and the mean diameter, expressed to the nearest 5 mm, is taken as the measurement of flow. The flow test is valid only if the concrete is cohesive and uniform at the conclusion of the test. BS 1881: Part 105 limits the use of the flow test to concrete having maximum aggregate size up to 20 mm and a measured flow diameter between 500 and 650 mm.

13.3 Factors affecting workability

Various factors known to influence the workability of a freshly mixed concrete are shown in figure 13.3. From the following discussion it will be apparent that a change in workability associated with the constituent materials is mainly affected by water content and specific surface of cement and aggregate.

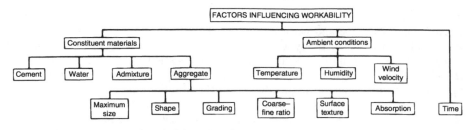

Figure 13.3 *Factors affecting workability of fresh concrete*

Cement and water

Typical relationships between the cement–water ratio (by volume) and the volume fraction of cement for different workabilities are shown in Chapter 17, figure 17.5. Hughes (1971) has shown that similar linear relationships exist, irrespective of the properties of the constituent materials. Workability is relatively insensitive to changes in only the cement content and, for practical purposes, may be considered dependent on only the water content for variations in the cement content of up to 10 to 20 per cent depending on the cement content, the effect of cement content being greater for the richer mixes.

For a given mix, the workability of the concrete decreases as the fineness of the cement increases as a result of the increased specific surface, this effect being more marked in rich mixtures. It should also be noted that finer cements improve the cohesiveness of a mix. With the exception of gypsum, the composition of cement has no apparent effect on workability. Unstable gypsum is responsible for *false set*, which can impair workability unless prolonged mixing or remixing of the fresh concrete is carried out. Variations in quality of water suitable for making concrete have no significant effect on workability.

Admixtures

The principal admixtures effecting improvement in the workability of concrete are water-reducing and air-entraining agents. The extent of the increase in workability is dependent on the type and amount of admixture used and the general characteristics of the fresh concrete.

Water-reducing admixtures are used to increase workability while the mix proportions are kept constant or to reduce the water content while maintaining constant workability. The former may result in a slight reduction in concrete strength. The water-reducing admixtures, including superplasticisers, are now more widely accepted as a workability aid. It should also be noted that the use of pulverised-fuel ash can also result in significant improvement in concrete workability.

Air-entraining agents are by far the most commonly used workability admixtures because they also improve both the cohesiveness of the plastic concrete and the frost resistance of the resulting hardened concrete. Two points of practical importance concerning air-entrained concrete are that for a given amount of entrained air, the increase in workability tends to be smaller for concretes containing rounded aggregates or low cement–water ratios (by volume) and, in general, the rate of increase in workability tends to decrease with increasing air content. However, as a guide it may be assumed that every 1 per cent increase in air content will increase the compacting factor by 0.01 and reduce the Vebe time by 10 per cent.

Aggregate

For given cement, water and aggregate contents, the workability of concrete is mainly influenced by the total surface area of the aggregate. The surface area is governed by the maximum size, grading and shape of the aggregate. Workability decreases as the specific surface increases, since this requires a greater proportion of cement paste to wet the aggregate particles, thus leaving a smaller amount of paste for lubrication. It follows that, all other conditions being equal, the workability will be increased when the maximum size of aggregate increases, the aggre-

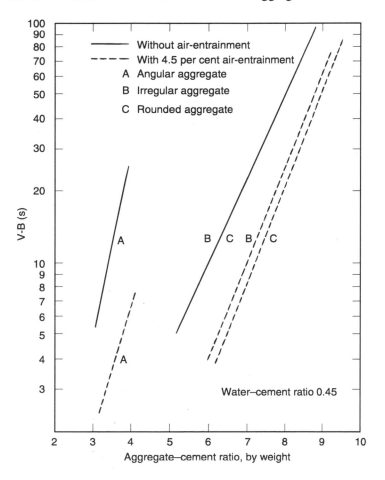

Figure 13.4 *Effect of aggregate shape on aggregate–cement ratio of concretes for different workabilities, based on Cornelius (1970)*

gate particles become rounded or the overall grading becomes coarser. However, the magnitude of this change in workability depends on the mix proportions, the effect of the aggregate being negligible for very rich mixes (aggregate–cement ratios approaching 2). The practical significance of this is that for a given workability and cement–water ratio the amount of aggregate which can be used in a mix varies depending on the shape, maximum size and grading of the aggregate, as shown in figure 13.4 and tables 13.2 and 13.3. The influence of air-entrainment (4.5 per cent) on workability is shown also in figure 13.4.

Several methods have been developed for evaluating the shape of aggregate, a subject discussed in Chapter 12. Angularity factors together with grading modulus and equivalent mean diameter provide a means of considering the respective effects of shape, size and grading of aggregate (see Chapter 17). Since the strength of a fully compacted concrete, for given materials and cement–water ratio, is not dependent on the ratio of coarse to fine aggregate, maximum economy can be obtained by using the coarse aggregate content producing the maximum workability for a given cement content (Hughes, 1960) (see figure 13.5). The use of optimum coarse aggregate content in concrete mix design is described in Chapter 17. It should be noted that it is the volume fraction of an aggregate, rather than its weight, which is important.

TABLE 13.2

Effect of maximum size of aggregate of similar grading zone on aggregate–cement ratio of concrete having water–cement ratio of 0.55 by weight, based on McIntosh (1964)

	Aggregate–cement ratio (by weight)					
Maximum aggregate size (mm)	Low workability		Medium workability		High workability	
	Irregular gravel	Crushed rock	Irregular gravel	Crushed rock	Irregular gravel	Crushed rock
9.5	5.3	4.8	4.7	4.2	4.4	3.7
19.0	6.2	5.5	5.4	4.7	4.9	4.4
37.5	7.6	6.4	6.5	5.5	5.9	5.2

TABLE 13.3

Effect of aggregate grading (maximum size 19.0 mm) on aggregate–cement ratio of concrete having medium workability and water–cement ratio of 0.55 by weight, based on McIntosh (1964)

	Aggregate–cement ratio	
Type of aggregate	Coarse grading	Fine grading
Rounded gravel	7.3	6.3
Irregular gravel	5.5	5.1
Crushed rock	4.7	4.3

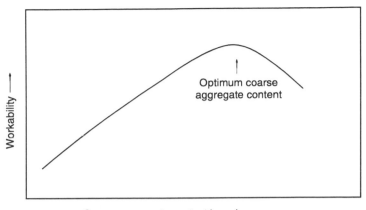

Figure 13.5 *A typical relationship between workability and coarse aggregate content of concrete, based on Hughes (1960)*

The effect of surface texture on workability is shown in figure 13.6. It can be seen that aggregates with a smooth texture result in higher workabilities than aggregates with a rough texture. Absorption characteristics of the aggregate also affect workability where dry or partially dry aggregates are used. In such a case workability drops, the extent of the reduction being dependent on the aggregate content and its absorption capacity.

Ambient conditions

Environmental factors that may cause a reduction in workability are temperature, humidity and wind velocity. For a given concrete, changes in workability are governed by the rate of hydration of the cement and the rate of evaporation of water. Therefore both the time interval from the commencement of mixing to compaction and the conditions of exposure influence the reduction in workability. An increase in the temperature speeds up the rate at which water is used for hydration as well as its loss through evaporation. Likewise wind velocity and humidity influence the workability as they affect the rate of evaporation. It is worth remembering that in practice these factors depend on weather conditions and cannot be controlled.

Time

The time that elapses between mixing of concrete and its final compaction depends on the general conditions of work such as the distance between the mixer and the point of placing, site procedures and general management. The associated reduction in workability is a direct result of loss of free water with time through evaporation, aggregate absorption and initial hydration of the cement. The rate of loss of workability is affected by certain characteristics of the constituent materials, for

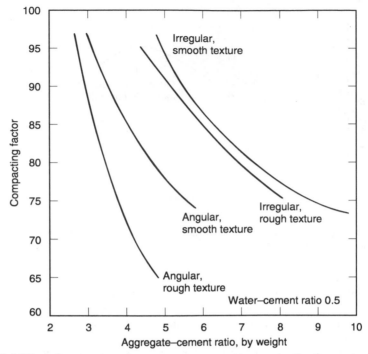

Figure 13.6 *Effect of aggregate surface texture on aggregate–cement ratio of concretes for different workabilities, based on Cornelius (1970)*

example, hydration and heat development characteristics of the cement, initial moisture content and porosity of the aggregate, as well as the ambient conditions.

For a given concrete and set of ambient conditions, the rate of loss of workability with time depends on the conditions of handling. Where concrete remains undisturbed after mixing until it is placed, the loss of workability during the first hour can be substantial, the rate of loss of workability decreasing with time as illustrated by curve A in figure 13.7. On the other hand, if it is continuously agitated, as in the case of ready-mixed concrete, the loss of workability is reduced, particularly during the first hour or so (see curve B in figure 13.7). However, prolonged agitation during transportation may increase the fineness of the solid particles through abrasion and produce a further reduction in workability. For concretes continuously agitated and undisturbed during transportation, the time intervals permitted (BS 5328) between the commencement of mixing and delivery on site are 2 hours and 1 hour respectively.

For practical purposes, loss of workability assumes importance when concrete becomes so unworkable that it cannot be effectively compacted, with the result that its strength and other properties become adversely affected. Corrective measures frequently taken to ensure that concrete at the time of placing has the desired workability are either an initial increase in the water content or an increase in the water content with further mixing shortly before the concrete is discharged. When this results in a water content greater than that originally intended, some reduction in strength and durability of the hardened concrete is to be expected unless the cement content is increased accordingly. This important fact is frequently over-

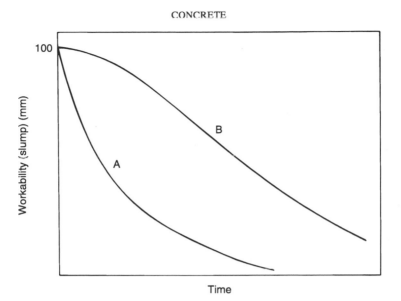

Figure 13.7 *Loss of workability of concrete with time: (A) no agitation and (B) agitated after mixing*

looked on site. It should be recalled that the loss of workability varies with the mix, the ambient conditions, the handling conditions and the delivery time. No restriction on delivery time is given in BS 8110: Part 1 but the concrete must be capable of being placed and effectively compacted without the addition of further water. For detailed information on the use of ready-mixed concrete the reader is advised to consult the work of Dewar (1973).

13.4 Stability

Apart from being sufficiently workable, fresh concrete should have a composition such that its constituent materials remain uniformly distributed in the concrete during both the period between mixing and compaction and the period following compaction before the concrete stiffens. Because of differences in the particle size and specific gravities of the constituent materials there exists a natural tendency for them to separate. This tendency is generally greater in high workability mixes. Flowing concrete is particularly prone to it and should be more carefully designed; as a guide a minimum of 450 kg m^{-3} combined fines (cement plus sand fractions less than 0.3 mm) should be used. Concrete capable of maintaining the required uniformity is said to be stable and most cohesive mixes belong to this category. For an unstable mix the extent to which the constituent materials will separate depends on the methods of transportation, placing and compaction. The two most common features of an unstable concrete are segregation and bleeding.

Segregation

When there is a significant tendency for the large and fine particles in a mix to become separated, segregation is said to have occurred. In general, the less cohesive the mix the greater the tendency for segregation to occur. Segregation is governed by the total specific surface of the solid particles including cement and the quality of mortar in the mix. Harsh, extremely wet and dry mixes as well as those deficient in sand, particularly the finer particles, are prone to segregation. As far as possible, conditions conducive to segregation such as jolting of concrete during transportation, dropping from excessive heights during placing and over-vibration during compaction should be avoided.

Blemishes, sand streaks, porous layers and honeycombing are a direct result of segregation. These features are not only unsightly but also adversely affect strength, durability and other properties of the hardened concrete. It is important to realise that the effects of segregation may not be indicated by the routine strength tests on control specimens since the conditions of placing and compaction of the specimens differ from those in the actual structure. There are no specific rules for suspecting possible segregation but after some experience of mixing and handling concrete it is not difficult to recognise mixes where this is likely to occur. For example, if a handful of concrete is squeezed in the hand and then released so that it lies in the palm, a cohesive concrete will be seen to retain its shape. A concrete which does not retain its shape under these conditions may well be prone to segregation and this is particularly so for wet mixes.

Bleeding

During compaction and until the cement paste has hardened there is a natural tendency for the solid particles, depending on size and specific gravity, to exhibit a downward movement. Where the consistency of a mix is such that it is unable to hold all its water some of it is gradually displaced and rises to the surface, and some may also leak through the joints of the formwork. Separation of water from a mix in this manner is known as bleeding. While some of the water reaches the top surface some may become trapped under the larger particles and under the reinforcing bars. The resulting variations in the effective water content within a concrete mass produce corresponding changes in its properties. For example, the strength of the concrete immediately underneath the reinforcing bars and coarse aggregate particles may be much less than the average strength and the resistance to permeation of fluids and ions in these areas is reduced. In general, the concrete strength tends to increase with depth below the top surface. The water which reaches the top surface presents the most serious practical problems. If it is not removed, the concrete at and near the top surface will be much weaker and less durable than the remainder of the concrete. This can be particularly troublesome in slabs which have a large surface area and are subjected to abrasive wear. On the other hand, removal of the surface water will unduly delay the finishing operation on the site.

The risk of bleeding increases when concrete is compacted by vibration although this may be minimised by using a correctly designed mix and ensuring that the concrete is not over-vibrated. Rich mixes tend to bleed less than lean mixes. The type of cement employed is also important, the tendency for bleeding to occur decreasing as the fineness of the cement or its alkaline and tricalcium aluminate (C_3A) content increases. Air-entrainment provides another very effective means of controlling bleeding in, for example, wet lean mixes where both segregation and bleeding are frequently troublesome.

14

Properties of Hardened Concrete

The properties of fresh concrete are important only in the first few hours of its history, whereas the properties of hardened concrete assume an importance which is retained for the remainder of the life of the concrete. The important properties of hardened concrete are strength, deformation under load, durability, permeability and shrinkage. In general, strength is considered to be the most important property and the quality of concrete is often judged by its strength. There are, however, many occasions when other properties are more important, for example, low permeability and low shrinkage are required for water-retaining structures. Although in most cases an improvement in strength results in an improvement of the other properties of concrete, there are exceptions. For example, increasing the cement content of a mix improves strength but results in higher shrinkage which in extreme cases can adversely affect durability and permeability. One of the primary objectives of this chapter is to help the reader to understand the factors which affect each of the important properties of hardened concrete.

14.1 Strength

The strength of concrete is defined as the maximum load (stress) it can carry. As the strength of concrete increases its other properties usually improve and since the tests for strength, particularly in compression, are relatively simple to perform concrete compressive strength is commonly used in the construction industry for the purpose of specification and quality control. Concrete is a comparatively brittle material which is relatively weak in tension.

Compressive strength

The compressive strength of concrete is taken as the maximum compressive load it can carry per unit area. Concrete strengths of up to 80 N mm^{-2} can be achieved by selective use of the type of cement, mix proportions, methods of compaction and curing conditions.

Concrete structures, except for road pavements, are normally designed on the basis that concrete is capable of resisting only compression, the tension being carried by steel reinforcement. In the UK a 150 mm cube is commonly used for determining the compressive strength. The standard method described in BS 1881: Part 116 requires that the test specimen should be cured in water at $20 \pm 2°C$ and crushed, by loading it at a constant rate of stress increase of between 12 and 24 N mm^{-2} min^{-1}, immediately after it has been removed from the curing tank.

Tensile strength

The tensile strength of concrete is of importance in the design of concrete roads and runways. For example, its flexural strength or modulus of rupture (tensile strength in bending) is utilised for distributing the concentrated loads over a wider area of road pavement. Concrete members are also required to withstand tensile stresses resulting from any restraint to contraction due to drying or temperature variation.

Unlike metals, it is difficult to measure concrete strength in direct tension and indirect methods have been developed for assessing this property. Of these the split cylinder test is the simplest and most widely used. This test is fully described in BS 1881: Part 117 and entails diametrically loading a cylinder in compression along its entire length. This form of loading induces tensile stresses over the loaded diametrical plane and the cylinder splits along the loaded diameter. The magnitude of the induced tensile stress f_{ct} at failure is given by

$$f_{ct} = \frac{2F}{\pi l d}$$

where F is the maximum applied load, and l and d are the cylinder length and diameter respectively.

The flexural strength of concrete is another indirect tensile value which is also commonly used (BS 1881: Part 118). In this test a simply supported plain concrete beam is loaded at its third points, the resulting bending moments inducing compressive and tensile stresses in the top and bottom of the beam respectively. The beam fails in tension and the flexural strength (modulus of rupture) f_{cr} is defined by

$$f_{cr} = \frac{FL}{bd^2}$$

where F is the maximum applied load, L the distance between the supports, and b and d are the beam breadth and depth respectively at the section at which failure occurs.

The tensile strength of concrete is usually taken to be about one-tenth of its compressive strength. This may vary, however, depending on the method used for measuring tensile strength and the type of concrete. In general the direct tensile strength and the split cylinder tensile strength vary from 5 to 13 per cent and the flexural strength from 11 to 23 per cent of the concrete cube compressive strength. In each case, as the strength increases the percentage decreases. As a guide, the modulus of rupture may be taken as 0.7 $\sqrt{}$ cube strength N mm^{-2} and the direct tensile strength as 0.45 $\sqrt{}$ cube strength N mm^{-2} although, where possible, values based on tests using the actual concrete in question should be obtained.

14.2 Factors influencing strength

Several factors which affect the strength of concrete are shown in figure 14.1. In this section their influence is discussed with particular reference to compressive strength. In general, tensile strength is affected in a similar manner.

Influence of the constituent materials

Cement

The influence of cement on concrete strength, for given mix proportions, is determined by its fineness and chemical composition through the process of hydration (see Chapter 12). The gain in concrete strength as the fineness of its cement particles increases is shown in figure 14 .2. The gain in strength is most marked at early ages and after 28 days the relative gain in strength is much reduced. At some later age the strength of concrete made with fine cements may not be very different from that made with normal cement (300 m^2 kg^{-1}).

The role of the chemical composition of cement in the development of concrete strength can best be appreciated by studying table 14.1 and figures 14.3 and 14.4. It is apparent that cements containing a relatively high percentage of tricalcium silicate (C_3S) gain strength much more rapidly than those rich in dicalcium silicate (C_2S), as shown in figure 14.3; however, at later ages the difference in the corresponding strength values is small. In fact there is a tendency for concretes made with low-heat cements eventually to develop slightly higher strengths (figure 14.4). This is possibly due to the formation of a better quality gel structure in the course of hydration.

Water

A concrete mix containing the minimum amount of water required for complete hydration of its cement, if it could be fully compacted, would develop the maximum attainable strength at any given age. A water–cement ratio of approximately 0.25 (by weight) is required for full hydration of the cement but with this water content a normal concrete mix would be extremely dry and virtually impossible to compact.

A partially compacted mix will contain a large percentage of voids and the concrete strength will drop. On the other hand, while facilitating placing and compaction, water in excess of that required for full hydration produces a somewhat porous structure resulting from loss of excess water, even for a fully compacted concrete. In practice a concrete mix is designed with a view to obtaining maximum compaction under the given conditions. In such a concrete, as the ratio of water to cement increases the strength decreases in a manner similar to that illustrated in figure 17.3 (see Chapter 17).

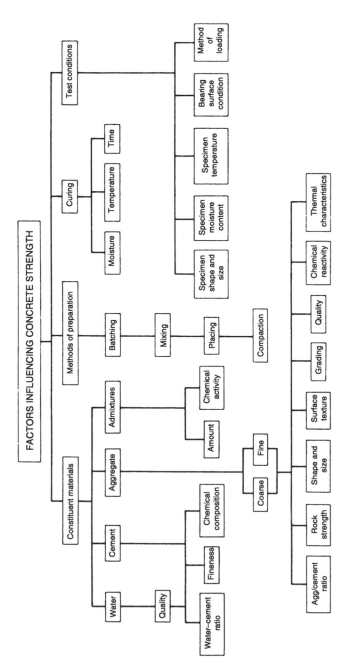

Figure 14.1 *Factors affecting strength of concrete*

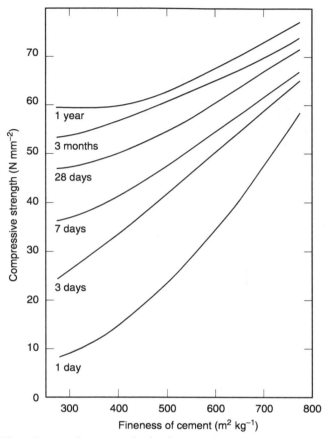

Figure 14.2 *Effect of cement fineness on the development of concrete strength, based on Bennett and Collings (1969)*

Aggregate

When concrete is stressed, failure may originate within the aggregate, the matrix or at the aggregate–matrix interface; or any combination of these may occur. In general the aggregates are stronger than the concrete itself and in such cases the aggregate strength has little effect on the strength of concrete.

TABLE 14.1
Chemical composition of various Portland cements with similar fineness

Cement	Compound composition (per cent)			
	C_3S	C_2S	C_3A	C_4AF
A	55	16	12	8
B	50	21	11	9
C	43	33	5	11
D	35	41	6	12

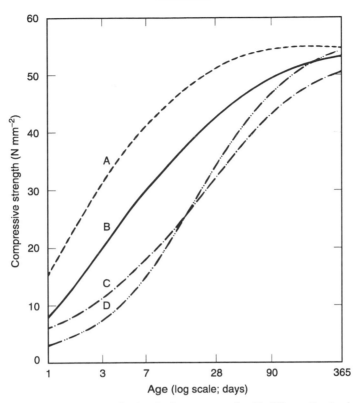

Figure 14.3 *Development of strength of typical concrete made with different Portland cements (see table 14.1)*

The bond (aggregate–matrix interface) is an important factor determining concrete strength. Bond strength is influenced by the shape of the aggregate, its surface texture and cleanliness. A smooth rounded aggregate will result in a weaker bond between the aggregate and matrix than an angular or irregular aggregate or an aggregate with a rough surface texture. The associated loss in strength however may be offset by the smaller water–cement ratio required for the same workability. Aggregate shape and surface texture affect the tensile strength more than the compressive strength. A fine coating of impurities, such as silt and clay, on the aggregate surface hinders the development of a good bond. A weathered and decomposed layer on the aggregate can also result in poor bond as this layer can readily become detached from the sound aggregate beneath.

The aggregate size also affects the strength. For given mix proportions, the concrete strength decreases as the maximum size of aggregate is increased. On the other hand, for a given cement content and workability this effect is opposed by a reduction in the water requirement for the larger aggregate. However, it is probable that beyond a certain size of aggregate there is no obvious advantage in further increasing the aggregate size except perhaps in some instances when larger aggregate may be more readily available. The optimum maximum aggregate size varies with the richness of the mix, being smaller for the richer mixes, and generally lies between 10 and 50 mm.

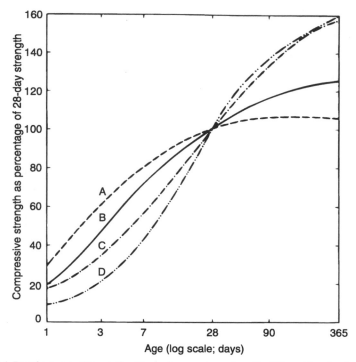

Figure 14.4 *Development of strength of typical concrete made with different Portland cements (see table 14.1)*

Concrete of a given strength can be produced with aggregates having a variety of different gradings provided due care is exercised to ensure that segregation does not occur. The suitability of a grading to some extent depends on the shape and texture of the aggregate. Aggregates which react with the alkali content of a cement adversely affect concrete strength. This is rarely a problem in the UK, even though some cases have been reported since 1976 (BRE, 1982).

Admixtures

As a general rule, admixtures can only affect concrete strength by changing the hydration processes and the air content of the mix and/or by enabling changes to be made to the mix proportions, most importantly to the water–cement ratio. Accelerating admixtures increase the rate of hydration thereby providing an increased early strength with little or no change at later ages unless the increased rate of heat evolution causes internal cracking in which case a lower strength will result. In contrast, with retarding admixtures the early strength of concrete is reduced owing to the delay in setting time. Provided no air is entrained, the concrete strength will be approximately the same as that of the control mix within a few days. Air entrainment in concrete will cause a reduction in strength at all ages and to achieve a required strength the mix cement content has to be increased. However, in practice the increased yield and improvement in workability are also taken into account in the mix design when using air-entraining admixtures and there is generally no significant change in concrete strength for the usual range of air content. Most water-reducing admixtures, including superplasticisers, do not have any significant effect on the hydration of the cement. Thus, when these admixtures are used to improve workability no significant change in strength should be expected, provided of course that the air content remains unchanged. On the other hand if the water content of the mix is reduced while maintaining workability an increase in strength corresponding to the new water–cement ratio will result.

It should be noted that the effect of particular admixtures in concrete depends on the precise nature of the admixtures themselves, the constituent materials and proportions of the mix and the ambient conditions (particularly temperature). In consequence, although for practical purposes the above mentioned guidelines apply, to obtain an accurate estimate of their effect on concrete strength in individual circumstances it is necessary to carry out trial mixes.

Influence of the methods of preparation

When concrete materials are not adequately mixed into a consistent and homogeneous mass, some poor quality concrete is inevitably the result. Even when a concrete is adequately mixed care must be taken during placing and compaction to minimise the probability of the occurrence of bleeding, segregation and honeycombing all of which can result in patches of poor quality concrete. A properly designed concrete mix is one that does not demand the impossible from site operatives before it can be fully compacted in its final location. If full compaction is not achieved the resulting voids produce a marked reduction in concrete strength.

Influence of curing

Curing of concrete is a prerequisite for the hydration of the cement content. For a given concrete, the amount and rate of hydration and furthermore the physical make-up of the hydration products are dependent on the time–moisture–temperature history.

Generally speaking, the longer the period during which concrete is kept in water, the greater its final strength. It is normally accepted that a concrete made with ordinary Portland cement and kept in normal curing conditions will develop about 75 per cent of its final strength in the first 28 days. This value varies with the nominal strength of concrete however, increasing as concrete strength increases. The gain in strength with age up to the time of loading can now be used only for estimating the static modulus of elasticity of concrete in structures (BS 8110: Part 2). The age factors for strength prescribed for this purpose are given in table 14.2.

The development of concrete strength under various curing conditions is shown in figure 14.5. It is apparent that concrete left in air achieves the lowest strength values at all ages owing to the evaporation of the free mixing water from the concrete. The gain in strength depends on a number of factors such as relative humidity, wind velocity and the size of structural member or test specimen. Figure 14.5 shows that both the increased hydration due to improvements in initial curing

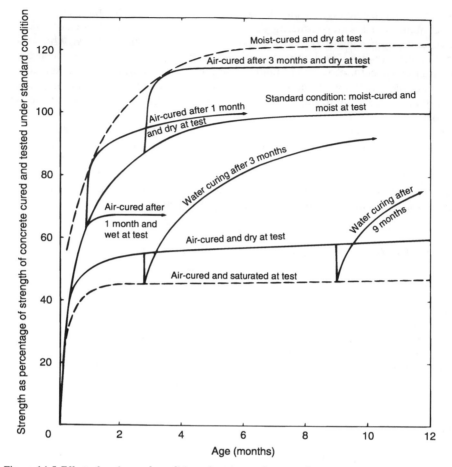

Figure 14.5 *Effect of curing and condition of concrete when tested on concrete strength, based on Gilkey (1937)*

TABLE 14.2
Age factors for strength of concrete

Characteristic strength (N mm⁻²)	Age at loading					
	7 days	28 days	2 months	3 months	6 months	1 year
20	0.68	1.00	1.10	1.15	1.20	1.25
25	0.66	1.00	1.10	1.16	1.20	1.24
30	0.67	1.00	1.10	1.17	1.20	1.23
40	0.70	1.00	1.10	1.14	1.19	1.25
50	0.72	1.00	1.08	1.11	1.15	1.20

(moist or water curing at normal temperature) and the condition of the concrete at the time of testing have a significant effect on the final apparent strength of concrete. It should also be noted that moist (or water) curing after an initial period in air results in a resumption of the hydration process and the concrete strength is further improved with time, although the optimum strength may not be realised.

The temperature at which concrete is cured has an important bearing on the development of its strength with time. The rate of gain in strength of concrete made with ordinary Portland cement increases with increase in concrete temperature at early ages (figure 14.6), although at later ages the concrete made and cured

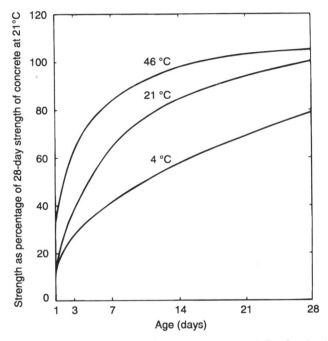

Figure 14.6 *Comparative compressive strength of concrete cast, sealed and maintained at different temperatures, based on Price (1951)*

at lower temperatures shows a somewhat higher strength. Figure 14.7 shows how a high temperature during the placing and setting of concrete can adversely affect the development of its strength from early ages. On the other hand, when the initial temperature is lower than the subsequent curing temperature, then higher temperatures during final curing result in significantly higher strengths (figure 14.8). A possible explanation for this behaviour is that a rapid initial hydration appears to form a gel structure (hydration product) of an inferior quality and this adversely affects concrete strength at later ages. Concretes made with other Portland cements would respond to temperature in a somewhat similar manner.

It has been suggested that the strength of concrete can be related to the product of age and curing temperature, commonly known as *maturity*. However, such relationships are dependent on a number of factors such as curing temperature his-

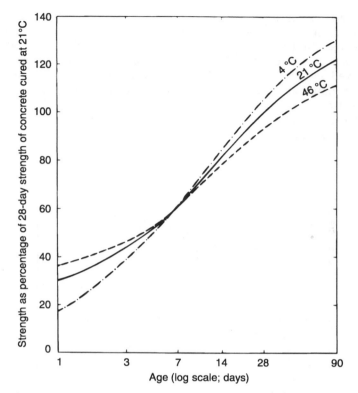

Figure 14.7 *Comparative compressive strength of sealed concrete specimens maintained at different temperatures for 2 hours after casting and subsequently cured at 21°C, based on Price (1951)*

tory, particularly the temperature at early ages (see figures 14.7 and 14.8), and are therefore limited in their general applicability for predicting concrete strength.

Influence of test conditions

The conditions under which tests to determine concrete strength are carried out can have a considerable influence on the strength obtained and it is important that these effects are understood if tests results are to be correctly interpreted.

Specimen shape and size

Three basic shapes used for the determination of compressive strength are the cube, cylinder and square prism. Each shape gives different strength results and furthermore for a given shape the strength also varies with size. Figure 14.9 shows the influence of specimen diameter and it can be seen that as the size decreases, the apparent strength increases. The measured strength of concrete is also affected by height–diameter ratio (figure 14.10). For height–diameter ratios less than 2

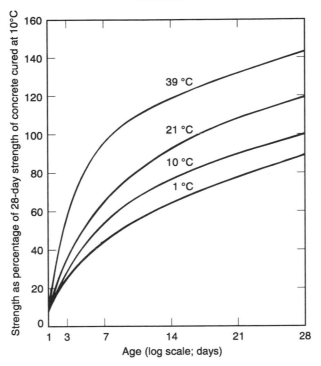

Figure 14.8 *Comparative compressive strength of concrete cast, sealed and maintained at 10°C for the first 24 hours and subsequently cured at different temperatures, based on Price (1951)*

strength begins to increase rapidly owing to the restraint provided by the machine platens. Strength remains sensibly constant for height–diameter ratios between 2 and 3 and thereafter shows a slight reduction. The relative influence of slenderness may be modified by several inherent characteristics of concrete such as strength, air-entrainment, strength of aggregate, degree of moist curing and moisture content at the time of testing.

BS 1881: Part 116 specifies the use of concrete cubes for determining compressive strength; 150 mm cubes are widely used in the UK for quality-control purposes. However, cored cylindrical specimens are used for measuring the compressive strength of concrete in *in situ* and precast members (BS 1881: Part 120). This standard gives a set of equations for converting the measured strength of a core into an equivalent *in situ* cube strength. It should be noted that the estimation of equivalent standard cube strengths from core strengths is no longer considered to be valid, because of large variations in the relationship between the two strengths arising from variations in site conditions including percentage of reinforcement, dimensions of members and the methods of compaction and curing. The effect on the measured strength of the variations in the height–diameter ratio of drilled cores is taken into account by using the correction factors given in table 14.3. The effect of the shape of the test specimen is taken into account by multiplying the cylinder (core) strength, for a cylinder (core) having a height–diameter ratio of 2, by 1.25 to obtain the equivalent cube strength. Since the relationship between height–diameter ratio and strength depends on the type of concrete, the

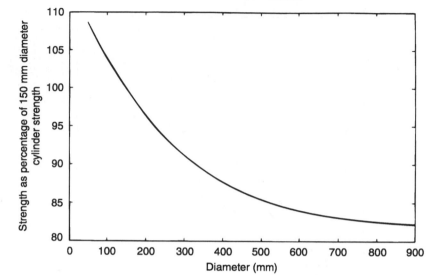

Figure 14.9 *Effect of specimen size on the apparent 28-day concrete compressive strength for specimens with a height–diameter ratio of 2 and aggregate whose maximum diameter is one-quarter of the specimen, based on Price (1951)*

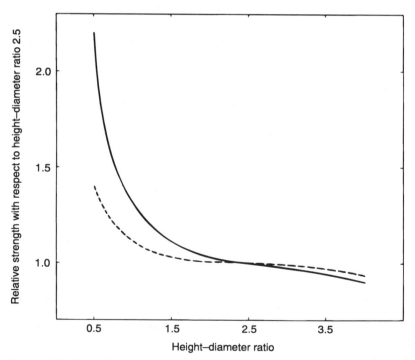

Figure 14.10 *Effect of height–diameter ratio on concrete compressive strength for specimens moist-cured at room temperature and tested wet*

use of one set of correction factors of the type given in table 14.3 can only be suitable for a limited range of concrete materials. It should be noted that the correction factors given in table 14.3 are most likely to favour low-strength concretes. The estimation of the equivalent *in situ* cube strength from compression tests performed on cored cylindrical specimens can be further influenced by a multitude of other factors and the reader is advised to consult the Concrete Society (1976) Technical Report on concrete core testing for strength, the British Standard Guide on Assessment of Concrete Strength in Existing Structures (BS 6089), Murphy (1984) and Munday and Dhir (1984).

TABLE 14.3

Correction factor for compression tests on cylinder (cores)

Specimen height–diameter ratio	Strength correction factor
2.00	1.00
1.75	0.97
1.50	0.92
1.25	0.87
1.00	0.80

Specimen moisture content and temperature

This should not be confused with the effect of moisture and temperature during curing. The strength of concrete can be influenced by the absence or presence of moisture and by temperature only when these conditions generate internal stresses which change the magnitude of external load required to bring about failure. Since the mode of failure in different strength tests is different, it follows that the influence of moisture content and temperature on the apparent strength varies.

In the case of compression tests, air-dry concrete has a significantly higher strength than concrete tested in a saturated condition (figure 14.5). The lower strengths of wet concrete can be attributed mainly to the development of internal pore pressure as the external load is applied.

The flexural strength of saturated concrete is greater however than that of concrete which is only partially dry owing to tensile stresses developed near the surface of the dry concrete by differential shrinkage. The initial drying is therefore critical with the apparent strength reaching a minimum value within the first few days; thereafter it begins to increase gradually as the concrete approaches a completely dry state. A thoroughly dry concrete specimen in which tensile stresses are not being induced has a higher apparent flexural strength than saturated concrete. The indirect tensile or split cylinder strength is lower for saturated concrete than for thoroughly dried concrete.

Since the influence of the moisture content on concrete strength varies with the type of test, standard strength tests (BS 1881: Part 116) should be performed on specimens in a saturated condition.

Method of loading

The compressive strength of concrete increases as the lateral confining pressure increases (figure 14.11).

The apparent strength of concrete is affected by the rate at which it is loaded. In general, for static loading, the faster the loading rate the higher the indicated strength. However, the relative effects of the rate of loading vary with the nominal strength, age and extent of moist curing. High-strength mature concretes cured in water are most sensitive to loading rate and particularly so for loading rates greater than 600 N mm^{-2} min^{-1}. BS 1881: Part 116 requires concrete in compression tests to be loaded at 12–24 N mm^{-2} min^{-1} and within this small range of loading rates variations in the measured strength of concrete will be insignificant. The standard rates of loading for flexural and split cylinder tests correspond to rates of increase of tensile stress of 1.2–6.0 N mm^{-2} min^{-1} and 1.2–2.4 N mm^{-2} min^{-1} respectively (BS 1881: Parts 117 and 118). For the flexural test the standard requires the use of the lower loading rates for low strength concretes and the higher loading rates for high strength concretes.

When loads on a structure are predominantly cyclic (repeated loading and unloading) in character, the effects of fatigue should also be considered. This kind of loading produces a reduction in strength. A reduction in strength of as much as 30 per cent of the normal static strength value can take place, although this depends on the stress–strength ratio, the frequency of loading and the type of concrete.

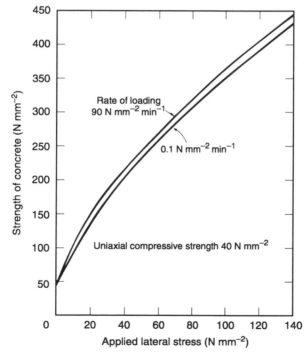

Figure 14.11 *Effect of lateral compression on concrete compressive strength*

Structural concrete is commonly subjected to sustained loads. It is probable that concrete can withstand higher loads if a constant load is maintained before loading to failure. Improvement in compressive strength can occur for sustained loads up to 85 per cent of the normal static strength although the actual gain in strength depends on the duration and magnitude of load, type of concrete and age. The increase in strength is probably due to consolidation of the concrete under sustained load and the redistribution of stresses within the concrete.

14.3 Deformation

Load-dependent deformation

Concrete deforms under load, the deformation increasing with the applied load and being commonly known as elastic deformation (figure 14.12). Concrete also continues to deform with time, under constant load; this is known as time-dependent deformation or creep (figure 14.13).

Elastic deformation

Unlike that for metals, the load–deformation relationship for concrete subjected to a continuously increasing load is non-linear in character. The non-linearity is most marked at higher loads. When the applied load is released, the concrete does not fully recover its original shape (see figure 14.12). Under repeated loading and unloading, the deformation at a given load level increases, although at a decreasing rate, with each successive cycle. All these characteristics of concrete indicate that it should be considered as a quasi-elastic material and when computing the elastic constants, namely, the modulus of elasticity and Poisson's ratio, the method employed should be clearly stated. It is only for simplicity and convenience that the elastic modulus is assumed to be constant in both concrete technology and design of concrete structures.

Modulus of elasticity

This is defined as the ratio of load per unit area (stress) to the elastic deformation per unit length (strain)

$$\text{modulus of elasticity } E = \text{stress/strain} = \sigma/\epsilon$$

The modulus is used when estimating the deformation, deflection or stresses under normal working loads. Since concrete is not a perfectly elastic material, the modulus of elasticity depends on the particular definition adopted (see figure 14.12).

The initial tangent modulus is of little value for structural applications since it has significance only for low stresses during the first load cycle. The tangent modulus is difficult to determine and is applicable only within a narrow band of stress levels. The secant modulus is easily determined and takes into account the total deformation at any one point. The method prescribed in BS 1881: Part 121

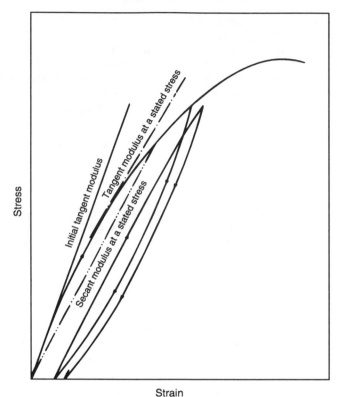

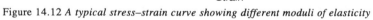

Figure 14.12 *A typical stress–strain curve showing different moduli of elasticity*

requires repeated loading and unloading before the specimen is loaded for determination of its secant modulus of elasticity from a stress–strain curve, which then
approaches a straight line for stresses up to one-third of the concrete cylinder
strength. The modulus of elasticity for most concretes, at 28 days, ranges from 15
to 40 kN mm^{-2}. As a guide, the modulus may be assumed to be $3.8\sqrt{}$(concrete
strength, N mm^{-2}) kN mm^{-2} for normal weight concrete. For structural design
purposes, the short-term elastic modulus values for normal weight concrete given
in BS 8110: Part 2 may be used.

Poisson's ratio

When concrete is subjected to axial compression, it contracts in the axial direction
and expands laterally. Poisson's ratio is defined as the ratio of the lateral strain to
the associated axial strain and varies from 0.1 to 0.3 for normal working stresses.
A value of 0.2 is commonly used.

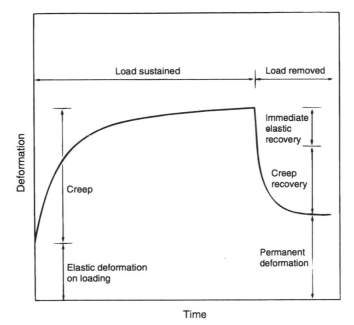

Figure14.13 *A typical illustration of deformation of concrete subjected to constant load*

Factors influencing the elastic behaviour of concrete

Concrete is a multiphase material and its resistance to deformation under load is dependent on the stiffness of its various phases, such as aggregate, cement paste and voids, and the interaction between individual phases. In general the factors which influence the strength of concrete also affect deformation although the extent of their influence may well vary. The modulus of elasticity increases with strength, although the two properties are not directly related because different factors exert varying degrees of influence on strength and modulus of elasticity. Although relationships between the two properties may be derived they are only applicable within the range of variables considered. In short there is no unique relationship between strength and modulus of elasticity.

The influence of stress level and rate of loading on both the axial and lateral strains is shown in figure 14.14 for a concrete which has been cured in water (20°C ± 1°C) and air-dried before loading in uniaxial compression. It should be noted that the slower the rate of loading the greater the strains at a given stress level. This is due to the fact that during loading both elastic and creep strains occur, the creep strains increasing with the duration of time for which the load acts. The variation of tangent modulus and Poisson's ratio, for the same concrete, with stress level and rate of loading are shown in figure 14.15. The general trends depicted in the figure apply to all concretes although the exact relationship may vary with the type of concrete and methods employed for measuring the properties.

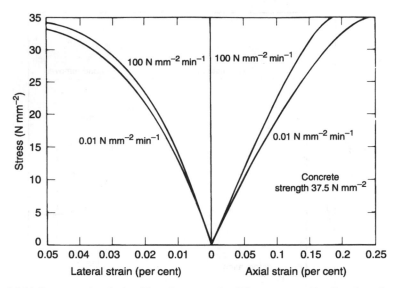

Figure 14.14 *Stress–strain relationships of concrete for different rates of loading, based on Dhir and Sangha (1972)*

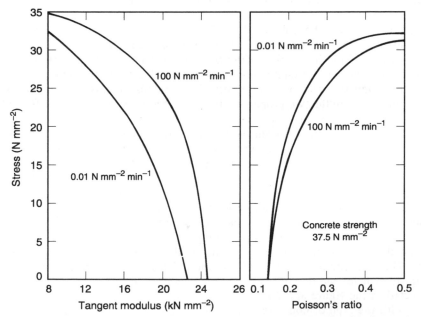

Figure 14.15 *Variation of tangent modulus and Poisson's ratio with stress level and rate of loading, based on Dhir and Sangha (1972)*

Creep deformation

When concrete is subjected to a sustained load it first undergoes an instantaneous deformation (elastic) and thereafter continues to deform with time (figure 14.13). The increase in strain with time is termed creep. It should be noted that after an initially high rate of creep the creep continues but at a continuously decreasing rate, except when the sustained load is large enough to cause failure, in which case the rate of creep increases before failure occurs. The removal of the sustained load results in an immediate reduction in strain and this is followed by a gradual decrease in strain over a period of time. The gradual decrease in strain is called creep recovery. Creep and creep recovery are related phenomena with somewhat similar characteristics. Creep is not wholly reversible and some permanent-strain remains after creep recovery is complete.

Since concrete in service is subjected to sustained loads for long periods of time the effect of creep strains, which normally exceed the elastic strains, must be considered in structural design. Failure to include the effects of creep strains may lead to serious underestimates of beam deflections and overall structural deformation and, in those cases where structural stability is involved, the provision of members with inadequate strength. In prestressed concrete, allowance must be made for the loss of tension in the prestressing tendons resulting from the shortening of a member under the action of creep. Creep strain can also be beneficial in that it can relieve local stress concentrations which might otherwise lead to structural damage. A classic example of this is the reduction of shrinkage stresses in restrained members.

Factors influencing creep

Both the type of concrete (as described by its ingredients, curing history, strength and age) and the relative magnitude of the applied stress with respect to concrete strength (stress–strength ratio) affect the creep strain. For a given concrete the creep strain is almost directly proportional to the stress–strength ratio for ratios up to about one-third. For the same stress–strength ratio, the creep strain increases as both cement content and water–cement ratio increase (figure 14.16) and decreases as the relative humidity and age at loading increase (figure 14.17).

The influence of the constituent materials of concrete on creep is somewhat complex. The different types of cement influence creep because of the associated different rates of gain in concrete strength. For example, concrete made with rapid-hardening Portland cement shows less creep than concrete made with ordinary Portland cement and loaded at the same age. Normal rock aggregates have a restraining effect on concrete creep and the use of large, high modulus aggregates can be beneficial in this respect.

The environmental conditions of concrete subjected to a sustained load can also have a marked effect on the magnitude of creep. For example, concrete cured under humid conditions and then loaded in a relatively dry atmosphere undergoes a greater creep strain than if the original conditions had been maintained. A concrete allowed to reach moisture equilibrium before loading, however, is not adversely affected. Creep decreases as the mass of concrete increases owing to the slower rate at which loss of water can take place within a large mass of concrete.

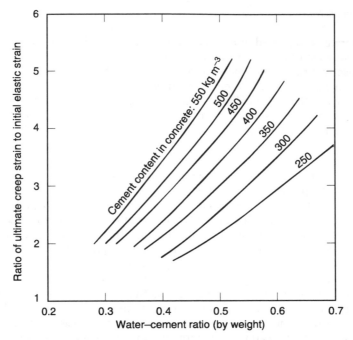

Figure 14.16 *Effect of cement content and water–cement ratio on ultimate creep strain of concrete at 50 per cent relative humidity loaded at an age of 28 days, based on CEB-FIP (1970)*

Several mathematical equations have been proposed for estimating creep strains. Both the rate of creep and its ultimate magnitude are dependent on a multitude of factors. These expressions can only be truly applicable to concretes similar to those for which they were designed. Nevertheless, creep strains are usually

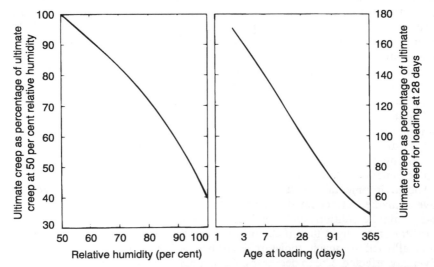

Figure 14.17 *Effect of relative humidity and age at loading on creep, based on Neville (1970)*

estimated for design purposes. It has been noted that, for a given concrete, creep strain depends on the stress–strength ratio. For practical purposes, concrete creep strain may be assumed to be directly proportional to the elastic deformation up to a stress–strength ratio of about two-thirds. On this assumption the ultimate creep strain may be estimated for a given water–cement ratio and cement content using figure 14.16. This creep value may be modified for different environmental moisture conditions and ages on loading using figure 14.17. It may be assumed that 50 per cent of the total estimated creep will take place within one month and 75 per cent within the first six months after loading.

Load-independent deformation

Concrete deformations also occur as a result of load-independent effects, such as various types of shrinkage as well as thermal expansion strain.

Shrinkage

Shrinkage of concrete is caused by the settlement of solids and the loss of free water from the plastic concrete (plastic shrinkage), by the chemical combination of cement with water (autogenous shrinkage) and by the drying of concrete (drying shrinkage). Where movement of the concrete is restrained, shrinkage will produce tensile stresses within the concrete which may cause cracking. Most concrete structures experience a gradual drying out and the effects of drying shrinkage should be minimised by the provision of movement joints and careful attention to detail at the design stage.

Plastic shrinkage

Shrinkage which takes place before concrete has set is known as plastic shrinkage. This occurs as a result of the loss of free water with, or without, significant settlements of solids in the mix. Since evaporation usually accounts for a large proportion of the water losses plastic shrinkage is most common in slab construction and is characterised by the appearance of surface cracks which can extend quite deeply into the concrete. Crack patterns associated with significant settlement (plastic settlement cracks) generally coincide with the line of the reinforcement. Preventive measures are usually based on methods of reducing water loss. This can be achieved in practice by making the mix more cohesive and by covering concrete with wet hessian or polythene sheets or by spraying it with a membrane curing compound.

Autogenous shrinkage

In a set concrete, as hydration proceeds, a net decrease in volume occurs since the hydrated cement gel has a smaller volume than the sum of the cement and water constituents. As hydration continues in an environment where the water content is constant, such as inside a large mass of concrete, this decrease in volume of the cement paste results in shrinkage of the concrete. This is known as autogenous shrinkage because, as the name implies, it is self-produced by the hydration of cement. However, when concrete is cured under water, the water taken up by cement during hydration is replaced from outside and furthermore the gel particles absorb more water, thus producing a net increase in volume of the cement paste and an expansion of the concrete. On the other hand if concrete is kept in a dry atmosphere water is drawn out of the hydrated gel and additional shrinkage, known as drying shrinkage, occurs.

Several factors influence the rate and magnitude of autogenous shrinkage. These include the chemical composition of cement, the initial water content, temperature and time. The autogenous shrinkage can be up to 100×10^{-6} of which 75 per cent occurs within the first three months.

Drying shrinkage

When a hardened concrete, cured in water, is allowed to dry it first loses water from its voids and capillary pores and only starts to shrink during further drying when water is drawn out of its cement gel. This is known as drying shrinkage and in some concretes it can be greater than 1500×10^{-6}, but a value in excess of 800×10^{-6} is usually considered to be undesirable for most structural applications. After an initial high rate of drying shrinkage concrete continues to shrink for a long period of time but at a continuously decreasing rate (see figure 14.18). For practical purposes, it may be assumed that for small sections 50 per cent of the total shrinkage occurs in the first year.

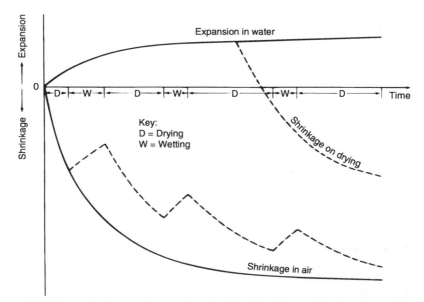

Figure 14.18 *Drying shrinkage and expansion characteristics of concrete*

When concrete which has been allowed to dry out is subjected to a moist environment, it swells. However, the magnitude of this expansion is not sufficient to recover all the initial shrinkage even after prolonged immersion in water. Concrete subjected to cyclic drying and wetting approaches the same shrinkage level as that caused by complete drying (figure 14.18). A test procedure for determining shrinkage is described in BS 1881: Part 5.

Factors affecting drying shrinkage

Several factors influence the overall drying shrinkage of concrete. These include the type, content and proportion of the constituent materials of concrete, the size and shape of the concrete structure, the amount and distribution of reinforcement and the relative humidity of the environment.

In general, drying shrinkage is directly proportional to the water–cement ratio and inversely proportional to the aggregate–cement ratio (see figure 14.19). Because of the interaction of the effects of aggregate–cement and water–cement ratios, it is possible to have a rich mix with a low water–cement ratio giving higher shrinkage than a leaner mix with a higher water–cement ratio (Dhir *et al.*, 1978). For a given water–cement ratio, shrinkage increases with increasing cement content.

Since the aggregate exerts a restraining influence on shrinkage the maximum aggregate content compatible with other required properties is desirable. When the

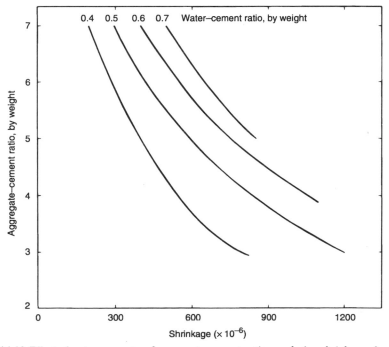

Figure 14.19 *Effect of water–cement and aggregate–cement ratios on drying shrinkage of concrete at 20°C and 50 per cent relative humidity, based on Lea (1970)*

aggregate itself is susceptible to large moisture movement, this can aggravate shrinkage (or swelling) of the concrete (BRE Digest 35, 1968; Dhir *et al.*, 1978) and may result in excessive cracking and large deflections of beams and slabs. The composition and fineness of cement can also affect its shrinkage characteristics. In general, shrinkage increases as the specific surface area of cement increases (table 14.4) although this effect is slight and is usually overshadowed by the effects of water–cement ratio and aggregate–cement ratio. Increases in dicalcium silicate (C_2S) content and ignition loss usually result in increased shrinkage. Tricalcium aluminate (C_3A) appears to influence the expansion of concrete under moist conditions. Nevertheless, the shrinkage characteristics of concrete cannot reliably be predicted from an analysis of the chemical composition of its cement. In general, admixtures which reduce the water requirement of concrete without affecting its other properties will reduce its shrinkage. Air-entrainment itself has no significant influence on shrinkage. Calcium chloride may considerably increase shrinkage.

The size and shape of a specimen affect the rate of moisture movement in concrete and this in turn influences the rate of volume change. Since drying begins from the surface, it follows that the greater the surface area per unit mass, the greater the rate of shrinkage. For a given shape, the initial rate of shrinkage is greater for small specimens although there will be little difference in the ultimate drying shrinkage, if this stage is ever reached for very large masses of concrete.

TABLE 14.4

Influence of the fineness of cement on drying shrinkage of concrete (aggregate–cement ratio 3) after 500 days, Bennett and Loat (1970)

Moist curing (days)	Specific surface area of cement ($m^2 \, kg^{-1}$)	Drying shrinkage $\times 10^{-6}$	
		Water–cement ratio = 0.375	Water–cement ratio = 0.450
1	280	460	520
	490	540	680
	740	540	690
28	280	380	520
	490	460	610
	740	420	580

The shrinkage of reinforced concrete is less than that of plain concrete owing to the restraint developed by the reinforcement. This restraint induces tensile stresses in the concrete which may be large enough to cause cracking.

The relative humidity and temperature of the environment have a significant effect on both the rate and magnitude of shrinkage in as much as they affect the movement of water in concrete. The duration of initial moist curing has little effect on ultimate shrinkage although it affects the initial rate of shrinkage.

Carbonation shrinkage

The process of carbonation is discussed in greater detail in Chapter 15. Carbonation shrinkage occurs as a result of the reaction between carbonic acid and calcium hydroxide in cement paste, producing calcium carbonate and water. There is an accompanying decrease in cement paste volume and permeability as well as an increase in the paste strength. Carbonation shrinkage can be of the same order of magnitude as drying shrinkage, but is sensitive to the relative humidity of concrete, with greatest effects occurring between 25 and 50 per cent relative humidity.

Wetting expansion

Concrete will expand when saturated owing to replenishment of the water lost in the self-desiccation process of hydration and water being drawn into the calcium silicate hydrates. However, the swelling is resisted by the stiffness of the paste, and is generally much smaller than the shrinkage, as a result of drying. In addition, the effects of drying shrinkage are reversible by re-wetting the concrete, although many drying and wetting cycles are required before the concrete will regain fully the volume losses of the initial drying cycle.

Thermal expansion

Concrete expands with increasing temperature, in common with most other materials. This phenomenon has greatest significance for mass concrete structures where the initial heat of hydration can result in high temperature gradients causing thermal stresses and cracking at areas of restraint. The coefficient of thermal expansion of concrete is mainly dependent on the moisture content of concrete, and varies between 10 and 20×10^{-6} per °C, with the maximum for a given concrete occurring around 70 per cent relative humidity.

15

Concrete Durability

In addition to its ability to sustain loads, concrete is also required to be durable during its service life. The durability of concrete can be defined as its resistance to deterioration processes that may occur as a result of interaction with its environment (external), or between the constituent materials or their reaction with contaminants present (internal). Alternatively, damage may be the result of corrosion of embedded steel reinforcement leading to the breakdown of the surrounding concrete. In general terms, premature deterioration is the result of physical or chemical phenomena occurring within or at the concrete surface, see figure 15.1.

The quality of the concrete in the near surface or cover zone plays an important role in influencing concrete durability. This interfaces with the environment and forms the first line of defence to external physical and chemical attack and, in the case of reinforced or prestressed concrete, offers protection to embedded steel. Ensuring the quality of concrete in this region and the provision of constituent materials of required chemical resistance in relation to the environmental conditions are critical factors with respect to the provision of concrete durability in structures.

The aims of this chapter are to review the main processes, both chemical and physical, that may lead to durability problems and premature deterioration in concrete structures and to consider means of providing durable concrete, capable of functioning under these conditions.

15.1 Permeation properties of concrete

Many processes leading to the deterioration of concrete involve the ingress of aggressive fluids into concrete from the surrounding environment, followed by physical or chemical processes attacking the concrete fabric or reinforcement, frequently leading to the build-up of expansive forces and disruption. Similarly, damage occurring internally within concrete is also influenced mainly by moisture movements. Clearly, the ability of concrete to limit these processes influences the durability and rate of deterioration.

Assessment of the resistance of concrete to potential damage can be made through measurement of the permeation properties, which define the ease with

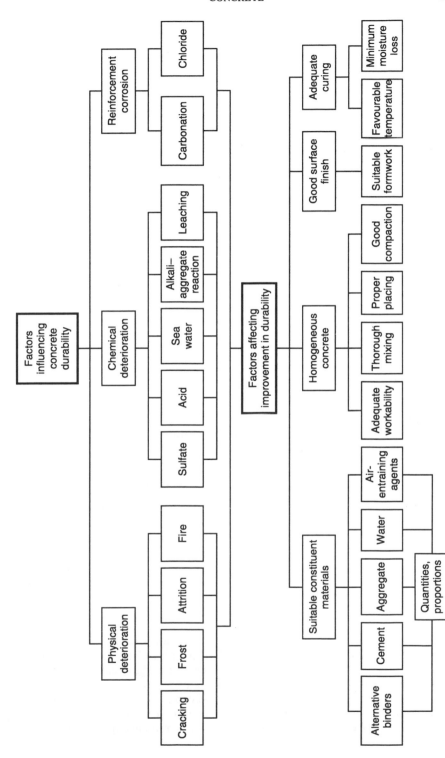

Figure 15.1 *Factors influencing concrete durability*

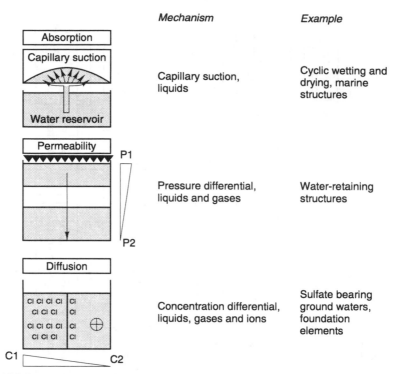

Figure 15.2 *Permeation properties of concrete*

which agents can move into or out of concrete (Dhir *et al.*, 1987). The permeation properties of concrete, see figure 15.2, embrace three distinct basic transport mechanisms which can operate in a semi-permeable medium such as concrete. These may be defined as follows:

Absorption. The process by which concrete takes in a liquid e.g. water or aqueous solution, by capillary attraction. The rate at which water enters is termed absorptivity (or sorptivity) and depends on the size and interconnection of the capillary pores in concrete, and the moisture gradient existing from the surface.

Permeability. The flow property of concrete which quantitatively characterises the ease by which a fluid passes through it, under the action of a pressure differential. This property depends on the pressure gradient and on the size and interconnection of the capillary pores in concrete.

Diffusion. The process by which a vapour, gas or ion can pass into concrete under the action of a concentration gradient. Diffusivity defines the rate of movement of the agent and is influenced by the concentration gradient from the concrete surface, the type of ingressing agent and any reaction with the hydrating cement paste, and size and interconnection of the capillary pores in concrete.

These properties depend mainly on factors including concrete mix proportions (cement content, water–cement ratio (figure 15.3)), workmanship and curing.

The significance of these individual, but related, mechanisms may vary depending on the exposure conditions to which concrete is subjected. For example, in dams and water-retaining structures, permeability is likely to be a significant factor with respect to serviceability. On the other hand, for chloride ingress leading to reinforcement corrosion, both diffusion and absorption processes are likely to be dominant. Under chemical attack from sulfates, diffusion alone may be important.

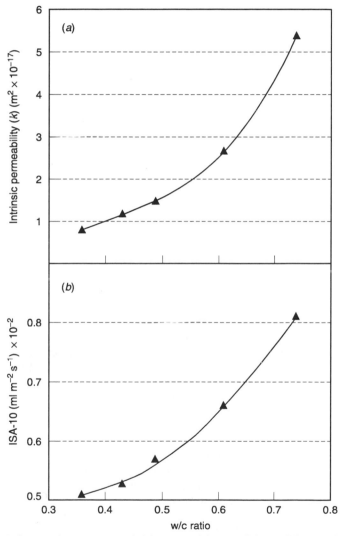

Figure 15.3 *Influence of water–cement (w/c) ratio on (a) permeability and (b) initial surface absorption of water by concrete in the first ten minutes*

15.2 Mechanisms of deterioration

Protective nature of concrete

In most cases, concrete provides very good protection to embedded steel owing to a number of physical and chemical phenomena (figure 15.4). It is only under specific conditions that this protection will break down and corrosion occur.

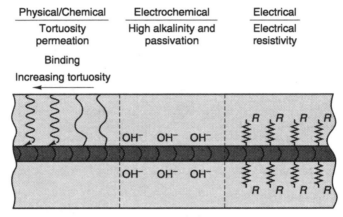

Figure 15.4 *Protective effects of concrete cover to embedded steel*

The physical characteristics (microstructure) of the concrete cover in combination with the chemical resistance, as influenced by the binder and its ability to bind and immobilise ingressing agents, without expansions, are the main factors influencing the effectiveness of concrete against corrosion initiation.

In addition, the high alkalinity of the concrete pore fluids (pH > 12.5), owing to calcium hydroxide produced from cement hydration and the soluble metal alkalis present in cement, provides conditions that leave embedded steel in a thermodynamically stable or passive and non-corroding condition.

Further protection may also be provided through the high electrical resistance of concrete, which is influenced largely by the moisture content and the constituent materials, and may limit the passage of corrosion currents through the medium.

Corrosion of reinforcement occurs normally as a result of the ingress of aggressive agents through the concrete cover and the destruction of the passive conditions. The two processes mainly responsible for this form of damage are the ingress of carbon dioxide (carbonation) and chloride ion. The damage occurring manifests itself on the concrete surface in the form of cracks, spalls or rust staining, as a result of expansive product formation on the reinforcement, although in some cases it is possible for significant loss of steel cross-section to occur without any visible damage at the concrete surface.

Carbonation of concrete

Carbonation of concrete is the process by which carbon dioxide present in air (on average, 0.03 per cent by volume) diffuses into concrete and chemically reacts with the alkali components, principally calcium hydroxide, produced through cement hydration, leading to the formation of calcium carbonate ($CaCO_3$) and causing an associated reduction in alkalinity (pH reduced to approximately 8.0). The chemical reactions associated with this process are given below:

Step 1   (a) $H_2O + CO_2 \rightarrow HCO_3^- + H^+$

              (b) $HCO_3^- \rightarrow CO_3^{2-} + H^+$

Step 2 $Ca(OH)_2 + 2H^+ + CO_3^{2-} \rightleftharpoons CaCO_3 + 2H_2O$ (neutralisation)

Carbonation continues from the concrete surface, as a penetrating front, throughout the life of the structure, with the depth achieved in a given concrete being approximately proportional to the square root of time.

The main factors associated with concrete controlling the rate of carbonation, and which also influence the permeation properties and pore fluid chemistry, include the concrete mix proportions, in particular the water–cement ratio and the cement content (figure 15.5). The workmanship, including how well the concrete has been cured, may also have some effect on the process.

The important environmental conditions influencing carbonation are the relative humidity, temperature and carbon dioxide concentration (figure 15.5). Maximum carbonation normally occurs for relative humidities in the region of 50–70 per cent. At lower relative humidities, there is insufficient water in the pores to promote the reactions, while at higher humidities, water in the pores inhibits diffusion of carbon dioxide. The rate of carbonation generally increases with increasing carbon dioxide concentration, and indoor conditions may, for example, owing to higher levels, be more severe than external exposures. For a given humidity, increases in carbonation rate may be expected with increasing temperature for the range found typically in practice.

Carbonation-induced corrosion

The effect of carbonation on concrete durability is significant only for concrete in which steel is embedded. Once the carbonation front has penetrated the cover depth, the reduced alkalinity leads to a breakdown of the passive conditions existing, and corrosion of the reinforcement may occur. The products of corrosion occupy a greater volume than that of the steel which has been affected, and the associated expansive forces normally cause breakdown of the concrete cover.

The rate at which corrosion resulting from this process occurs, depends mainly on the environmental conditions, with the influence of the concrete mix proportions being less important (Dhir *et al.*, 1992). The rate of corrosion due to carbonation becomes significant when the relative humidity in the environment is greater than 75 per cent, and increases with relative humidity above this level to near saturation. For environments of relative humidity less than 75 per cent, corrosion occurring is unlikely to be of practical significance.

From the foregoing it is clear that conditions resulting in optimum rates of carbonation are unlikely to promote corrosion, while under conditions for which corrosion might occur, significant carbonation is unlikely. Therefore, cyclic or seasonal wetting and drying are likely to represent the most serious exposure with respect to this form of attack. The provision of a concrete cover that is greater than

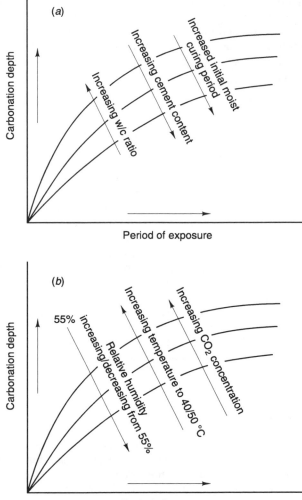

Figure 15.5 *Influence on carbonation of concrete of (a) concrete characteristics and (b) environmental conditions*

the anticipated depth of carbonation, during the life of the structure, is necessary if the risk of corrosion from the effects of carbonation is to be avoided.

Chloride ingress

Soluble chlorides present in de-icing salt or occurring in sea water can enter concrete by both absorption or diffusion processes along interconnected pores or else through cracks in the concrete surface. For plain concrete, the presence of chlorides is not generally a cause for concern in the UK, although salt weathering at the surface does cause problems in some other regions. However, for concrete containing reinforcement, the presence of chlorides is potentially very serious and this has been responsible for the damage of many structures over the last 25 years.

Chloride present in concrete may exist in one or more of three forms (figure 15.6):

1. free in the concrete pore fluids;
2. physically adsorbed to the pore walls;
3. chemically bound within the cement hydrates.

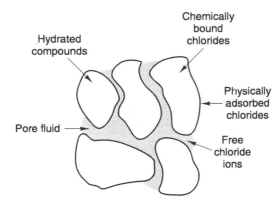

Figure 15.6 *Different forms of chloride present in concrete (Tuutti, 1982)*

The condition of chlorides present in concrete is critical with respect to the onset of corrosion. Those present in the free form represents the most serious threat, since they are capable of further penetration into concrete at depth and of breaking down the passive film surrounding the reinforcement, thereby initiating corrosion. Chlorides existing in the chemically bound or physically adsorbed states are in these forms, unavailable for contribution to chloride transportation or corrosion processes, although change between forms is possible.

As with carbonation, the rate at which chloride ingress into concrete occurs is influenced by a number of factors, which depend mainly on the characteristics of concrete and the environment (figure 15.7).

The main factors associated with concrete, which again influence the physical and chemical resistance, include the cement content, water–cement ratio and binder type (Dhir *et al.*, 1990). Similarly, factors relating to concrete manufacture and workmanship may also have some influence on the ingress of chloride ions into concrete.

The humidity and temperature of the environment have a significant influence on chloride ingress. If the exposure is subject to periodic wetting and drying, which is the most common type of exposure, then it is likely that chloride will enter concrete under both absorption when the concrete is dry and diffusion once the pores become filled. Chlorides tend to move faster under the action of absorption, although this effect generally occurs only in the outer few millimetres of the concrete surface. The rate of chloride ingress also increases with temperature, and

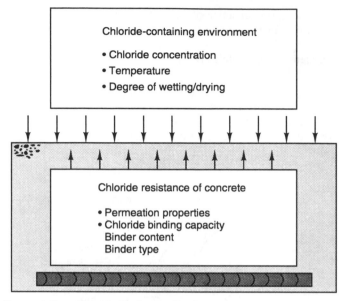

Figure 15.7 *Factors influencing chloride ingress in concrete*

this form of damage may therefore be more of a problem in hot climates (Dhir *et al.*, 1993).

Adequate protection to reinforcement can generally be achieved by the use of properly designed concrete mixes, producing dense impermeable concrete with adequate concrete cover to the reinforcement. Materials such as PFA and GGBS enhance the resistance of the concrete cover to chloride ingress and offer an economic option for increasing service life (figure 15.8). In addition, careful attention should be given to all aspects of detailing in design and in particular to the prevention of water collection on a structure. Where appropriate, protective coatings and linings can also be provided to the concrete surface and/or reinforcement.

Chloride-induced corrosion

The data available suggest that before corrosion due to chlorides can initiate, a certain quantity, or chloride threshold level, must be present at the site of the reinforcement. There is uncertainty about the precise level causing corrosion, with chloride values of between 0.1 to 0.4 per cent by weight of cement often quoted. On achievement of this condition, the propagation of corrosion thereafter depends on the continued availability and passage of chlorides and oxygen to the corroding site. This form of corrosion is normally characterised by small anodic pits, with serious local losses in steel cross-section often occurring. The impact that this has depends on a number of factors, including the type of element under exposure, the section geometry, reinforcement bar size and cover depth. Prestressed concrete structures, in particular, may be at greater risk owing to the small diameter bars and the stressed condition.

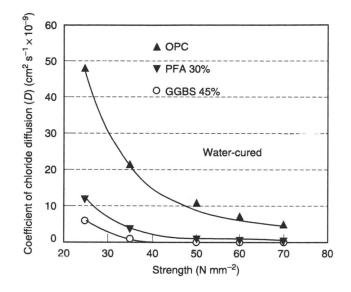

Figure 15.8 *Influence of binder type on chloride diffusion in concrete*

If significant quantities of chloride have been introduced into concrete at the mixing stage via the constituent materials used (such as water, aggregates, or admixtures containing calcium chloride), little can be done thereafter to rectify the situation. The use of chloride-based admixtures is no longer permitted in current practice, and limits on the total chloride content of mixes are prescribed in BS 8110: Part 1 (1985), see table 15.1, to ensure adequate durability from the constituent materials used. However, it is probable that chloride already introduced in this way in older structures may be responsible for future damage.

TABLE 15.1
Limits of chloride in concrete (BS 8110: Part 1:1985)

Type or use of concrete	Maximum total chloride content expressed as a percentage of chloride ion by mass of cement (inclusive of PFA or GGBS when used)
Prestressed concrete	0.1
Heat-cured concrete containing embedded metal	
Concrete made with cement complying with BS 4027 or BS 4248	0.2
Concrete containing embedded metal and made with cement complying with BS 12, BS 146, BS 1370, BS 4246 or combinations with GGBS or PFA	0.4

15.3 Chemical attack

Sulfate attack

Most sulfate solutions react with calcium hydroxide, $Ca(OH)_2$, and calcium aluminate, C_3A, to form calcium sulfate (gypsum) and calcium sulfoaluminate compounds (ettringite). Calcium, sodium and magnesium sulfates occur widely in soils (particularly clays), ground water and sea water, and contact with concrete is possible in these locations. The compounds formed have limited solubility in water and their volume is greater than the volume of the compounds of cement paste from which they originate. This increase in volume within the hardened concrete contributes to the build-up of internal stresses and the breakdown of its structure.

The intensity and rate of sulfate attack depend on a number of factors, including type of sulfate involved (magnesium sulfate is the most aggressive), its concentration and the continuity of supply to concrete. The permeation characteristics (microstructure) and presence of cracks also affect the severity of the attack. Similarly, the type of cement is a very important factor and the resistance of various cements to sulfate attack increases in the following order: normal Portland cement, Portland blastfurnace cement and low-heat Portland cement, sulfate-resisting Portland cement and supersulfated cement, and finally high-alumina cement.

Owing to their inherent ability to fix the $Ca(OH)_2$ liberated by hydration of Portland cements, pozzolanas such as PFA and GGBS can be employed to effect improvement in the resistance of concrete to sulfate attack.

To protect concrete from sulfate attack, every effort should be made to produce an impermeable concrete. Requirements given in BS 8110: Part 1 (1985), which considers the cement type, water–cement ratio and minimum cement content for concrete exposed to various degrees of sulfate attack, are reproduced in table 15.2. More recently, the Building Research Establishment has produced similar guidelines, which also take account of the cement type and sulfate concentration and type (BRE, 1991).

The number of construction joints in a structure should be minimised, since these can be particularly prone to attack. For concrete structures housed in sulfate-bearing soils, protective linings such as the various proprietary self-adhesive membranes, or protective coatings, such as bitumens, tars and epoxy resins, may also be applied on exterior surfaces; although the long-term effectiveness of some coatings cannot always be assured.

Another form of external attack, not usually found in the UK but widely reported in regions with climate and geological environments normally associated with desert terrain, is that known as *salt weathering* (Fookes and Collis, 1976). This is typified by a physical disintegration of the concrete surface resulting from the entry of soluble salts (sulfates and/or chlorides) into the pore spaces within the concrete matrix and the pores in the aggregates, and their subsequent crystallisation within these pores. Internal stresses induced in the concrete by the volume changes of the salt crystals, associated with moisture and temperature changes, can lead to cracking and subsequent disintegration of the surface concrete. Progressive deterioration of the surface, which commonly occurs within a short distance of the ground surface, might normally be expected. This form of attack is unaffected by

TABLE 15.2
Concrete for use in sulfate exposure condition (BS 8110: Part 1: 1985)

Class	Concentration of sulfates expressed as SO_3			Type of cement	Dense, fully compacted concrete made with 20 mm nominal maximum size aggregates complying with BS 882 or BS 1047	
	In soil		*In ground water* ($g\ l^{-1}$)			
	Total SO_3 (%)	*SO_3 in 2:1 water:soil extract (g l^{-1})*			*Cement* content not less than (kg m^{-3})*	*Free water–cement* ratio not more than*
1	Less than 0.2	Less than 1.0	Less than 0.3	BS 12 cements combined with PFA BS 12 cements combined with GGBS	—	—
2	0.2 to 0.5	1.0 to 1.9	0.3 to 1.2	All cements listed in 6.1.2.1 (BS 8110) BS 12 cements combined with PFA BS 12 cements combined with GGBS	330	0.50
				BS 12 cements combined with minimum 25% or maximum 40% PFA BS 12 cements combined with minimum 70% or maximum 90% GGBS	310	0.55
				BS 4027 cements (SRPC) BS 4248 cements (SSC)	280	0.55
3	0.5 to 1.0	1.9 to 3.1	1.2 to 2.5	BS 12 cements combined with minimum 25% or maximum 40% PFA BS 12 cements combined with minimum 70% or maximum 90% GGBS	380	0.45
				BS 4027 cements (SRPC) BS 4248 cements (SSC)	330	0.50
4	1.0 to 2.0	3.1 to 5.6	2.5 to 5.0	BS 4027 cements (SRPC) BS 4248 cements (SSC)	370	0.45
5	over 2	over 5.6	over 5.0	BS 4027 cements (SRPC) and BS 4248 cements (SSC) with adequate protective coating	370	0.45

* Inclusive of PFA, GGBS, SRPC (sulfate-resisting Portland cement) and SSC (super-sulfated cement).

the type of cement and can only be completely eliminated by the provision of an adequate protective lining or surface coating.

External forms of protection together with the use, where appropriate, of sulfate-resistant cements will not be fully effective in the prevention of sulfate attack if significant quantities of sulfates have already been introduced into the concrete, via its constituent materials, during mixing. Limits on the soluble sulfate content of concrete mixes are prescribed in BS 8110: Part 1 (1985) to minimise, in a practical manner, the possible effects of *internal* sulfates (Fookes and Collis, 1976).

Acid attack

As with sulfates, acid conditions may be found in soils and ground waters. These may be organic by nature and the result of plant decay or dissolved carbon dioxide, or may be due to industrial or agricultural effluent (Kay, 1992). Exposure of concrete to these conditions is likely to result in damage. However, unlike sulfate attack, where expansive product formation occurs, in this case disintegration and destruction of the hydrated cement paste are likely.

The mechanism of attack is that of reaction between the acid and hydrate compounds and calcium hydroxide, resulting in the formation of calcium salts associated with the acid. The rate of damage is controlled by the solubility of the calcium salt produced, being more rapid with increasing solubility (Comité Euro-International du Beton, 1992), with the pH of the attacking solution tending to be of less importance. The fact that the calcium salt solubility is the principal factor controlling the rate of damage means that more rapid deterioration tends to occur when conditions of flowing water exist, rather than static.

Guidance on the provision of concrete for exposure to acid conditions is given in BRE Digest 363 (1991). Factors considered include the concrete application, the contaminants in contact with concrete, the pH and whether the water is static or flowing. Recommendations are given in relation to the type of cement to be used under such conditions. The reduced calcium hydroxide content and lower porosity of well cured PFA and GGBS concretes have been found to be beneficial in reducing the rate of attack (BRE Digest 363, 1991). However, it appears that the permeation of concrete is of less importance for acid attack, compared with other damaging processes (Comité Euro-International du Beton, 1992).

Sea water attack

For concrete exposed to sea water, there are a number of phenomena that may lead to durability problems. In terms of chemical attack, both sulfates and chlorides are present, however where there is a combination, the expansive sulfate reactions tend to be reduced and damage to reinforcement as a result of corrosion is the main durability consideration. In addition, frost damage, abrasion due to the action of waves, salt crystallisation, marine growth and biological attack are other factors that may lead to deterioration of concrete.

As indicated in figure 15.9 (Comité Euro-International du Beton, 1992), for structures exposed to sea water, it is possible for division to be made into zones, where particular forms of damage are likely. In the upper zones, chloride-induced corrosion and frost damage only are likely. In addition to these, abrasion damage is possible in the splash zone and chemical damage, e.g. due to sulfate and biological attack, become possible sources of damage in the tidal and submerged zones.

Alkali–aggregate reactions

Certain natural aggregates can react chemically with the alkalis present in Portland cement. This leads to the formation of a gel, which in the presence of moisture

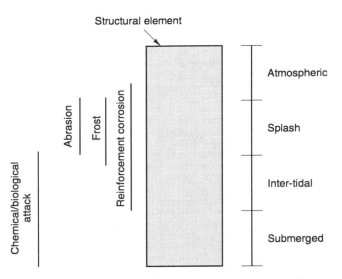

Figure 15.9 *Damaging effect of sea water at different locations on structure (based on Comité Euro-International du Beton, 1992)*

swells, leads to the build-up of expansive forces and causes disintegration of concrete. The process is accelerated with increasing environmental temperature.

The most common form of alkali–aggregate reaction (AAR) is alkali–silica reaction (ASR) which is associated with siliceous rocks containing silica in its active (amorphous) form, such as cherts, siliceous limestones and certain volcanic rocks. The minimum alkali content of cement required to produce enough alkali silicate gel (expansive product) to damage concrete is 0.6 per cent of soda equivalent. In concrete, an alkali limit of 3 kg m^{-3} based on the certified average alkali level supplied by the cement manufacturer is recommended in BS 8110: Part 1 (1985).

The simplest preventative measure for this type of deterioration is not to use alkali-reactive aggregate with cements having a high alkali content. Adverse effects of the alkali–silica reaction can be minimised by using suitable proportions of either PFA or GGBS in the concrete mix, although there is some uncertainty relating to their alkali contribution to concrete. Minimising the risk of ASR can be achieved by following one of a number of routes: ensuring the environment is dry, limiting the alkali content of cement and concrete, or using aggregate of proven record. For further guidance on minimising the risk of ASR in concrete, the reader is referred to Concrete Society (1987) and Hobbs (1988).

The second type of expansive alkali–aggregate reaction is known as alkali–carbonate reaction (ACR), which occurs with certain argillaceous dolomitic or magnesium containing limestones, although not all such reactive rocks necessarily produce deleterious expansions. The environmental factors affecting the ACR and the preventative measures are essentially the same as for the alkali–silica reaction, but there is some doubt about the effectiveness of pozzolanic materials in controlling the alkali–carbonate reaction. The reactive carbonate rocks are relatively more rare than the reactive silica rocks, and this form of reaction has not been reported in the UK.

Leaching

Calcium hydroxide, $Ca(OH)_2$, in hardened cement paste dissolves readily in water. Thus, if concrete in service absorbs or permits the passage of water through it, the calcium hydroxide in the hardened cement paste is removed, or leached out. Leaching can seriously impair the durability of concrete. Hydraulic structures in which water may pass through cracks, areas of segregated or porous concrete, or lines along poor construction joints suffer from this sort of attack. Concrete may also absorb rain or ground water, and the presence of carbon dioxide in such waters enhances the process of leaching. Free carbon dioxide and organic acids are found in acid peaty ground water.

A homogeneous and dense concrete with low permeability significantly reduces the effectiveness of the leaching action. The use of pozzolanas (e.g. pulverised-fuel ash) can also be helpful, as they reduce the permeation of concrete and fix $Ca(OH)_2$, thus lowering the soluble lime present. Care should be taken in selecting and proportioning the constituent materials and in curing the concrete, to ensure that leaching is minimised.

15.4 Physical attack

Damage to concrete can occur owing to expansion or contraction of elements under restraint, caused by the nature of concrete itself (e.g. drying shrinkage, cracking), or through variations in environmental conditions (e.g. cyclic freezing and thawing, wetting and drying), or as a result of direct physical interaction with the environment (e.g. cavitation effects of fast moving water or abrasive effects of traffic on industrial floors). These conditions may lead to deterioration of concrete itself, or else damage to the surface, leaving it vulnerable to attack from other aggressive media. Although perhaps less widely addressed than some of the chemical deterioration processes, this form of damage has in many cases also led to the need for repairs.

Cracking in concrete

Damage to concrete structures may arise through cracking in members that are in a condition of restraint, as a result of a number of different effects or processes which cause a build-up of stress. The control of cracking in concrete is an important aspect of design, in terms of both the structure and durability requirements.

In general, the basic mechanisms causing cracking in concrete structures can be divided into three groups as follows (Comité Euro-International du Beton, 1992):

1. dimensional changes due to phenomena occurring within concrete, e.g. moisture movements or temperature changes;
2. expansion of embedded steel or material within concrete, leading to disruption and breakdown;
3. displacements caused by external influences, e.g. applied loading, differential settlement.

A detailed consideration of means of preventing such damage is outside the scope of this book. For further details, the reader should refer to Concrete Society (1986). In general, the prevention of cracking can be achieved through effective structural design, good detailing and construction practices, and material selection in relation to the environmental conditions.

Frost and de-icing action

Deterioration of concrete due to the action of frost may be brought about by the disruptive action of freezing in relatively young concrete, or by alternate freezing and thawing of the free water in mature concrete.

In concrete that has not set but has been exposed to low temperatures, the mix water will freeze, with an associated increase in volume. Water is then not available for cement hydration and setting is delayed. When thawing occurs, cement hydration may begin and, provided concrete is revibrated, it should set and harden without loss of strength. If setting of concrete prior to freezing has occurred, then expansion of water on freezing disrupts the weak paste, concrete cannot be vibrated and there is a resulting reduction in strength. With increasing concrete maturity and degree of cement hydration, there is a decrease in susceptibility to early frost damage. These effects mean that care is needed for concreting in cold weather conditions.

Damage to mature concrete may occur if it is exposed to conditions of frost during service life. If concrete is in a condition of restraint, has a moisture content near saturation and is subjected to repeated cycles of freezing and thawing, then the effects of ice growth in the capillaries and the associated development of hydraulic and osmotic pressures will result in the build-up of expansive forces within the concrete pore system, leading to its disruption and breakdown. The use of de-icing salts, through their effects on freezing temperature, may also influence freeze–thaw behaviour, leading to differential expansions between parts of structures, and they may also increase the potential for scaling.

The resistance of concrete to freezing and thawing can be improved through the production of concrete of suitable permeation characteristics. This can be achieved through the use of a mix with the lowest water–cement ratio compatible with sufficient workability for placing and compacting into a homogeneous mass. Durability can be further improved by using air-entrainment, with air contents of between 3 and 6 per cent of the volume of concrete being adequate for most applications. The use of air-entrained concrete is particularly suitable for roads where salt is used for de-icing. BS 8110: Part 1 (1985) recommends the use of entrained air for concrete characteristic strengths below 50 N mm^{-2}, where it is likely to be exposed to freezing and thawing actions while wet and its surfaces subject to the effects of de-icing salts. The level of air contents required varies for different maximum size of aggregate used, ranging from an average of 7 per cent for 10 mm to 4 per cent for 40 mm nominal maximum aggregate sizes. It is also important that, wherever possible, provision is made for adequate drainage of exposed concrete surfaces – dry concrete surface is not affected by frost. The type of cement used has little effect on the process, although during the very early stages of hydration, the use of a high early strength cement can be beneficial. Damage to a structure resulting from the expansion and contraction of concrete should be minimised by providing joints which permit movement without restraint.

Physical attrition of concrete

In some situations, concrete may be prone to damage as a result of direct physical interaction between the environment and concrete surface.

Abrasion/erosion

This type of damage to concrete is normally caused by interaction between objects or material in contact with the concrete surface. Examples where this type of damage may occur include industrial floors, as a result of impact or wearing action, or in hydraulic structures through water-borne solids moving at speed and acting directly on concrete surfaces.

Factors including finish of concrete and curing, both of which have a significant influence on the quality of the concrete near surface, are important in relation to abrasion resistance (Dhir *et al.*, 1991). While the properties of the concrete surface are important in preventing damage, once signs have occurred, the type of aggregate and the aggregate/paste bond are factors influencing subsequent deterioration (Campbell-Allen and Roper, 1991). Under conditions where abrasion/erosion is likely, aggregates known to have good resistance to this type of action should be used.

Cavitation

Under certain conditions of hydraulic flow, e.g. changes in slope or geometry in a channel, the formation of cavities or pits as a result of the action of flowing water on the concrete providing the wearing surface may occur. This is caused by water vapour under a high-energy condition having repeated impact with the concrete surface, and is referred to as cavitation. Once signs of deterioration have initiated at the concrete surface, development of damage at depth and distance from this point is likely, with serious breakdown of the concrete surface eventually resulting.

The best way to minimise cavitation is through effective hydraulic design and the prevention of regions of turbulent flow. Resistance to cavitation generally increases with concrete strength and, where risks exist, medium to high strength concrete should be used. The selection of aggregates to achieve a good bond is important, since damage is likely to result from aggregate pull-out. Some advantage in minimising cavitation damage may also be possible through the use of smaller sized aggregate (Campbell-Allen and Roper, 1991).

Effects of heat/fire damage

Damage to concrete under high temperatures occurs through the effects on both hydrated cement paste and aggregate. Factors influencing damage include the temperature reached during the heat exposure, characteristics of the concrete, and the loading conditions during the period of temperature rise. The extent of damage tends to vary with depth of concrete and is most severe at the concrete surface. In addition to concrete, the effects of high-temperature exposure may also damage reinforcement. A summary of the possible effects that can occur both during heating and cooling are given in table 15.3 (Concrete Society, 1990).

An example of the effect of temperature on compressive strength of concrete is given in figure 15.10. For temperatures of up to about 300°C, the strength of concrete is not significantly reduced. However, after repeated cycles up to this temperature, progressive strength loss may be expected (Campbell-Allen and Roper, 1991). At temperatures greater than 500°C, a significant reduction in compressive strength may occur. Similar influences on the modulus of elasticity are also observed at these temperatures (Concrete Society, 1990).

In addition to the effects on the engineering properties of concrete, the concrete colour may change as a result of heat exposure. This normally coincides with significant strength loss and is due to the presence of ferrous salts in the aggregate or sand, and is more prominent in concrete with siliceous aggregates (Concrete Society, 1990).

Damage arising from the effects of high temperatures normally occurs in the form of spalling. This may occur explosively, shortly after exposure to heat, or over longer periods and less violently. Further details on fire damage to concrete and its repair may be obtained from Technical Report 53 produced by the Concrete Society (1990).

TABLE 15.3

Effects on reinforced concrete of fire attack (Concrete Society, 1990)

Stage	Probable effects
1. Rise in temperature	Surface crazing
2. Heat transfer to interior concrete	Loss of concrete strength, cracking and spalling
3. Heat transfer to reinforcement (accelerated if spalling occurs)	Reduction of yield strength; possible buckling and/or deflection increase
On cooling	
4. Reinforcement cools	Recovery of yield strength appropriate to maximum temperature attained: any buckled bars remain buckled
5. Concrete cools	Cracks close up; reduction in strength until normal temperature is reached; deflection recovery incomplete for severe fire; further deformations and cracking may result as concrete absorbs moisture from the atmosphere

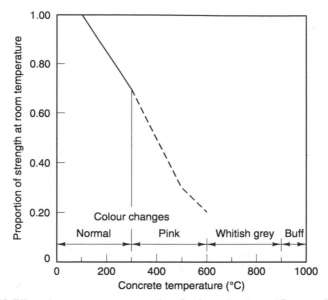

Figure 15.10 *Effect of temperature on strength and colour of concrete (Concrete Society, 1990)*

15.5 Production of durable concrete structures

A summary of the important factors in the design and specification of durable concrete is given in figure 15.11. As indicated, the exposure effects are critical and
have a significant influence on the choice of binder, with resulting interdepend-
ency on other factors, including workmanship and curing which influence the permeation properties and therefore the cover required.

The traditional route to specifying concrete for durability, for example that followed in BS 8110 (1985), prescribes parameters including concrete grade,
water–cement ratio and cement content, and in some cases type, in relation to the
environment. As an alternative approach, provision of durable concrete can be
made through a performance-based specification, where concrete is tested for
durability to ensure achievement of a required performance with respect to a standard test procedure.

Exposure conditions

Exposure effects are critical to the integrity of structures, and the provision of
durable structures depends on how accurately the exposure conditions can be
defined. The important factors influencing the degree of aggressivity include:

1. humidity/availability of moisture;
2. aggressive substances present;
3. temperature.

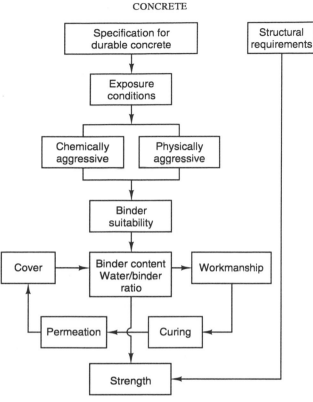

Figure 15.11 *Factors influencing specification for concrete durability*

On the basis of these, various national and international codes of practice offer guidance. One of the more recent is given in EuroCode 2 (DD ENV 1992-1-1, 1992), and is summarised in table 15.4. This classifies the range of exposure conditions likely to be encountered in practice, which is then used to establish the concrete material requirements. The difficulty lies in interpreting the classification and in deciding on the exposure class for a given structure under consideration.

Binder suitability

The ability of concrete to withstand any exposure condition is largely influenced by the characteristics of the binder, in that the latter influences the permeation characteristics of the paste phase, the chemical binding capacity and resistance to aggressive agents. For example, the use of PFA and GGBS may effectively improve resistance to several forms of chemical attack, but may adversely affect performance with respect to others.

TABLE 15.4
Classification of exposure conditions – EuroCode 2 (DD ENV 1992-1-1,1992)

Exposure class	Examples of environmental conditions
1. Dry environment	Interior of buildings for normal habitation or offices
2. Humid environment (a) without frost	Interior of buildings where humidity is high (e.g. laundries) Exterior components Components in non-aggressive soil and/or water
(b) with frost	Exterior components exposed to frost Components in non-aggressive soil and/or water and exposed to frost Interior components when the humidity is high and exposed to frost
3. Humid environment with frost and de-icing salts	Interior and exterior components exposed to frost and de-icing agents
4. Sea water environment (a) without frost	Components completely or partially submerged in sea water, or in the splash zone Components in saturated salt air (coastal area)
(b) with frost	Components partially submerged in sea water or in the splash zone and exposed to frost Components in saturated salt air and exposed to frost

The following classes may occur alone or in combination with the above classes:

5. Aggressive chemical environment (a)	Slightly aggressive chemical environment (gas, liquid or solid) Aggressive industrial atmosphere
(b)	Moderately aggressive chemical environment (gas, liquid or solid)
(c)	Highly aggressive chemical environment (gas, liquid or solid)

Binder content and water–binder ratio

Having established a suitable binder type, the quantity required to provide the necessary protection against a given rate and magnitude of aggression will be required. Similarly, because of the dependence of permeation on the binder type, content and water–binder ratio, these require to be considered collectively. Table 15.2 shows how the binder parameters may be varied for concrete exposed to sulfate attack, depending on exposure severity and binder type. Other factors that may be considered at this stage include the mixing water and the use of chemical admixtures.

Additional considerations

Whilst a concrete of a given binder content and water–binder ratio may be deemed to have the required durability potential, the effect of several other factors should be considered carefully and, where necessary, the binder content and water–binder ratio revised.

Workmanship

This should take account of the degree of compaction and the quality of the finish that can be achieved on site.

Curing

Curing determines the rate and magnitude of hydration of the binder paste, with some binders being particularly sensitive to curing. Thus the degree and duration of curing, particularly at the early ages, can greatly influence the hydration products and the microstructure of the binder, and thereby alter both the binding capacity and permeation characteristics. Most codes specify the degree of curing required, however they give no guidance as to its effect on binder ability to resist deterioration.

Permeation

Besides chemical binding, adequate permeation properties are a key factor in determining the binder durability characteristics, and therefore the provision of adequate permeation properties should be achieved. However, little guidance is given on these within National Standards.

Cover

The thickness of cover is critical to the protection afforded by concrete to the embedded reinforcement. Increasing this will increase the chemical and physical barrier against the environment. The thickness of cover should be optimised with the binder quality, content and water–binder ratio. It should be noted that, from structural design considerations, the cover depth should ideally be as low as possible.

Strength

For the binder type and water–binder ratio required for durability as outlined above, a corresponding strength of concrete is also given or will be achieved. This should be compared with the strength value required in the structural design, and the higher value adopted.

Performance specification approach

This represents a recent development in specifying concrete for durability. With this method, concrete is considered with respect to the relevant deterioration process, the required design life and the criteria that define the end of that life.

This approach employs a test method which directly assesses the resistance of concrete to a standardised deterioration process. It differs from the conventional approach in that no reference to material contents or bulk properties (e.g. cement content, water–cement ratio and strength) is made, but only a specified performance to an agreed test method.

Before such an approach can be fully implemented, a number of factors require to be established, including the relationship between the test and real performance, the sampling and test procedure, and the test precision. Such tests may be carried out on cements, or other constituent materials, concrete test specimens or the actual structure itself.

16

Assessment of *in situ* Concrete Quality

In situ testing of concrete structures may be required when data are needed to allow engineering decisions to be based on the condition of a structure. The reasons for this may include preliminary testing to gather data prior to structural renovation, or diagnostic testing to assess reasons for deterioration or distress, regular structural inspection, or testing to resolve disputes over concrete quality. In addition, testing may often accompany a change of ownership of the building, and is essential after fire damage to assess the structural capability of concrete.

16.1 Planning a test programme

The first stage of any structural assessment should be a consideration of all available documented records from the time of construction, particularly those of concrete mix proportions, material sources, cube test results, and the temperature and humidity at the time of casting. Such information may be difficult to obtain for older buildings, and in such cases, people who were involved with the original design or construction may be helpful, although word-of-mouth accounts can be extremely unreliable and contradictory.

Prior to specifying or carrying out any test, it is necessary to complete a thorough visual inspection of the structure to establish which areas require testing and to identify the appropriate tests (Institution of Structural Engineers, 1980; Bungey, 1989, 1992). Difficulties relating to access to the structure and any dangers associated with particular test methods should also be assessed at this time, bearing in mind current health and safety regulations. Many important concrete details relating to quality of workmanship, structural integrity and material failure or distress can also be identified at this stage. The main visual features which are indicative of specific problems are as follows:

1. segregation/bleeding at shutter joints (mix constituent problems);
2. plastic shrinkage cracking (mix constituent problems);
3. honeycombing (low standards of workmanship);
4. excessive deflection or flexural cracking (structural inadequacy);
5. distortion of doors or windows (long-term deflections caused by creep, thermal or structural movement);
6. surface cracking, spalling or staining (material deterioration due to reinforcement corrosion, sulfate attack, frost action or alkali–aggregate reaction).

Typical visual features relating to specific problems are summarised in table 16.1. The information gained from these symptoms assists in the selection of the correct type of test for further investigation. The inspecting engineer should note the position, frequency and seriousness of deflections, cracks and surface defects such as spalling and discoloration, and assess the likely difficulties in access for testing or repair (FIP, 1986). It is important therefore, that the engineer is able to identify the causes of different cracks, and the reader is referred to a publication by the Comité Euro-International du Beton (1992) for further guidance.

Photography also assists in the presentation of the preliminary inspection report, which should contain as much detail as possible from existing documentation, such as the original concrete mix proportions and position and quantity of reinforcement, as well as the number, position, estimated cause and seriousness of any

TABLE 16.1

Symptoms associated with common problems in concrete structures

Structural problem	*Symptoms*
Early age effects	
Plastic settlement	Cracking over reinforcement, at tops of columns and at depth changes
Rapid drying (plastic shrinkage)	Diagonal and random cracking
Structural deficiency	Deflections, cracking and spalling at areas of highest stress
Frost damage	Cracking, spalling and erosion of section
Thermal effects	Cracking and spalling at areas of restraint
Physical damage	Cracking, spalling and erosion
Fire damage	Cracking and spalling, colour changes (pink at > 300°C)
Long-term effects	
Frost damage	Cracking, spalling and erosion of section
Thermal effects	Cracking and spalling at areas of restraint
Shrinkage	Cracking, mostly of thin walls and slabs
Physical damage	Cracking, spalling and erosion
Reinforcement corrosion	Cracking, and spalling, rust staining
Chemical attack	Cracking, spalling and erosion
Alkali–aggregate reaction	Cracking and spalling of section
Creep	Deflection, cracking and spalling at areas of highest stress

symptoms of distress. The report should be circulated and discussed with all parties involved in the planning of a test programme.

It is important that the objectives of a test programme are clearly defined at the outset and are agreeable to all parties concerned, particularly when there is a dispute over concrete quality. Furthermore, all parties involved in a dispute should be aware of the variability which can be found in concrete structures as a result of varying concrete supply and differences in compaction and curing regimes.

In choosing the test locations and the frequency of testing, engineers should ensure that the test programme will provide sufficient data to allow sensible conclusions to be reached, whilst keeping costs and structural damage to a minimum. It is also important that the test types and procedures are agreed, particularly for tests not covered by a National Standard. Some indication of the accuracy of the tests to be used should also be given at this stage, normally by reference to a standard work on the subject, such as that by Bungey (1992).

16.2 Test methods

Concrete structures can be assessed for strength, comparative quality, local integrity, potential durability and to determine the causes of deterioration. However, it is of the utmost importance that the testing system employed should provide the information required to answer appropriate questions about the condition of the structure. The engineer should also advise his client of any limitations of the proposed tests. For example, a survey using ultrasonic pulse velocity techniques can identify areas of poor compaction, but clients should be warned that favourable results are not necessarily indicative of fully satisfactory concrete.

Current British Standards for testing *in situ* concrete are listed in table 16.2, although there are many other methods available which are not covered by these standards. Any methods selected for a testing programme should be chosen primarily for their ability to yield the information necessary for a diagnostic analysis of the structure to be made accurately, bearing in mind the preliminary information that may be available from the visual survey, table 16.1. The choice of test method should reflect any initial diagnosis, and a guide to choosing methods to obtain further information is given in table 16.3. Since all *in situ* tests have limitations, it is a common practice for tests to be used in combination, but when selecting tests attention should be paid to:

1. the required amount of access to use the selected method;
2. the degree of damage caused by the test;
3. the cost of the tests;
4. how much data will be gained; and
5. the test reliability.

A combination of test methods may lead to higher confidence in all results, particularly when similar trends are found with different tests. However, test combinations do increase overall cost, and extra tests should be restricted to those which are cheap, non-destructive and quick. It may also be necessary to use one type of test to ascertain which areas require more detailed testing, such as location of steel with a covermeter prior to corrosion testing, or the use of ultrasound to identify areas of poor compaction which may warrant taking cores for strength testing.

TABLE 16.2
British Standards for testing in situ *concrete*

British Standard	Subject
BS 1881	Testing concrete
Part 5	Methods for testing for properties other than strength
Part 116	Determination of cube strength
Part 117	Determination of tensile strength
Part 118	Determination of flexural strength
Part 119	Compressive strength of beams broken in flexure
Part 120	Compressive strength of concrete cores
Part 121	Modulus of elasticity of concrete
Part 124	Chemical analysis of concrete
Part 201	Non-destructive tests for hardened concrete
Part 202	Surface hardness testing by rebound hammer
Part 203	Measurement of ultrasonic pulse velocity
Part 204	Use of electromagnetic covermeters
Part 205	Radiography measurements in concrete
Part 206	Determination of concrete strain
Part 207	Assessment of strength from surface tests
Part 208	Initial surface absorption test
Part 210	Permeability tests
Part 211	Determination of corrosion potential
BS 6089	Measurement of strength *in situ*

16.3 Testing for strength

A selection of common methods for estimating *in situ* strength of concrete is given in table 16.4. These fall into two main categories of partially destructive and destructive tests. Further details of these can be obtained from relevant literature

TABLE 16.3
Test methods used to assess different concrete properties

Concrete property	Methods
Mix constituents	Chemical analysis
Detection of reinforcement	Covermeter testing, X- and γ-radiography
Strength	Cores, pull-out and internal fracture tests, break-off and pull-off tests, penetration resistance, ultrasonic pulse velocity and rebound hammer if calibrated
Cracking	Ultrasonic pulse velocity, crack microscope
Compaction and honeycombing	Ultrasonic pulse velocity, γ-radiography, cores
Density	γ-radiography
Concrete microstructure	ISAT, modified Figg and permeability tests
Moisture content	Electrical resistivity, microwave absorption
Abrasion resistance and soundness	Rebound hammer, abrasive wear tests, infra-red thermography

TABLE 16.4

Description, merits and limitations of in situ strength tests

Test method	Recommended number of replicates	Prediction accuracy (±, %)	Advantages	Disadvantages	Cost
IN SITU TESTS					
Penetration resistance	*The system fires a steel alloy probe into the concrete, and the depth of penetration is measured (normally 20–40 mm)*				
	3	20	Quick, easy access	Slender members may crack, calibration and safety precautions required	Moderate
Internal fracture	*An expanding 6 mm anchor bolt is fixed in the concrete to a depth of 20 mm. The bolt is pulled against a reaction tripod of 80 mm diameter until a peak load, which corresponds to internal fracturing of the concrete, is observed*				
	6	30	Quick, simple test	Complex failure mode, variable results, power supply required	Low
Pull-out	*The test comprises a compressed steel split ring, expanded into a 25 mm groove, which is undercut in an 18 mm hole at a depth of 25 mm. This is pulled out and the peak load recorded*				
	4	20 (10 for a specific mix)	Reliability, compressive failure mode	Time consuming, destructive, skilled operator and power supply needed	High
Pull-off	*The test consists of a circular metal disc bonded to the surface of the concrete by adhesive. A direct tensile force is applied with a hand-operated jack, which pulls off the disc and an accompanying portion of concrete. The peak load is recorded*				
	6	15	Easy test	Bonding requires careful preparation of concrete surface	Low–moderate
Break-off	*The test measures the force required to break off a 55 mm diameter by 70 mm core drilled in the concrete with a load cell in an enlarged slot*				
	5	20	Quick test	Coring equipment/power/water supplies required, destructive test	Moderate
LABORATORY TESTS					
Cores	*Cores provide a sample which can be used for laboratory strength, chemical composition, durability and permeation testing properties*				
Standard (100–150 mm)	3	5	Accurate strength prediction, other parameters measured accurately	Destructive to concrete and steel, laboratory delay	Moderate
Small (<100 mm)	9	10	Less destructive than standard cores, other parameters measured accurately	Laboratory delay, strength prediction less accurate than standard cores	Moderate

(Concrete Society, 1976; Munday and Dhir, 1984; Samarin and Dhir, 1984; Bungey, 1989, 1992).

The accuracy of all *in situ* strength tests depends on the nature of concrete constituent materials, degree of bleeding and segregation that has occurred within the test concrete, the variability of the test method used, the number of replicate tests performed at each location and the reliability of the test calibration.

Some of the partially destructive *in situ* strength tests covered by BS 1881: Part 207 are illustrated in figure 16.1, and the main points to note are given in table 16.4. Whilst these types of test give only an indication of the surface strength of concrete, and are not as accurate as testing cores taken from the structure, considerably less damage is caused and results are obtained instantly.

The most accurate way to assess *in situ* strength of concrete is to cut and test concrete cores (BS 1881: Part 120). Whilst the numbers of cores cut may be limited by considerations of cost, structural damage, access difficulties and delays in laboratory sample preparation and testing, cores taken from a structure can also be used to test concrete in a variety of other ways, including assessment of the potential and residual durability of the concrete (see Chapter 15) and chemical analysis.

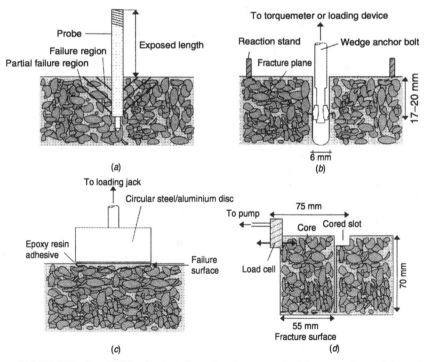

Figure 16.1 *BS 1881: Part 207* In situ *tests for estimating strength: (a) penetration resistance test; (b) internal fracture test; (c) pull-off test; (d) break-off test*

16.4 Testing for comparative concrete quality

For rapid surveying of large areas of concrete, where core cutting or *in situ* strength testing would be uneconomical, *in situ* comparative testing can be used. The test methods in this category are not usually calibrated against a physical parameter such as strength, but yield fast results of a comparative nature and cause little or no damage to the concrete surface. Areas of poor compaction, cracking, internal voidage and delamination can be identified.

The available test methods are described in table 16.5, and a selection of those covered by BS 1881 are shown in figure 16.2. The reader is referred to relevant literature for further details (Concrete Society, 1988; Bungey, 1989, 1991, 1992; Gilberton and Lees, 1991; Parrott, 1991).

The most commonly used comparative tests are the ultrasonic pulse velocity and the surface hardness methods, shown in figure 16.2, (a) and (b) respectively.

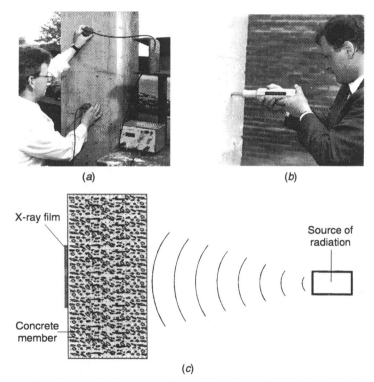

Figure 16.2 *BS 1881 tests for comparative quality of concrete: (a) ultrasonic pulse velocity test equipment (BS 1881: Part 203); (b) Schmidt hammer testing of concrete for surface hardness (BS 1881: Part 202); (c) radiography test equipment (BS 1881: Part 205)*

Surface hardness

The surface hardness test (BS 1881: Part 202), figure 16.2(b), is based on the principle that the rebound height of an elastic mass depends on the hardness of the surface from which it rebounds. The test is carried out with a standard rebound or Schmidt hammer. The results are sensitive to mix constituent materials including cement type and content, coarse aggregate type and the characteristics of the member under test, including the degree of compaction, age, curing regime, surface under test, member mass, carbonation, moisture content, temperature and state of stress (Bungey, 1989). For this reason, whilst the equipment is generally supplied

TABLE 16.5

Description, merits and limitations of in situ tests for comparative concrete quality

Test method	Main applications	Advantages	Disadvantages	Cost
Surface hardness	*The test is based on the fact that rebound height of an elastic mass depends on the hardness of the surface from which it rebounds*			
	Uniformity of surface zone	Quick, cheap, simple test	Strength correlations require calibration	Low
Ultrasonic pulse velocity	*The test measures the velocity of high-frequency ultrasonic pulses. Most concretes have pulse velocities of 3.6–5.0 km s^{-1}*			
	Uniformity of interior	Well-established, simple method	Access to opposite faces required for best results.	Moderate
Crack measurement	*Available methods include graduated magnifying glasses, templates, feeler gauges*			
	Width, age of cracks	Cheap, quick, low cost	Only measures surface crack widths	Low
Radiography	*The test photographs the concrete interior, indicating variations in density, location of voids, poor compaction and reinforcement*			
	Location of poor compaction, voids, reinforcement	Portable, gives photo of variations in internal density	Depth of penetration < 500 mm; radiation – safety precaution	Moderate
Radiometry	*The test directs γ-rays at the concrete, and measures the intensity of the radiation emerging, which is indicative of concrete density*			
	Comparative density	Portable and simple	Extensive safety/ transport precautions	Moderate–high
Thermography	*Measures surface temperature differentials as concrete heats or cools, and indicates delamination, moisture, ducts or voids*			
	Location of major ducts, voids, moisture, delamination	No contact with surface, hand-held front-end equipment	Comparative application, complex equipment, cost	Moderate–high
Radar	*Radar techniques can be used to find reinforcement bars, voids, ducts and areas of wet concrete*			
	Location of voids, cracks, steel and moisture	No contact with the concrete surface, fast survey	Expensive, specialised equipment, difficult interpretation of results	High
Screed test	*The surface is indented by controlled hammer blows, and the depth of the indentation measured*			
	Screed soundness	Well-established quick, simple, suitable for quality control	Affected by many factors, not related to strength	Low

with charts to convert rebound height into strength of concrete, high accuracy can only be expected if the apparatus is calibrated against a range of laboratory concrete samples made with the same materials as the structure under test. Without such a calibration, the test is more useful for indicating areas of lower strength of an unspecified value, which may be indicative of unsatisfactory mix proportioning or compaction. Such areas may thereafter warrant further and more expensive testing.

When using the equipment, it is advisable to define a test grid to reduce operator variability. Since the method is very sensitive to local surface variation, particularly to aggregate particles below the surface of the concrete, it is necessary to take up to 25 readings in an area up to 300 mm square and average the results. The test surface should preferably be formed, smooth, clean and dry, but trowelled surfaces may be tested if rubbed smooth with the carborundum stone which is usually supplied with the apparatus.

Ultrasonic pulse velocity

The ultrasonic pulse velocity test (BS 1881: Part 203), figure 16.2(a), measures the velocity of a high-frequency ultrasonic pulse passed through a concrete member between two transducers, with higher velocity taken to reflect higher density and hence a better quality of concrete. The range of pulse velocities found in concrete is 3.6–5.0 km s^{-1}. To carry out the test under optimum conditions, access is required to opposite sides of the test member (direct method), but it is possible to obtain usable results with a semi-direct method (pulse passing between adjacent faces) or an indirect method (pulse passing between transducers placed on the same face), although the amplitude of the received signal decreases in both cases, particularly the latter.

The measured pulse velocity of *in situ* concrete can be affected by many factors including the temperature at time of test, stress history if 50 per cent of the cube strength has been exceeded (Bungey, 1980), moisture condition (wet concrete has up to 5 per cent higher pulse velocity than dry concrete) and the presence of reinforcement, which generally has a higher pulse velocity than concrete (5.1–5.9 km s^{-1}). In addition, in heavily reinforced areas, there is a high possibility of poor compaction, which will introduce further uncertainty into the results.

The most reliable application of the test is determination of concrete uniformity and the location of areas of poor compaction and internal voidage, although it is possible, with calibration as for the Schmidt hammer, to estimate concrete strength. The test is quick, and large areas of concrete can be covered in a relatively short time. It is advisable to prepare a test grid to ensure readings are taken at known stations. The main feature of good test practice is a good acoustic coupling between the transducers and the concrete surface. This is usually provided using grease, petroleum jelly or liquid soap, with thicker mediums being most suited to rough faces.

Radiography

Radiography testing of concrete structures is covered by BS 1881: Part 205, and is shown schematically in figure 16.2(c). This technique is used to provide a photograph of the interior of concrete members in terms of the relative density, thereby indicating areas of poor compaction, voidage and reinforcing steel. Since γ-radiation is used to activate radiation-sensitive film behind the member under test, radiation safety requirements are necessary and the test should only be carried out by specialists.

16.5 Testing for durability

Table 16.6 briefly describes the test types which can be used for *in situ* durability testing, and some examples are shown in figure 16.3. Durability tests for concrete are divided conveniently into those which predict future potential durability and those which identify causes of current deterioration, although the same tests can often be used for both purposes.

In recent years, *in situ* durability tests have received increasing attention (Sadegzadeh and Kettle, 1986; Dhir *et al.*, 1987; British Cement Association, 1988; Concrete Society, 1988; Vassie, 1991). There has been a corresponding move in some sectors of the construction industry to specify concrete by parameters which are indicative of durability, as well as the standard parameters such as water–cement ratio, cement content and grade. However, the testing of durability remains an area largely uncovered by British Standards, although this is likely to change in the future.

The most commonly encountered concrete durability problem with concrete structures is corrosion of reinforcing steel due to the ingress of deleterious external agencies from the environment (Chapter 15). Of great importance to this is the provision of an adequate depth of concrete cover to the steel reinforcement, and poor workmanship at the time of construction can often lead to this provision being too low. In the case of dispute, cover provision can be measured using simple, battery-operated electromagnetic field generators which can locate and size embedded steel and provide a digital readout converted from eddy current or magnetic induction effects. Detailed guidance for the use of such equipment is given in BS 1881: Part 204.

The prediction of future potential durability can be determined by measuring the resistance of concrete to the ingress of external agencies by the use of permeation methods such as intrinsic permeability, absorption, or ionic/gaseous diffusion tests, which are sensitive to the microstructure of concrete. Few of these can be applied *in situ* and cores are generally cut for laboratory preparation and subsequent testing. However, cores can be tested in other ways, including chemical analysis for chloride content, sulfate content, carbonation, cement type and content, aggregate type and grading, alkali reactivity and original water content (BS 1881: Part 124).

Whilst permeability index tests such as the modified Figg test (Concrete Society, 1988) can be applied *in situ* (figure 16.3(a)), the results are affected by the residual moisture content of the concrete, and specimens should ideally be dried in a standard manner prior to test. Although guidance is given on pre-condi-

TABLE 16.6
Durability tests

Test method	Damage	Information provided	Advantages	Disadvantages	Cost
Abrasion resistance	*This gives guidance on the ability of concrete to resist mechanical wear*				
	Minor	Cause/risk of deterioration	Direct test of abrasion	Not directly related to wearing effects	Moderate
Surface absorptivity	*This is measured using the Initial Surface Absorption Test (ISAT) and has close correlation with several aspects of durability*				
	Moderate	Cause/risk of deterioration, corrosion	Easy, quick test	Influence of concrete moisture condition	Moderate
Permeability	*The coefficient of permeability of concrete cannot be measured in situ but permeability index tests can be used*				
	Moderate	Cause/risk of deterioration, corrosion	Easy test, quick	Affected by moisture condition of concrete	Moderate
Chemical analysis	*Laboratory with specialised equipment and skilled operators*				
	Moderate	Cause/risk of deterioration, corrosion	Concrete composition	Sample size/type governed by type of test	High
Entrained air content	*Performed by magnification of a polished sample. Discrete voids are counted manually or by imaging computer*				
	Moderate	Cause/risk of deterioration	The only available method	Laboratory-based, coring required	High
Cover measurement	*Hand-held devices locate embedded steel accurately, giving information about cover depth and the bar diameter*				
	None	Cause/risk of corrosion	Simple, cheap, fast	Calibration, depth of penetration 300 mm	Low
Carbonation depth	*Guidance on the risk of reinforcement corrosion by spraying a broken section with an indicator solution such as phenolphthalein in alcohol, which is colourless below pH 9 and purple above*				
	Minor	Cause/risk of corrosion	Easy, cheap and accurate	Damage, location-specific results	Low
Half-cell potential	*Electrochemical activity can be picked up by comparing the electrode potential of embedded reinforcement with that of a half-cell on the concrete surface and plotting as an isopotential contour map, from which areas of concern can be identified*				
	Minor	Corrosion activity and risk	Corrosion risk identified using an isopotential contour map	Corrosion rates/degree not determined	Moderate–high
Resistivity	*The test passes an A/C electrical current between outer electrodes, whilst measuring the voltage drop between two inner electrodes*				
	Minor	Corrosion activity and risk	Quick to perform	Corrosion rates not determined	Moderate–high

tioning in BS 1881: Part 5, the standard allows *in situ* measurements to be taken after two days without rain, irrespective of the actual moisture content of the concrete, and the viability of such non-standard procedures is disputed (Dhir *et al.*, 1987). In the laboratory, the most reliable and economical test method is the Initial Surface Absorption Test (ISAT), figure 16.3(b), which is at present covered by BS 1881: Part 5 (soon to be dealt with in BS 1881: Part 208). The test is sensitive to changes in the microstructure of concrete arising from differences in mix proportions, curing and binder and aggregate type. The results of this test have been correlated with many aspects of durability, including concrete strength and resistance to carbonation, chloride diffusion, freeze–thaw attrition and abrasion (Dhir *et al.*, 1994).

The assessment of corrosion risk can be determined *in situ* by a half-cell potential method (Vassie, 1991), figure 16.3(c), which detects the presence of electrochemical activity associated with reinforcement corrosion. Interpretation of the results is normally carried out by plotting an isopotential contour map from which areas where there is a risk of corrosion can be identified, although the actual corrosion rates cannot be determined. In areas of risk, resistivity tests (Millard, 1991), figure 16.3(d), which measure the ease with which an electrical current can flow through concrete, may also provide an indication of the likelihood of significant corrosion occurring.

16.6 Interpretation of results and reporting

The interpretation and reporting of *in situ* test results may be considered in four phases as follows:

1. calculation from measured values;
2. assessment of result variability;
3. calibration against concrete properties;
4. reporting.

Calculation from measured values

The methods employed for calculation of specific parameters from directly measured values will depend to a large extent on the test method employed. Most tests have well-defined procedures.

Assessment of result variability

The number of test replicates and test stations used has a significant bearing on the ease with which the variability of concrete within members and between members can be assessed. For confidence, the variability of any tests employed should be studied. The results can then be analysed using various statistical tools such as graphical and numerical methods. In general, variation in the hardened concrete due to differences in the supplied product will tend to be random and arise from differences in materials, batching, transport and handling techniques.

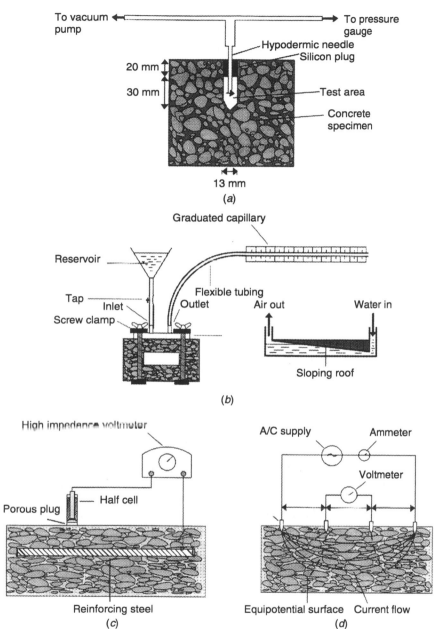

To vacuum pump

To pressure gauge

Hypodermic needle

Silicon plug

20 mm

30 mm

Test area

Concrete specimen

13 mm

(a)

Graduated capillary

Reservoir

Flexible tubing

Tap

Inlet

Outlet

Air out

Water in

Screw clamp

Sloping roof

(b)

High impedance voltmeter

A/C supply

Ammeter

Voltmeter

Half cell

Porous plug

Reinforcing steel

Equipotential surface   Current flow

(c)

(d)

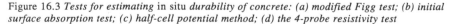

Figure 16.3 *Tests for estimating* in situ *durability of concrete: (a) modified Figg test; (b) initial surface absorption test; (c) half-cell potential method; (d) the 4-probe resistivity test*

Calibration against concrete properties

The accuracy with which test results can be calibrated against a particular desired concrete property is controlled by the variability of the particular test method, the operator skill and the variability of the concrete under test. It should be noted that differences between the laboratory conditions under which calibrations are obtained and site conditions can vary, and that this can affect the accuracy of calibration. For example, calibration of permeation tests such as the ISAT (figure 16.3(b)) should be carried out with concrete moisture conditions similar to those which are expected on site, since the results are significantly affected by the residual moisture content of concrete.

Reporting

The final report should contain enough details to allow correct and appropriate decisions for remedial action to be made (Tomsett, 1991). A typical report should have the following elements:

1. An *introduction* which gives the reasons for carrying out the testing.
2. A *background* description of the available documentation and the visual survey as well as conclusions relating to preliminary diagnosis and the choice of test drawn from these. Any photographic work undertaken should be presented here.
3. A description of the *test methods,* including details of all locations tested, test types, adopted methodology, expected result variability, calibration and interpretation details.
4. A *results* section, giving all results and interpretation, as well as drawing attention to areas where the structure is substandard and may require further testing or remedial action.
5. *Recommendations for further testing.* If a series of non-destructive tests have shown a particular area to be suspect, it may be necessary to take cores for both strength testing and/or more detailed laboratory analysis before specifying an expensive repair and maintenance programme.
6. *Conclusions* should summarise the findings of the study concisely.
7. *Recommendations for rehabilitation* should specify the nature and extent of the repair and reconstruction programme, and should include information such as member position, access details, size of the affected area, specified repair materials and the time schedule for the repairs. Sufficient details on all aspects of the proposed repairs should be given to allow contractors to prepare accurate tenders for the work.

17

Concrete Mix Design and Quality Control

Concrete mix design can be defined as the procedure by which, for any given set of conditions, the proportions of the constituent materials are chosen so as to produce a concrete with all the required properties for the minimum cost. In this context the cost of any concrete includes, in addition to that of the materials themselves, the cost of the mix design, of batching, mixing and placing the concrete and of site supervision.

Quality control refers in the first instance to the supervision exercised on site to ensure both that the materials used in the production of the concrete are of the required quality and that the stated mix proportions are adhered to as closely as is possible with the available site facilities. It also refers to the control testing of the properties of the concrete in both its fresh and hardened state to ensure that it conforms with the design requirements and, where it deviates significantly from this, the taking of the necessary corrective action. The analysis of the results of such testing to determine the compliance, or otherwise, with the concrete specification depends on the particular contractual conditions governing compliance. Different standards and codes of practice have slightly different requirements in this respect. Specific details of these, and of the different *producer* and *purchaser* risks associated with each are outside the scope of this text.

Mix design, quality control and the overall cost of the finished concrete product are all interdependent and this will become increasingly apparent as the factors affecting mix design are discussed.

In this chapter the basic requirements for any concrete and the way in which these may be incorporated into concrete mix design are described. The essentials of quality control associated with the production of concrete on site have been considered earlier. Some of the statistical methods of quality control used for the interpretation of control tests on the hardened concrete are described here.

17.1 Required concrete properties

The basic requirements for concrete are conveniently considered at two stages in its life.

In its hardened state (in the completed structure) the concrete should have adequate durability, the required strength and also the desired surface finish.

In its plastic state, or the stage during which it is to be handled, placed and compacted in its final form, it should be sufficiently workable for the required properties in its hardened state to be achieved with the facilities available on site. This means that

(1) the concrete should be sufficiently fluid for it to be able to flow into and fill all parts of the formwork, or mould, into which it is placed;
(2) it should do so without any segregation, or separation, of the constituent materials while being handled from the mixer or during placing;
(3) it must be possible to fully compact the concrete when placed in position; and
(4) it must be possible to obtain the required surface finish.

If concrete does not have the required workability in its plastic state, it will not be possible to produce concrete with the required properties in its hardened state. The dependence of both durability and strength on the degree of compaction has been noted earlier. Segregation results in variations in the mix proportions throughout the bulk of the concrete and this inevitably means that in some parts the coarser aggregate particles will predominate. This precludes the possibility of full compaction since there is insufficient mortar to fill the voids between the coarser particles in these zones. This results in what is descriptively known as honeycombing on the surface of the hardened concrete with reduced durability and strength as well as unacceptable surface finish.

The means by which each of the required concrete properties may be achieved are now considered.

Durability

Adequate durability of exposed concrete can frequently be obtained by ensuring full compaction, an adequate cement content and a low water–cement ratio, all of which contribute to producing a dense, impermeable concrete. The choice of aggregate is also important particularly for concrete wearing surfaces and where improved fire resistance is required. Aggregate having high shrinkage properties should be used with caution in exposed concrete. For particularly aggressive environments additional precautions may be necessary. For example, in marine areas and for concrete subjected to road de-icing salts, the use of PFA and GGBS as partial replacement of Portland cement will increase the resistance of concrete to the ingress of chlorides which corrode the reinforcing steel. In addition, sulfate-resisting cements can be used in sulfate-bearing soils and freeze–thaw attrition of concrete can be resisted by the use of air-entraining admixtures.

Durability is not a readily measured property of the hardened concrete. However, for a correctly designed concrete mix any increase in the water content, or reduction in the cement content, on site, with the associated reduction in durability, will be accompanied by an increase in the water–cement ratio and a corresponding reduction in concrete strength. The latter can be determined quite easily using control specimens and for this reason the emphasis in control testing is on the determination of concrete strength.

Strength

Although durability is sometimes the overriding criterion, the strength of the concrete is frequently an important design consideration, particularly in structural applications where the load-carrying capacity of a structural member may be closely related to the concrete strength. This will usually be the compressive strength although occasionally the flexural or indirect tensile strength may be more relevant.

The strength requirement is generally specified in terms of a *characteristic strength* (BS 8110: Part 1) coupled with a requirement that the probability of the strength falling below this shall not exceed a certain value. Typically this may be 5 per cent or a 1 in 20 chance of a strength falling below the specified characteristic strength, this generally being the 28-day strength.

An understanding of the factors affecting concrete strength on site, and of the probable variations in strength, is essential if such specifications are to have any real meaning at the mix design stage. A histogram showing the frequency of cube compressive strengths for a road contract is shown in figure 17.1. What might appear to be large variations in strength can be seen, with actual cube strengths ranging from 8.5 to 34.0 N mm^{-2}. Some of the differences between individual cube strengths may be attributed to testing errors including sampling, preparation and curing of the test cubes, the testing machine itself and to the actual test procedure. Differences in strength can also occur owing to variations in the quality of the

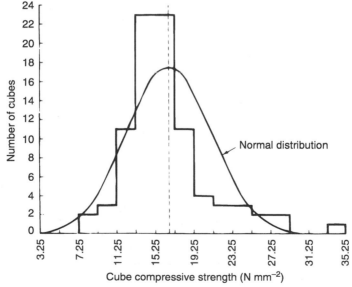

Figure 17.1 *Histogram for cube compressive strength*

cement but the principal factor affecting the strength is the quantity of water, or more specifically the water–cement ratio, in the concrete mix.

If the proportions of aggregate and cement and also the quality of the aggregate are maintained constant, the water–cement ratio can be controlled very effectively at the mixer by adding just sufficient water to give the required workability. Once a suitable mix has been obtained the workability can be assessed quite satisfactorily by an experienced mixer operator, with periodic control tests of the workability. However, human error will inevitably result in some variation in the water–cement ratio either side of the desired value. Any variation in mix proportions or significant changes in the aggregate grading will affect the quantity of water needed to maintain the required workability and this too will result in variations in the water–cement ratio and hence in concrete strength.

All these factors tend to give water–cement ratios which are as likely to be greater as they are to be less than the target value. The actual water–cement ratios tend therefore to have a *normal* or gaussian distribution about the mean, or target, value. The relationship between water–cement ratio and concrete strength is nonlinear. Nevertheless over a limited range the relationship will be approximately linear and it might be expected that concrete strengths will also tend to have a normal distribution. This can be seen in figure 17.1 in which a typical bell-shaped normal distribution curve is shown.

Design mean strength

The assumption of a normal distribution of concrete strengths forms the basis of mix design and statistical quality control procedures for satisfying the strength requirement. For a normal distribution, the probability of a strength lying outside specified limits either side of the mean strength can be determined. These limits (figure 17.2) are usually expressed in terms of the standard deviation s defined by

$$s = \left[\frac{\Sigma (f_c - f_{cm})^2}{n - 1} \right]^{\frac{1}{2}} = \left[\frac{\Sigma (f_c)^2 - (\Sigma f_c)^2/n}{n - 1} \right]^{\frac{1}{2}} \text{ N mm}^{-2}$$

where f_c is an observed strength, f_{cm} is the best estimate of the mean strength, equal to $(\Sigma f_c)/n$ and n is the number of observations. The probabilities of a strength lying

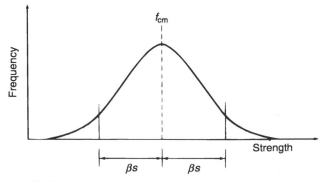

Figure 17.2 *Normal distribution curve*

TABLE 17.1
Probability values

Probability of an observed strength lying outside the range $(f_{cm} \pm \beta s)$	β	Probability of an observed strength being less than $(f_{cm} - \beta s)$
1 in 50	2.33	1 in 100
1 in 20	1.96	1 in 40
1 in 10	1.64	1 in 20

outside the range $(f_{cm} \pm \beta s)$ for different values of β are given in table 17.1, in which the probabilities of strengths falling below the lower limit $(f_{cm} - \beta s)$ are also given.

If the specified characteristic strength f_{cu} is the strength below which not more than 1 in 20 of the population of strengths shall fall, it follows that

$$f_{cu} = f_{cm} - 1.64s$$

or

$$f_{cm} = f_{cu} + 1.64s$$

Hence if the standard deviation likely to be obtained on site can be assessed, the mean strength for which the concrete must be designed can be determined. The

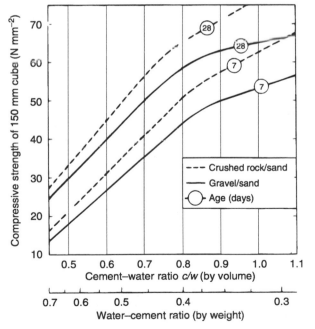

Figure 17.3 *Cube compressive strength relationships for concrete made with ordinary Portland cement*

free water–cement ratio required to give this mean concrete strength can then be estimated using curves such as those in figure 17.3.

The effect of site control on the mean strength f_{cm} required to give a characteristic strength f_{cu} is shown in figure 17.4. The poorer the control the higher the mean strength required and this will mean a lower water–cement ratio. This in turn will require a smaller aggregate–cement ratio, or larger cement content per unit volume, for a given workability and a consequent increase in the total cost of materials.

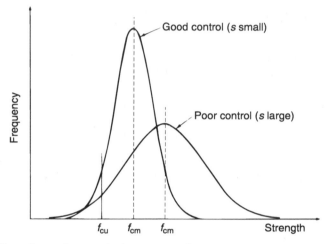

Figure 17.4 *Effect of control on required mean strength*

Workability

Suitable workabilities for different types of work are given in table 17.2, workability in this context referring to those qualities measured by the slump, compacting factor and Vebe tests. In this table the bracketed values are the least reliable although the associated tests can generally be used for control purposes once the actual value corresponding to a suitable workability has been determined.

For a given water–cement ratio, the principal factors affecting workability are the shape and grading of the coarse and fine aggregates and the aggregate–cement ratio. The choice of suitable concrete mix proportions must take all these factors into account.

Specification of concrete mixes

The interdependence of water–cement ratio and strength, and of type and grading of aggregate, aggregate–cement ratio, water–cement ratio and workability has been noted earlier. When specifying concrete mixes it is essential that this inter-

TABLE 17.2

Workability requirements

Type of work	Vebe time (s)	Compacting factor	Slump (mm)	Workability
Heavily reinforced sections with vibration. Simply reinforced sections without vibration	(3)	0.92	(25–100)	Medium
Simply reinforced sections with vibration. Mass concrete without vibration	6	(0.86)	(10–50)	Low
Mass concrete and large sections with vibration. Road slabs vibrated using power-operated machines	12	(0.80)	—	Very low

dependence is recognised if meaningful specifications and compliance requirements are to be achieved.

Full details of methods for specifying essentially two basic types of concrete mix, namely prescribed mixes and design mixes, and for testing for compliance are given in BS 5328. For both types, the permitted types of cement and aggregate, the nominal maximum size of aggregate and requirements relating to the production of concrete are specified. Other requirements, excluding those enumerated below, may also be given (where appropriate). In this connection useful guidelines for hot climates are contained in ACI 305 'Hot-weather concreting' (1977). The fundamental difference between the two types of mix is that for *prescribed* mixes the mix proportions (cement, coarse and fine aggregates) are specified and tested for compliance, whereas for *designed* mixes the concrete grade (characteristic concrete strength) is specified and tested for compliance. For both types of mix, concrete strength may be the overriding criterion and therefore *prescribed* mixes must always have an ample strength margin to ensure that the required strength will be achieved. For this reason greater economy will generally result from using a *designed* mix.

17.2 Concrete mix design

A number of different approaches to mix design have been proposed, one of the most recent and widely used being that described in 'Design of Normal Mixes' (DOE, 1988). A more general approach proposed by Hughes (1971) considers the optimum coarse aggregate content, this being the volume fraction of coarse aggregate which enables a given strength and desired workability to be achieved with the minimum cement content. Hughes' approach which forms the basis of the simplified mix design procedure described here is considered to be a useful introduction to the DOE (1988) method, which is outlined briefly later.

Mix design (Hughes' approach)

The basic steps in designing a concrete mix are outlined below. These are followed by illustrative examples in which concrete mixes are designed to meet specific requirements.

(1) *Durability*. For the degree of exposure determine the type of cement, the minimum cement content and maximum water–cement ratio, being guided by relevant specifications.

(2) *Strength.* Where a maximum water–cement ratio has been obtained in step 1, if this is a characteristic value determine the associated concrete strength using relationships such as those shown in figure 17.3. Compare this with the specified characteristic strength and accept the larger value as the required characteristic strength f_{cu}. If the maximum water–cement ratio obtained in step 1 is the highest permitted target *mean* value, this is not considered until step 4.

(3) Determine the required mean concrete strength from

$$f_{cm} = f_{cu} + \beta s$$

using an appropriate value of β. The standard deviation s should be based on recent test results for conditions similar to those which will occur on site. Where no such data are available, conservative values should be adopted. For example, when $\beta = 1.64$, values of $\beta s = 0.67 f_{cu}$ for $f_{cu} \leqslant 22.5$ N mm^{-2} and $\beta s = 15.0$ N mm^{-2} for $f_{cu} \geqslant 22.5$ N mm^{-2} might be used.

(4) Determine the cement–water ratio c/w (by volume) associated with the mean strength f_{cm} using, for example, figure 17.3. Compare this with the durability cri-

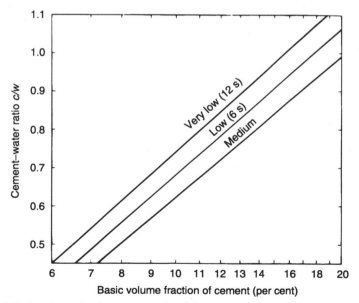

Figure 17.5 *Basic volume fraction of cement for different warkabilities, based on Hughes (1971)*

terion where this is related to the target *mean* water–cement ratio and adopt the larger cement–water ratio c/w (by volume) for design purposes.

(5) *Workability.* Determine the *basic* volume fraction of cement required to give the desired workability (see table 17.2), for the cement–water ratio c/w obtained in step 4, using figure 17.5.

(6) Using table 17.3 and the gradings of the available coarse and fine aggregates, determine the corresponding grading moduli G_a and G_b respectively, and the equivalent mean diameter D_b of the fine aggregate.

TABLE 17.3
Aggregate grading parameters

BS 410 test sieve (mm)	37.50	20.0	10.0	5.00	2.36	1.18	0.60	0.30	0.15	0.075
Grading modulus ($\times 0.90$ mm^{-1})	1/4	1/2	1	2	4	8	16	32	63	
Equiv. mean diameter ($\times 0.90$ mm)			8	4	2	1	1/2	1/4	1/8	

(7) Using figure 17.6 obtain the corrections to the *basic* volume fraction of cement required when (a) the coarse aggregate is crushed rock; (b) the fine aggregate is crushed rock; (c) the coarse aggregate grading modulus G_a differs from 0.46; or (d) the fine aggregate grading modulus G_b is greater than 16 or less than 9. Hence determine the corrected volume fraction of cement c. Compare this with the corresponding durability requirement, step 1, and accept the higher value as the required volume fraction of cement c. Where the durability requirement is the overriding factor, increase the cement–water ratio c/w accordingly so that the workability remains unchanged, using figure 17.5.

(8) Using the volume fraction of cement c and the volume ratio c/w determine the volume fraction of water w.

(9) Evaluate the fine-to-coarse size ratio $G_a D_b$. The optimum coarse aggregate content depends on the relative sizes of the coarse and fine aggregate particles, more space between the coarse aggregate particles being required to accommodate the fine aggregate particles as the relative size of the latter increases. It has been shown (Hughes, 1968) that a satisfactory measure of the ratio of fine-to-coarse aggregate size is given by $G_a D_b$. Using table 17.4, an estimate of the solids fraction of loose coarse aggregate a_b is obtained. Hence determine the recommended volume fraction of coarse aggregate a, using figure 17.7.

TABLE 17.4
Solids fraction of loose coarse aggregate

Aggregate size	Solids fraction a_b	
	20–10 mm	*20–5 mm*
Irregular gravel	0.56	0.58
Crushed rock	0.46	0.48

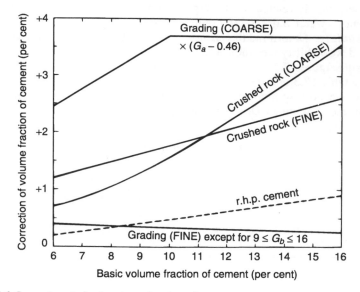

Figure 17.6 *Corrections to basic volume fraction of cement*

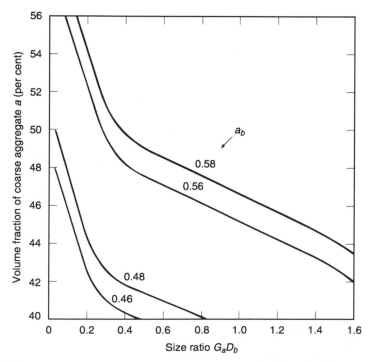

Figure 17.7 *Recommended volume fraction of coarse aggregate, based on Hughes (1971)*

(10) The sum of the volume fractions a, c, w and the volume fraction of fine aggregate b must be unity. The volume fraction of the fine aggregate b is equal therefore to $1 - (a + c + w)$.

(11) Determine batch quantities per cubic metre using the specific gravities of the constituent materials, the values for the aggregates being those for saturated surface-dry aggregates.

(12) Trial mixes in the laboratory are generally recommended so that any necessary preliminary adjustments to the mix proportions can be made before full-scale trial mixes under site conditions are carried out.

Trial mixes

The necessity for trial mixes arises, for example, since the relationships between concrete strength and water–cement ratio (see figure 17.3) depend to some extent on the local aggregates and the source of cement. Experience in a particular locality of the quality and characteristics of both cement and aggregates can be of considerable value at the mix design stage.

Segregation of the water and cement fines from the concrete can occur with lean concretes having a high workability. The cohesiveness of such mixes can be improved, and hence the tendency for wet segregation to occur reduced, by reducing the water content and/or increasing the proportion of fine to coarse aggregate. Either of these causes some reduction in workability and some reduction in the volume fraction of the combined fine and coarse aggregates is generally necessary if the workability is to be maintained. Alternatively the use of an air-entraining agent can be beneficial in such cases, only minor adjustments being necessary to allow for the volume fraction of the entrained air. Typically this will include some reduction in the volume fractions of both water and fine aggregate.

High-strength concrete

For very high-strength concretes a skew distribution of concrete strength might be expected, as opposed to a normal distribution, as the upper limit of the attainable strength for the materials used is approached. The actual water–cement ratio might still be expected, however, to follow a normal distribution as discussed earlier. In such cases it is recommended that at step 4, the cement–water ratio c/w associated with the specified characteristic strength f_{cu} should be obtained and increased by an appropriate factor. It is recommended that this factor should be based upon the control ratio suggested by Erntroy (1960).

Example 1

Characteristic cube strength   30 N mm^{-2} at 28 days (structural)
Type of work   simply reinforced with vibration
Degree of exposure   mild
Nominal cover to reinforcement   25 mm
Cement   ordinary Portland

TABLE 17.5
Aggregate grading

BS 410 test sieve (mm)	37.5	20.0	10.0	5.00	2.36	1.18	0.60	0.30	0.15	0.075
				per cent passing						
Irregular gravel	100	98	46	4	0					
Natural sand			100	96	78	60	36	8	3	
				per cent retained						
Irregular gravel		2	52	42	4					
Natural sand				4	18	18	24	28	5	3

Coarse aggregate   irregular gravel (20–5 mm)
Fine aggregate   natural sand
Specific gravities of the coarse and fine aggregates and cement are 2.54, 2.60 and 3.14 respectively
Site control (previous data)   standard deviation $s = 5.5$ mm^{-2}

The grading of the aggregates is given in table 17.5 in the form in which this is frequently reported, that is, in terms of the percentage passing the standard sieve ·sizes. The aggregate grading moduli G_a and G_b and equivalent mean diameter D_b are calculated using the percentages retained between successive sieve sizes, the required values being the differences between the percentages passing the corresponding sieve sizes (see table 17.5).

Mix design

(1) Durability (BS 8110: Part 1)
   Maximum *mean w/c* by weight = 0.65 (min. *c/w* by vol. = 0.49)
   Minimum *mean* cement content = 275 kg m^{-3}
(2) —
(3) Strength $f_{cm} = f_{cu} + \beta s = 30 + 1.64(5.5) = 39.2$ N mm^{-2}
(4) From figure 17.3 *c/w* by volume = 0.595
   Durability requirement = 0.49   O.K.
(5) Workability, table 17.2   low workability appropriate
   From figure 17.5   basic volume fraction of cement = 8.5 per cent
(6) Using table 17.3

Gravel (G_a)			Sand (G_b)			Sand (D_b)		
$2 \times \frac{1}{4}$	=	$\frac{1}{2}$	4×1	=	4	4×8	=	32
$52 \times \frac{1}{2}$	=	26	18×2	=	36	18×4	=	72
42×1	=	42	18×4	=	72	18×2	=	36

$$4 \times 2 \quad = \quad \dfrac{8}{76.5} \qquad 24 \times 8 \quad = \quad 192 \qquad\qquad 24 \times 1 \quad = \quad 24$$

$$28 \times 16 \quad = \quad 448 \qquad\qquad 28 \times \tfrac{1}{2} \quad = \quad 14$$

$$5 \times 32 \quad = \quad 160 \qquad\qquad 5 \times \tfrac{1}{4} \quad = \quad 1\tfrac{1}{4}$$

$$3 \times 63 \quad = \quad \dfrac{189}{1101} \qquad\qquad 3 \times \tfrac{1}{8} \quad = \quad \dfrac{\tfrac{3}{8}}{179.7}$$

$G_a = 0.90\,(76.5) \times 10^{-2} = 0.689\ \text{mm}^{-1}$, $G_b = 0.90\,(1101) \times 10^{-2} = 9.91\ \text{mm}^{-1}$;

$D_b = 0.90\,(179.7) \times 10^{-2} = 1.62\ \text{mm}$

(7) From figure 17.6 (a) —
  (b) —
  (c) $3.2\,(0.689 - 0.46) = +0.74$ per cent

Corrected volume fraction $c = 8.5 + 0.74 = 9.3$ per cent
Durability requirement $= 275\ \text{kg m}^{-3}$

$$\equiv \frac{275 \times 100}{(\text{s.g. cement})\ 1000}$$

$$= 8.8\ \text{per cent O.K.}$$

Therefore volume fraction of cement $c = 0.093$

(8) From step 4, $c/w = 0.595$
  Therefore volume fraction of water $w = 0.093/0.595 = 0.156$
(9) Using step 6, $G_a D_b = 0.689 \times 1.62 = 1.12$
  From table 17.4 $a_h = 0.58$
  From figure 17.7 volume fraction of coarse aggregate $a = 0.460$
(10) $a + c + w = 0.709$
  Therefore volume fraction of fine aggregate $b = 0.291$
(11) Batch quantities

water	0.156×1000	=	$156\ \text{kg m}^{-3}$
cement	$0.093 \times 1000\,(3.14)$	=	$292\ \text{kg m}^{-3}$
sand	$0.291 \times 1000\,(2.60)$	=	$757\ \text{kg m}^{-3}$
gravel	$0.460 \times 1000\,(2.54)$	=	$1168\ \text{kg m}^{-3}$
			$2373\ \text{kg m}^{-3}$

Adjustment to mix proportions

The foregoing example illustrates the basic mix design procedure. Some of the possible corrections to the mix proportions thus obtained are now considered.

Let trial mixes for the concrete designed in the example show that the desired mean strength f_{cm} is being achieved but the workability is too low, the Vebe time being 8 s and the compacting factor 0.83. Referring to figure 17.5, to increase the workability by the required amount, namely, about one-half of the interval between successive degrees of workability, the volume fraction of cement should be increased by about 0.4 per cent. Hence c is increased from 0.093 to 0.097, w is increased accordingly from 0.156 to 0.163 to maintain the same c/w ratio, a remains unchanged at 0.460 and b, the sand content, is reduced to 0.280 to maintain the relationship $a + b + c + w = 1$. The corresponding batch quantities are obtained as before.

Alternatively let the workability be satisfactory but the estimated mean strength f_{cm} be only 35 N mm^{-2}. From figure 17.3, to increase the strength by 4 N mm^{-2} requires an increase of 0.045 in the c/w ratio, giving a final value of $c/w = 0.640$. From figure 17.5, the corresponding basic volume fraction of cement is 9.1 per cent, that is, an increase of 0.6 per cent over the value previously obtained. The corrected volume fractions are therefore $c = 0.099$, $w = 0.155$, $a = 0.460$ (as before) and $b = 0.286$. The change in w is very small and may frequently be neglected in which case w and a will be unchanged and the increase in c, namely 0.006, will be reflected by a corresponding decrease in b. The required corrections to the batch quantities will be an increase in cement content of 19 kg m^{-3} and a reduction in sand content of 16 kg m^{-3}.

Where both workability and strength need minor adjustments, similar considerations treating each correction in turn enable the required adjustments to mix proportions to be obtained.

Example 2

All data as for example 1 except that the degree of exposure will be severe and the nominal cover to the reinforcement will be 40 mm.

Mix design

    (1)   Durability (BS 8110: Part 1)
           Maximum *mean w/c* by weight = 0.55, (min. *c/w* by vol. = 0.58)
           Minimum *mean* cement content = 325 kg m^{-3}
    (2)   —
    (3)   Strength $f_{cm} = 39.2$ N mm^{-2} (as for example 1)
    (4)   From 17.3 c/w by volume = 0.595
           Durability requirement = 0.58 O.K.
    (5)   (as for example 1)
    (6)   (as for example 1)
    (7)   From example 1.
           Corrected volume fraction $c = 8.5 + 0.74 = 9.3$ per cent
           Durability requirement = 325 kg m^{-3} = 10.4 per cent
           Therefore required volume fraction of cement $c = 0.104$
           By inspection, a *basic* volume fraction of cement = 9.6 per cent
           requires, from figure 17.6, a correction given by
                    $3.6 (0.689 - 0.46) = + 0.83$ per cent
           giving a corrected volume fraction $c = 9.6 + 0.83 = 10.4$ per cent
           as required from durability considerations.
    (5R)   Workability, table 17.2 low workability appropriate

For a basic volume fraction of cement = 9.6 per cent, from
figure 17.5 c/w by volume = 0.66 > 0.58 O.K.

(4R) From figure 17.3 strength f_{cm} = 46 N mm^{-2}

(3R) Strength $f_{cu} = f_{cm} - \beta s = 46 - 1.64 (5.5) = 37$ N mm^{-2}
This is the required *characteristic* strength to meet the durability
requirements given in (1), for concrete for which figure 17.3 is directly
applicable.

(8) From step 5R, c/w = 0.66
Therefore volume fraction of water w = 0.104/0.66 = 0.157

(9) From example 1
Volume fraction of coarse aggregate a = 0.460

(10) $a + c + w$ = 0.721
Therefore volume fraction of fine aggregate b = 0.279

(11) Batch quantities

water	0.157 × 1000	=	157 kg m^{-3}
cement	0.104 × 1000 (3.14)	=	326 kg m^{-3}
sand	0.279 × 1000 (2.60)	=	725 kg m^{-3}
gravel	0.460 × 1000 (2.54)	=	1168 kg m^{-3}

2376 kg m^{-3}

Mix design (DOE method)

This method uses previously established data to obtain mix proportions for an initial trial mix. These mix proportions are obtained for a particular characteristic strength and workability and depend on the type of cement, the type and maximum size of the coarse aggregate and the type and grading of the fine aggregate to be employed. The design method is split into five stages, each dealing with a particular aspect of the design and having as its end result either one of the main mix parameters or final proportions of one constituent.

In stage one consideration of the characteristic strength together with the expected strength variability leads to the target mean strength and hence the free water–cement ratio (by weight) required to achieve this. Stage two provides a value for the free water content required to achieve the specified workability. Combination of the results of stages one and two provides the cement content as stage three. The total aggregate content is then obtained from consideration of the expected plastic density of the mix in stage four and stage five completes the procedure giving values for the proportions of fine and coarse aggregate in the total.

A trial mix to these proportions can then be made and tested and, if necessary, the results obtained used with the data given in the DOE publication to modify the mix proportions to obtain closer compliance with the specified values of strength and workability. Provision is made in the method for consideration to be given to durability requirements (specified in terms of a maximum free water–cement ratio and minimum cement content), and the method is suitable for design in terms of compressive or indirect tensile strength, with or without air-entraining admixtures.

17.3 Statistical quality control

It is necessary to check that the desired quality of concrete is being obtained on site and for this purpose control testing is required. Tests on the hardened concrete are usually performed on 150 mm cubes prepared from samples of the concrete used on site. The importance of correct sampling and preparation, curing and testing of the control cubes cannot be overemphasised. It is usual for sets of two control specimens to be made from each sample, one set being tested at 28 days and/or at 7 days or earlier in some cases, for example, when accelerated curing techniques are used. The early test results enable estimates to be made of the probable 28-day concrete strengths once a relationship between the strengths at the two ages has been established. Any necessary remedial action can then be taken at a much earlier stage than would otherwise be possible.

The control testing may indicate that the site concrete has the desired mean strength but a greater standard deviation than had been assumed at the design stage. In these circumstances, adjustment of the mix proportions is required so as to increase the mean strength. If, on the other hand, the standard deviation is less than had been assumed some economy can be achieved by reducing the mean strength.

It is possible also that a change in mean strength, or standard deviation, can occur unintentionally during the course of construction for some reason or other. It is important that any such changes are detected as soon as possible so that the necessary action can be taken and the reasons for the changes determined at the earliest opportunity. This requires a continuous assessment of the strengths of the control specimens. It is here that statistical methods of quality control are invaluable, generally incorporating visual aids in the form of control charts.

TABLE 17.6
*Control chart data**

Cube reference number	Estimated 28 day strength f_c	$(f_c - f_{cm})$	Cumulative sum of (3) CUSUM M	Range of adjacent results in (2)	(5) minus design mean range	Cumulative sum of (6) CUSUM SD	Mean of group of previous four results
(1)	(2)	(3)	(4)	(5)	(6)	(7)	(8)
1	37.5	− 0.5	− 0.5				
2	42.0	+ 4.0	+ 3.5	4.5	− 1.0	− 1.0	
3	42.5	+ 4.5	+ 8.0	0.5	− 5.0	− 6.0	
4	29.0	− 9.0	− 1.0	13.5	+ 8.0	+ 2.0	37.8
5	42.0	+ 4.0	+ 3.0	13.0	+ 7.5	+ 9.5	38.9
6	41.0	+ 3.0	+ 6.0	1.0	− 4.5	+ 5.0	38.7
7	47.5	+ 9.5	+15.5	6.5	+ 1.0	+ 6.0	39.9
8	31.5	− 6.5	+ 9.0	16.0	+10.5	+16.5	40.5
9	40.0	+ 2.0	+11.0	8.5	+ 3.0	+19.5	40.0
10	38.0	0	+11.0	2.0	− 3.5	+16.0	39.3
11	39.0	+ 1.0	+12.0	1.0	− 4.5	+11.5	37.2
12	35.5	− 2.5	+ 9.5	3.5	− 2.0	+ 9.5	38.2
13	35.5	− 2.5	+ 7.0	0	− 5.5	+ 4.0	37.1

(continued on p. 289)

TABLE 17.6 (continued)

Cube reference number	Estimated 28 day strength f_c	$(f_c - f_{cm})$	Cumulative sum of (3) CUSUM M	Range of adjacent results in (2)	(5) minus design mean range	Cumulative sum of (6) CUSUM SD	Mean of group of previous four results
(1)	(2)	(3)	(4)	(5)	(6)	(7)	(8)
14	34.0	− 4.0	+ 3.0	1.5	− 4.0	0	36.1
15	45.5	+ 7.5	+10.5	11.5	+ 6.0	+ 6.0	37.7
16	36.5	− 1.5	+ 9.0	9.0	+ 3.5	+ 9.5	37.9
17	41.5	+ 3.5	+12.5	5.0	− 0.5	+ 9.0	39.4
18	37.5	− 0.5	+12.0	4.0	− 1.5	+ 7.5	40.3
19	42.0	+ 4.0	+16.0	4.5	− 1.0	+ 6.5	39.4
20	37.0	− 1.0	+15.0	5.0	− 0.5	+ 6.0	39.5
21	40.5	+ 2.5	+17.5	3.5	− 2.0	+ 4.0	39.3
22	35.5	− 2.5	+15.0	5.0	− 0.5	+ 3.5	38.8
23	34.0	− 4.0	+11.0	1.5	− 4.0	− 0.5	36.8
24	35.5	− 2.5	+ 8.5	1.5	− 4.0	− 4.5	36.4
25	42.5	+ 4.5	+13.0	7.0	+ 1.5	− 3.0	36.9
26	36.0	− 2.0	+11.0	6.5	+ 1.0	− 2.0	37.0
27	31.0	− 7.0	+ 4.0	5.0	− 0.5	− 2.5	36.3
28	38.5	+ 0.5	+ 4.5	7.5	+ 2.0	− 0.5	37.0
29	44.5	+ 6.5	+11.0	6.0	+ 0.5	0	37.5
30	37.5	− 0.5	+10.5	7.0	+ 1.5	+ 1.5	37.9
31	34.0	− 4.0	+ 6.5	3.5	− 2.0	− 0.5	38.6
32	42.0	+ 4.0	+10.5	8.0	+ 2.5	+ 2.0	39.5
33	37.5	− 0.5	+10.0	4.5	− 1.0	+ 1.0	37.7
34	43.5	+ 5.5	+15.5	6.0	+ 0.5	+ 1.5	39.2
35	42.0	+ 4.0	+19.5	1.5	− 4.0	− 2.5	41.2
36	39.5	+ 1.5	+21.0	2.5	− 3.0	− 5.5	40.6
37	27.0	−11.0	+10.0	12.5	+ 7.0	+ 1.5	38.0
38	39.0	+ 1.0	+11.0	12.0	+ 6.5	+ 8.0	36.9
39	39.5	+ 1.5	+12.5	0.5	− 5.0	+ 3.0	36.3
40	47.5	+ 9.5	+22.0	8.0	+ 2.5	+ 5.5	38.2
41	42.0	+ 4.0	+26.0	5.5	0	+ 5.5	41.9
42	38.0	0	+26.0	4.0	− 1.5	+ 4.0	41.7
43	32.5	− 5.5	+20.5	5.5	0	+ 4.0	40.0
44	36.0	− 2.0	+18.5	3.5	− 2.0	+ 2.0	37.1
45	30.0	− 8.0	+10.5	6.0	+ 0.5	+ 2.5	34.1
46	41.0	+ 3.0	+13.5	11.0	+ 5.5	+ 8.0	34.9
47	28.5	− 9.5	+ 4.0	12.5	+ 7.0	+15.0	33.9
48	42.5	+ 4.5	+ 8.5	14.0	+ 8.5	+23.5	35.5
49	36.5	− 1.5	+ 7.0	6.0	+ 0.5	+24.0	37.1
50	37.5	− 0.5	+ 6.5	1.0	− 4.5	+19.5	36.2
51	37.0	− 1.0	+ 5.5	0.5	− 5.0	+14.5	38.4
52	37.0	− 1.0	+ 4.5	0	− 5.5	+ 9.0	37.0
53	35.0	− 3.0	+ 1.5	2.0	− 3.5	+ 5.5	36.6
54	39.5	+ 1.5	+ 3.0	4.5	− 1.0	+ 4.5	37.1
55	31.5	− 6.5	− 3.5	8.0	+ 2.5	+ 7.0	35.8
56	33.5	− 4.5	− 8.0	2.0	− 3.5	+ 3.5	34.9
57	42.5	+ 4.5	− 3.5	9.0	+ 3.5	+ 7.0	36.8
58	33.0	− 5.0	− 8.5	9.5	+ 4.0	+11.0	35.1
59	36.0	− 2.0	−10.5	3.0	− 2.5	+ 8.5	36.3
60	34.0	− 4.0	−14.5	2.0	− 3.5	+ 5.0	36.4

Characteristic strength f_{cu} = 30.0 N mm⁻², design standard deviation s = 5.0 N mm⁻², design mean strength f_{cm} = 38.0 N mm⁻², design mean range = $1.13s$ = 5.5 N mm⁻².

Shewart control charts

The standard control chart for concrete strength comprises a central horizontal line corresponding to the mean concrete strength f_{cm} with pairs of lines, either side of and equidistant from this central line, corresponding to the limits $f_{cm} \pm \beta s$. With only one pair the lower limit will be the characteristic strength f_{cu}.

A convenient visual presentation of the variation of the test results from the mean strength is obtained by plotting, on this control chart, the individual test results as these become available. For $\beta = 1.64$, if more than three results in any forty consecutive results fall below the lower limit, the degree of control is immediately suspect and the need for some remedial action is indicated. BS 5328 requires that each test result shall be the average of the results obtained from a set of two control specimens at any age.

Typical test results are given in table 17.6 together with the specified characteristic strength and design data. The associated control chart is shown in figure 17.8. In this case it would appear that adequate control is being exercised.

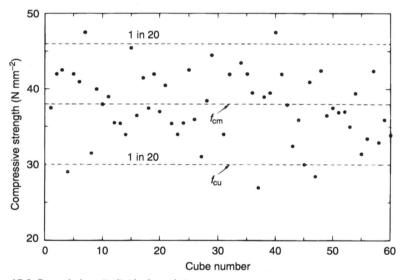

Figure 17.8 *Control chart (individual results)*

Unfortunately this type of control chart is relatively insensitive to changes in control and a reduction in mean strength during construction may not become apparent until possibly thirty or forty results are available following such a reduction. Furthermore no real indication is given of the actual mean strength at any time, although of course when forty or more results are available average values may be calculated.

Another similar control chart is also in current use where, instead of individual results, the mean strengths of groups of consecutive results are plotted. In this case the upper and lower limits, corresponding to $f_{cm} \pm \beta s$ in the standard control chart,

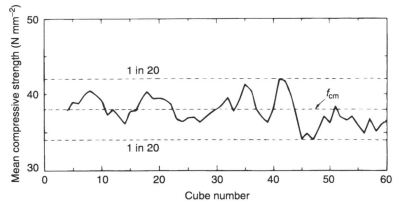

Figure 17.9 *Control chart (groups of four consecutive results)*

are $f_{cm} \pm (\beta s \sqrt{N})$ where N is the number of results in each group. In table 17.6, column 8, the mean strengths of groups of four results are given and the associated control chart is shown in figure 17.9. This modified chart is more sensitive than the standard chart to changes in control since each of the plotted points is based on more information, namely, the previous four results, than for the standard control chart. Examination of figure 17.9 does suggest that some reduction in mean strength may have occurred in the latter stages of the work covered by these test results, although no positive indication of any change in control is given.

Cumulative sum control charts

One of the deficiencies of the Shewart control charts is that full use is not being made of all the information available at the time a decision regarding a change in control has to be made. The application of *cumulative sum* techniques in the concrete industry in the form of cusum control charts enables all or any part of the available information to be used. The greater visual impact of these control charts and their greater sensitivity to changes in control enable such changes to be detected at a comparatively early stage. Cusum charts for mean strength and standard deviation are considered here.

Mean strength (CUSUM M)

To construct a cusum control chart for mean strength, the cumulative sum of the algebraic differences between the assumed or design mean strength and each test result is plotted as successive results become available. The mean slope over any portion of the cusum plot is then directly proportional to the difference between the actual and assumed mean strengths. An estimate of the actual mean strength f_{cm} may be obtained from

$$\bar{f}_{cm} = f_{cm} + \frac{(\text{change in CUSUM M over } n \text{ results})}{n}$$

Typical calculations are shown in table 17.6, columns 3 and 4, and the associated control chart is shown in figure 17.10. It is clear from this form of presentation that a reduction in the mean strength has probably occurred following result 42. An estimate of the actual mean strength between results 42 and 60 is given by $\bar{f}_{cm} = 38 + (-40.5/18) = 36$ N mm^{-2}. In practice, the reason for any such reduction in mean strength should be sought. The standard deviation is required before any decision to change the mix proportions is made.

Standard deviation (CUSUM SD)

A cusum control chart for standard deviation is obtained by plotting the cumulative sum of the differences between the design mean range for successive results, equal to $1.13s$, and the observed range. Typical calculations are shown in table 17.6, columns 5, 6 and 7, and the associated control chart in figure 17.10. In this case, an estimate of the actual standard deviation $\bar{s}$ may be obtained from

$$\bar{s} = s + \frac{(\text{change in CUSUM SD over } n \text{ results})}{1.13n}$$

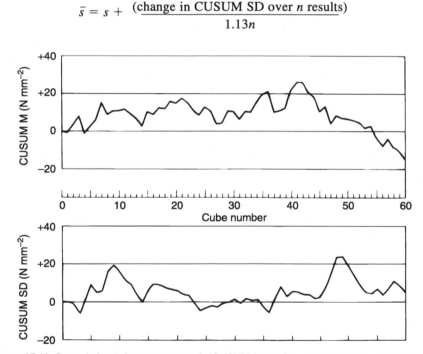

Figure 17.10 *Control charts for mean strength (CUSUM M) and standard deviation (CUSUM SD)*

It may be noted at this stage that all the test results in table 17.6 meet the requirements for compliance with characteristic strength given in BS 5328. For Grade 30 concrete (characteristic strength 30 N mm^{-2}) these are

(a) the average strength determined from any group of four consecutive test results shall be at least 33 N mm^{-2}, and

(b) the strength determined from any test result shall be at least 27 N mm^{-2}.

In figure 17.10 it is seen that, despite apparent local variations, the standard deviation remains close to the assumed value. In the circumstances, the fact that the actual mean strength, after result 42, is 2 N mm^{-2} lower than assumed indicates that some adjustment to the mix proportions must be considered if the characteristic strength requirements are to continue to be satisfied.

Local variations in the slope of the cusum plot, owing to random variations of the test results, may be expected even when there is no real change in the associated control parameter. Such local variations can be seen in figure 17.10. In general the smaller the number of results used the less reliable will be the estimated value of the parameter in question. To standardise the decision-making process and reduce the probability of making either unnecessary or delayed adjustments to concrete mix proportions, some form of transparent V-mask (figure 17.11) is used

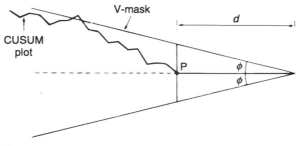

Figure 17.11 *V-mask*

in practice. On this mask, lines are marked on each side of and at an angle ϕ to the horizontal. A point P, distance d from the apex of the V, is placed on each point as it is plotted on the cusum chart. A change in control from that assumed is indicated whenever the cusum plot intersects one of the lines forming the V (see figure 17.11). The magnitudes of d and ϕ are so chosen that, for a given standard deviation (allowance being made for the random variations of results referred to above) and different departures of the control parameter from its assumed value, an acceptable number of results is to be expected before a change of control is indicated. The actual statistical methods used for evaluating appropriate values of d and ϕ are outside the scope of this book.

References

ACI Committee 305 (1977) Hot weather concreting, *J. Amer. Concr. Inst.*, Vol. 74, pp. 317–332..

Bährner, V. (1940) *Report on consistency tests on concrete made by means of the Vebe consistometer,* Report No. 1, Svenska Cemenfor, Stockholm.

Bennett, E.W. and Collings, B.C. (1969) High early strength concrete by means of very fine Portland cement, *Proc. Inst. Civ. Engrs.*, Vol. 43, pp. 443–452.

Bennett, E.W. and Loat, D.R. (1970) Shrinkage and creep of concrete as affected by the fineness of Portland cement, *Mag. Concr. Res.*, Vol. 22, pp. 69–78.

British Cement Association (1988) *The Diagnosis of Alkali–Silica Reaction*, BCA, London.

BS 12: 1991 Specification for Portland cement.

BS 146: 1991 Portland-blastfurnace cement.

BS 410: 1986 Specification for test sieves.

BS 812: 1973–1990 Testing aggregates (in several parts).

BS 882: 1992 Specification for aggregates from natural sources for concrete.

BS 915: Part 2: 1983 Specification for high alumina cement.

BS 1014: 1992 Specification for pigment for Portland cement and Portland cement products.

BS 1047: 1983 Specification for air-cooled blastfurnace slag aggregate for use in construction.

BS 1370: 1979 Specification for low heat Portland cement.

BS 1881: 1970–1972 Testing of concrete (in many parts).

BS 3148: 1980 Methods of test for water for making concrete (including notes on the suitability of the water).

BS 3797: Part 2: 1990 Specification for lightweight aggregates for concrete.

BS 3892 Pulverized-fuel ash; Part 1: 1993 Specification for pulverized-fuel ash for use as a cementitious component in structural concrete; Part 2: 1984 Specification for pulverized-fuel ash in grouts and for miscellaneous use in concrete.

BS 4027: 1991 Specification for sulfate-resisting Portland cement.

BS 4246: 1991 Specification for low heat Portland–blastfurnace cement.

BS 4248: 1974 Specification for supersulfated cement.

BS 4550 Methods of testing cement; Part 0: 1978 General introduction; Part 1: 1978 Sampling; Part 2: 1978 Chemical tests; Part 3: 1978 Physical tests; Part 4: 1978 Standard coarse aggregate for concrete cubes; Part 5: 1978 Standard sand for concrete cubes; Part 6: Standard sand for mortar cubes.

BS 5075 Concrete admixtures; Part 1: 1982 Specification for accelerating admixtures, retarding admixtures and water-reducing admixtures; Part 2: 1982 Specification for air-entraining admixtures; Part 3: 1985 Specification for superplasticizing admixtures.

BS 5328: 1991 Concrete.

BS 6089: 1981 Guide to assessment of concrete strength in existing structures.

BS 6588: 1991 Specification for Portland pulverized-fuel ash cement.

BS 6610: 1991 Specification for pozzolanic cement with pulverized-fuel ash as pozzolana.

BS 6699: 1992 Specification for ground granulated blastfurnace slag for use with Portland cement.

BS 8110 Structural use of concrete; Part 1: 1985 Code of practice for design and construction; Part 2: 1985 Code of practice for special circumstances.

Building Research Establishment (1968) *Shrinkage of Natural Aggregates in Concrete*, Digest 35, HMSO, London.

Building Research Establishment (1969) *Lightweight Aggregate Concretes – 3: Structural Applications*, Digest 111, HMSO, London.

Building Research Establishment (1970) *Lightweight Aggregate Concretes*, Digest 123, HMSO, London.

Building Research Establishment (1982) *Alkali – Aggregate Reaction in Concrete*, Digest 258, HMSO, London.

Building Research Establishment (1991) *Sulphate and Acid Resistance of Concrete in the Ground*, Digest 363, HMSO, London.

Bungey, J.H. (1980) The validity of ultrasonic pulse velocity testing of in-place concrete for strength. *NDT International* (IPC Press), December, pp. 296–300.

Bungey, J.H. (1989) *Testing of Concrete in Structures,* Surrey University Press/Chapman and Hall, Glasgow.

Bungey, J.H. (1991) Ultrasonic testing to identify alkali–silica reaction in concrete, *British Journal of Non-Destructive Testing,* Vol. 33, No. 5, May, pp. 227–231.

Bungey, J.H. (1992) *Testing Concrete in Structures – a guide to equipment for testing concrete in structures,* CIRIA Technical Note 143, Construction Industry Research and Information Association.

Campbell-Allen, D. and Roper, H. (1991) *Concrete Structures: Materials, Maintenance and Repairs,* Longman Scientific and Technical Press, London.

CEB-FIP (1970) *Recommendations for an International Code of Practice for Reinforced Concrete,* Cement and Concrete Association, London.

Comité Euro-International du Beton (1992) *Durable Concrete Structures Design Guide,* Thomas Telford, London.

Concrete Society (1976) *Concrete core testing for strength,* Tech. Report 11, Concrete Society, London.

Concrete Society (1980) *Guide to chemical admixtures for concrete,* Tech. Report 18, Concrete Society, London.

Concrete Society (1986) *Non-structural cracks in concrete,* Tech. Report 22, Concrete Society, London.

Concrete Society (1987) *Alkali–silica reaction: minimising the risk of damage. Guidance notes and model specification clauses,* Tech. Report 30, Concrete Society, London.

Concrete Society (1988) *Permeability of concrete, a review of testing and experience,* Tech. Report 31, Concrete Society, London.

Concrete Society (1990) *Assessment and repair of fire-damaged concrete structures,* Tech. Report 33, Concrete Society, London.

Cornelius, D.F. (1970) Air-entrained concretes: a survey of factors affecting air content and a study of concrete workability, *Road Research Laboratory Report LR 363,* Ministry of Transport, London.

Cusens, A.R. (1956) The measurement of the workability of dry concrete mixes, *Mag. Concr. Res.,* Vol. 8, pp. 23–30.

DD ENV 1992-1-1: 1992. EuroCode 2: Design of Concrete Structures, Part 1. General rules and rules for buildings.

Department of Environment (1988) *Design of Normal Concrete Mixes,* HMSO, London.

Department of Transport (1991) *Specification for Highway Works,* HMSO, London.

Dewar, J.D. (1973) The workability of ready-mixed concrete, *RILEM Seminar, Fresh Concrete: Important Properties and their Measurement,* University of Leeds.

Dhir, R.K. (1986) Pulverised-fuel ash, Chapter 7 in *Concrete Technology and Design, Vol. 3: Cement Replacement Materials;* ed. R.N. Swamy, Surrey University Press, pp. 197–255.

Dhir, R.K. and Sangha, C.M. (1972) A study of the relationships between time, strength, deformation and fracture of concrete, *Mag. Concr. Res.,* Vol. 24, pp. 197–208.

Dhir, R. K. and Yap, A.W.F. (1984a) Superplasticized flowing concrete: durability properties, *Mag. Concr. Res.,* Vol. 36, No. 127, pp. 99–111.

Dhir, R. K. and Yap, A.W.F. (1984b) Superplasticized flowing concrete: strength and deformation properties, *Mag. Concr. Res.,* Vol. 36, No. 129, pp. 203–215.

Dhir, R.K., Munday, J.G.L., Gribble, C.D. and Tharmabala, T. (1978) A new approach to the study of concrete shrinkage, *Proceedings of the International Conference on Materials for Developing Countries,* Bangkok, pp. 507–521.

Dhir, R.K., Hewlett, P.C. and Chan, Y.N. (1987) Near-surface characteristics of concrete: assessment and development of in-situ test methods, *Mag. Concr. Res.,* Vol. 39, No. 141, December, pp. 183–195.

Dhir, R.K., Jones, M.R., Ahmed, H.E.H. and Seneviratne A.M.G. (1990) Rapid estimation of chloride diffusion coefficient in concrete, *Mag. Concr. Res.,* Vol. 42, No. 152, September, pp. 177–185.

Dhir, R.K., Hewlett, P.C. and Chan, Y.N. (1991) Near surface characteristics of concrete: abrasion resistance, *RILEM, Materiaux et Constructions,* Vol. 24, March, pp. 122–128.

Dhir, R.K., Jones, M.R. and McCarthy, M.J. (1992) PFA concrete: carbonation – induced reinforcement corrosion rates, *Proceedings of the Institution of Civil Engineers, Structures and Buildings,* Vol. 94, No. 3, August, pp. 335–342.

Dhir, R.K., Jones, M.R. and Elghaly, A.E. (1993) PFA concrete: exposure temperature effects on chloride diffusion, *Cement and Concrete Research,* Vol. 23, No. 25, pp. 1105–1114.

Dhir, R.K., Jones, M.R., Byars, E.A. and Shaaban, I.G. (1994) Predicting concrete durability from its absorption, *ACI/CANMET International Conference,* Nice, France, May, pp. 1177–1194.

DIN 1048: Part 1: 1978 Testing methods for concrete, fresh concrete and hardened concrete in separately made test specimens (Berlin).

Erntroy, H.C. (1960) *The variation of works test cubes,* Research Report 10, Cement and Concrete Association, London.

FIP (1986) *Inspection and Maintenance of Reinforced and Prestressed Concrete Structures,* Thomas Telford, London.

Fookes, P.G. and Collis, L. (1976) Cracking and the Middle East, *Concrete,* Vol. 10, No. 2, pp. 14–19.

Gilberton, A. and Lees, T. (1991) Variation in test results for cement content, water content and water/cement ratio of concrete units, *Concrete,* Vol. 25, No. 4, May/June, pp. 46–49.

Gilkey H.J. (1937) Moist curing of concrete, *Engng. News Rec.,* Vol. 119, pp. 630–633.

Glanville, W.H., Collins, A. R. and Matthews, D.D. (1947) *The grading of aggregates and workability of concrete,* Road Research Tech. Paper No. 5, HMSO, London.

Goetz, H. (1969) The mode of action of concrete admixture, *Proceedings of the Symposium on the Science of Admixtures,* Concrete Society, London, pp. 1–5.

Harrison, T.A. and Spooner, D.C. (1986) *The Properties and Use of Concretes made with Composite Cements,* Int. Tech. Note 10, Cement and Concrete Association, London.

Hewlett, P.C. (Ed.) (1988) *Cement Admixtures: Uses and Applications,* 2nd edition, Longman Group, London.

Hobbs, D.W. (1988) *Alkali–silica Reaction in Concrete,* Thomas Telford, London.

Hughes, B.P. (1960) Rational concrete mix design, *Proc. Inst. Civ. Engrs.,* Vol. 17, pp. 315–332.

Hughes, B.P. (1966) A laboratory test for determining the angularity of aggregate, *Mag. Concr. Res.,* Vol. 18, pp. 147–152.

Hughes, B.P. (1968) The rational design of high quality concrete mixes, *Concrete,* Vol. 2, pp. 212–222.

Hughes, B.P. (1971) The economic utilization of concrete materials, *Proceedings of the Symposium on Advances in Concrete,* Concrete Society, London.

Institution of Structural Engineers (1980) *Appraisal of Existing Structures,* London.

Kay, T. (1992) *Assessment and Renovation of Concrete Structures,* Longman Scientific and Technical Press, London.

Lea, F.M. (1970) *The Chemistry of Cement and Concrete,* Edward Arnold, London.

McCurrich, L.H., Hardman, M.P. and Lammiman, S.A. (1979) Chloride-free accelerators, *Concrete,* Vol. 13, No. 3, pp. 29–32.

McIntosh, J.D. (1964) *Concrete Mix Design,* Cement and Concrete Association, London.

Millard, S.G. (1991) Reinforced concrete resistivity measurement techniques, *Proceedings of the Institution of Civil Engineers,* Vol. 91, Part 2, March, pp. 71–88.

Munday, J.G.L. and Dhir, R.K. (1984) *Assessment of in-situ Concrete Quality by Core Testing,* ACI Special Publication SP-82, ed. V.M. Malhotra, American Concrete Society, October, pp. 393–410.

Murphy, W. (1984) The interpretation of tests on the strength of concrete in structures, *In situ/Nondestructive testing of Concrete,* Special Publication SP-82, ed. V.M. Malhotra, American Concrete Institute, pp. 377–392.

Neville, A.M. (1986) *Properties of Concrete,* Pitman, London.

Neville, A.M. (1970) *Creep of Concrete: Plain, Reinforced and Prestressed,* North-Holland, Amsterdam.

Orchard, D.F. (1979) *Concrete Technology,* Vols. 1 and 2, Applied Science Publishers, Barking.

Parrott, L.J. (1991) *A Review of Methods to Determine the Moisture Conditions in Concrete,* British Cement Association, London.

Price, W.H. (1951) Factors influencing concrete strength, *J. Amer. Concr. Inst.,* Vol. 47, pp. 417–432.

Rixom, M.R. (1978) *Chemical Admixtures for Concrete,* Spon, London.

Sadegzadeh, M. and Kettle, R.J. (1986) Indirect and non-destructive methods for assessing abrasion resistance of concrete, *Mag. Concr. Res.,* Vol. 38, No. 137, December, pp. 183–190.

Samarin, A. and Dhir, R.K. (1984) *Deformation of in-situ Concrete Strength Rapidly and Confidently by Non-destructive Testing,* ACI Special Publication SP-82, ed. V.M. Malhotra, American Concrete Institute, October, pp. 77–94.

Short, A. and Kinniburgh, W. (1978) *Lightweight Concrete,* Applied Science Publishers, Barking.

Smith, M.R. and Collis, L. (Eds) (1993) *Aggregates,* Engineering Geology Special Publication No. 9, Geological Society, UK.

Spence, I.M., Ramsay, D.M. and Dhir, R.K. (1974) A conspectus of aggregate strength and the relevance of this factor as the basis for a physical classification of crushed rock aggregate, *Advances in Rock Mechanics,* National Academy of Sciences, Washington, pp. 79–84.

Taylor, W.T. (1965) *Concrete Technology and Practice,* Angus and Robertson, Melbourne.

Tomsett, H.N. (1991) Making concrete decisions based on NDT, *British Journal of NDT,* Vol. 33, No. 6, June, pp. 282–285.

Tuutti, K. (1982) *Corrosion of Steel in Concrete,* S-100 44, Swedish Cement and Concrete Research Institute, Stockholm.

Vassie. P.R. (1991) *The half-cell potential method of locating corroding reinforcement in concrete structures,* Application Guide No. 9, Transport and Road Research Laboratory, Crowthorne, Berks.

Further reading

Day, K.W. (1995) *Concrete Mix Design, Quality Control and Specification,* Spon/Chapman and Hall, London.

Malier, Y. (Ed.) (1992) *High Performance Concrete: From Material to Structure,* Spon, London.

Mays, G. (Ed.) (1992) *Durability of Concrete Structures: Investigation, Repair, Protection,* Spon, London.

Mehta, P.K. (1986) *Concrete: Structure, Properties and Materials*, Prentice-Hall International, Englewood Cliffs, New Jersey.

Ramachandran, V.S., Feldman, R.F. and Beaudoin, J.J. (1981) *Concrete Science: Treatise on Current Research*, Heyden, London.

Shah, S.P. and Ahmad, S.H. (Eds.) (1994) *High Performance Concretes and Applications*, Edward Arnold, London.

Soroka, I. (1979) *Portland Cement Paste and Concrete*, Macmillan, London.

Swamy, R.N. (Ed.) (1991) Blended cements in construction, *Proceedings of the International Conference on Blended Cements in Construction*, Sheffield, UK, Elsevier Applied Science, London.

Teychenne, D.C., Franklin, R.E. and Erntroy, H.C. (1975) *Design of Normal Concrete Mixes*. Revised Edition (1988) Teychenne, D.C., Nicolls, J.C., Franklin, R.E., Hobbs, D.W. *et al.* DOE/BRE, London.

Vazquez, E. (Ed.) (1990) Admixtures for concrete: improvement of properties, *Proceedings of the International RILEM Symposium*, Spain, Chapman and Hall, London.

IV
BITUMINOUS MATERIALS

Introduction

Most publications introducing bitumen commence with references to applications of natural seepages of bitumen in the Middle East, that were used several thousands of years BC, many of which still exist, and discuss the origins of the term 'bitumen', usually attributing it to Sanskrit. The early translators did not appreciate the difference between the chemical composition of bitumen and tar or pitch, and many of these confusions still remain in the minds of the general public. The historical references support the durability of bitumen but tend to imply that bitumen is simply a black sticky substance, used as an adhesive and waterproofer. An implication of this is that the uses of bitumen are empirically based.

Empiricism has certainly been the main basis of developments until relatively recently, but scientific principles are now being applied to the subject and research has shown how the basic laws of physics can be used to explain observed performance and hence be applied to design.

The civil engineer now requires material properties to be expressed in mathematical terms so that analysis and design can be carried out, usually with the aid of computers. The following chapters attempt to present some of the basic principles so that these can be applied to application and design, rather than referring to the use of standard specifications or handbooks, without understanding the principles.

This Part of the book deals with bituminous binders, the properties of bitumen–aggregate mixes and design of such mixes. Since the many applications of bitumen make use of the properties of the binder and the properties of the various mixtures with aggregates, the applications are referred to under the appropriate property. To avoid repetition, this approach has been adopted instead of dividing the applications into groups such as waterproofing, mechanical protection and so on, because such groupings involve different types of materials and different properties.

18

Bituminous Binders

18.1 Scope

The term 'bituminous binder' includes 'coal tar', 'natural asphalt' and bitumen obtained from petroleum crude (petroleum bitumen). Tars with suitable properties are now not widely available. The availability and production of natural asphalts, which contain appreciable amounts of mineral matter and thus present transport and handling problems, are very small in comparison with those of petroleum bitumens, on which attention will be concentrated. Although the chemical composition of tar is very different from that of bitumen, the basic principles underlying the physical properties and applications are common to all bituminous binders, although the numerical properties are different. For example, the consistency or viscosity of tar is more susceptible to temperature changes than is the viscosity of bitumen and as the temperature drops, tar is more liable to fracture than is bitumen.

Bitumens suitable for many purposes, including road and aircraft pavements and industrial applications (of which there are many hundreds), are manufactured from petroleum. Bitumens used for many applications including road and aircraft pavements are called 'penetration grade bitumens', 'asphaltic bitumens' or 'asphalt cement': the latter term being used in North America. The terms 'straight-run' and 'residual' are sometimes found in the literature, but these refer to manufacturing processes, now practically obsolescent, and these terms are not relevant to the user of the product.

Petroleum bitumens are complex mixtures of high molecular weight hydrocarbons, defined as being completely soluble in carbon disulphide. They are thus also soluble in most other hydrocarbons, but are relatively inert chemically and can therefore be used in protective coatings. They are, for example, resistant to hydrochloric acid, to dilute sulphuric acid and to alkalis. They have useful properties as adhesives and binders, as waterproofers and as insulators, and they are flexible, durable, non-toxic, tasteless and odourless when cold.

For most applications of bitumen, the main interest lies in the engineering properties, including such factors as the response to applied stress and the change in response when temperature changes. The chemical composition and chemical properties are generally of secondary importance except to the manufacturer.

'Penetration grade bitumens' are semi-solid at ambient temperatures and require to be heated to render them fluid enough for application by mixing or spraying. The heating requirements can be reduced by using them in the form of 'cutbacks' or 'liquid asphalts' (North America), which consist of bitumen fluxed or diluted with solvent or in the form of 'emulsions' or 'emulsified asphalts' (North America), which are dispersions of bitumen in water. Cutbacks and emulsions are mainly used for spray applications, such as surface dressing, for temporary repairs and for lightly trafficked surfacings.

The emphasis will be on the properties of 'penetration grade bitumens', since the use of 'cutbacks' or 'emulsions' may be regarded as a means of applying penetration grade bitumens at lower temperature, but they can only be used when the system is sufficiently porous to permit setting or curing by loss of the volatiles and are thus of lower durability than penetration grade bitumens (see section 18.5 and Chapter 19, section 19.6).

18.2 Characterisation

Since an almost infinite number of grades or hardness of 'penetration grade bitumens' can be produced, it is necessary to adopt convenient methods of describing them. The most obvious need is for a measurement of hardness or grade but since the consistency of bitumen is temperature dependent, the consistency becoming softer as the temperature rises, it is necessary either to measure the consistency at a fixed temperature or to determine the temperature at which a fixed consistency occurs.

Most of the tests used for characterising bitumen are of an empirical nature. It is, therefore, essential that the tests should always be carried out under the same conditions. To ensure that the results of tests carried out by different people at different times and in different places are strictly comparable, most bitumen tests have been standardised by organisations such as the British Standards Institution (BS) and the Institute of Petroleum (IP) in the UK, the American Society for Testing Materials (ASTM) and the Association of State Highway and Transportation Officials (AASHTO) in North America, and other similar bodies. The methods developed by these organisations should always be used.

Full descriptions of the tests are deliberately not given here, in order to discourage testing without reference to an appropriate standard. The latest version must always be fully complied with; dates are therefore not given.

Penetration

The first method of defining consistency, by measuring at a fixed temperature, is employed in the 'penetration test', covered by BS 2000-49, IP 49, ASTM D 5 and AASHTO T 49. In this test, the principles of which are illustrated in figure 18.1, a needle, of standardised dimensions, is allowed to penetrate into a sample of the bitumen, under a known load, at a fixed temperature, for a given time. The amount of indentation, in units of 0.1 mm, is termed the 'penetration'. The normal conditions of test are 100 g load, temperature $T = 25°C$ and a time $t = 5$ seconds. These are assumed to have been used unless otherwise stated, but tests at other temperatures can also be carried out and sometimes under other loading conditions. Penetrations can vary from close to zero for the soluble portion of some natural bitumens to over about 700×0.1 mm for road oils used for treating gravel roads.

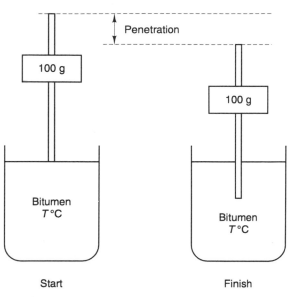

Figure 18.1 *The penetration test*

Softening point (SP)

The second method of defining hardness or consistency is to determine the temperature at which a fixed consistency occurs. This is used in the 'softening point (ring and ball)' test, which is illustrated in figure 18.2 and is standardised in BS 2000-58, IP 58, ASTM D 36 and AASHTO T 53. The test is carried out by recording the temperature, $T_{R\&B}$ or SP, at which the bitumen softens sufficiently to allow a steel ball, placed on a sample of the bitumen, to fall a fixed distance, when the bath temperature is raised at a specified rate. No change in the physical state of the bitumen occurs and the softening point must *not* be treated as a melting point, but merely as one point on a plot of consistency versus temperature. In the North American tests the temperature-controlled bath is not stirred during heating, as it is in the British versions, so the softening point occurs at a higher temperature than in the former. The temperature difference according to Krom (1950) is usually about 1.5°C. Softening points can vary from about 25° to 120°C.

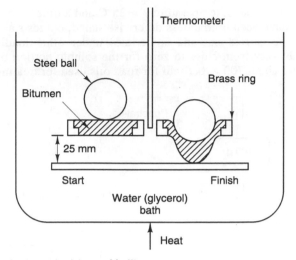

Figure 18.2 *The softening point (ring and ball) test*

Temperature susceptibility

Bitumens vary considerably in the change in hardness or consistency with changes in temperature, or their temperature susceptibility, and it is necessary to define this characteristic, which depends on the rheological type or flow characteristics of the bitumen. The most widely used method is that developed by Pfeiffer and van Doormaal (1936) and which is discussed in detail by Heukelom (1973). This is the 'Penetration Index' (PI) which is defined by the following equation, where A is the slope of the line in a plot of the logarithm of the penetration versus the temperature:

$$50A = \frac{20 - PI}{10 + PI} = \frac{50 \, d(\log_{10} \text{pen})}{dT} = 50 \, \frac{\log_{10} \text{pen} \, T_1 - \log_{10} \text{pen} \, T_2}{T_1 - T_2}$$

where $T_1 > T_2$.

The value of A for many bitumens varies between about 0.015 and 0.06 and the numbers have very little significance to the user. It was therefore decided to arrange the equation in the above form so that a normal bitumen at the time would have a PI of zero and other bitumens would be rated above or below the 'standard', which was 200 pen_{25} manufactured from crude oil from Panuco, Mexico. PIs for paving grade bitumens are usually in the range -1 to $+1$, tars as low as -2 and industrial grades $> +2$.

For ease of calculation the equation can be rewritten in the form:

$$PI = \frac{500(\log_{10} \text{pen} \, T_1 - \log_{10} \text{pen} \, T_2) + 20(T_2 - T_1)}{(T_2 - T_1) + 50(\log_{10} \text{pen} \, T_2 - \log_{10} \text{pen} \, T_1)}$$

The penetration index can thus be determined from measurements of penetration at two different temperatures, T_1 and T_2.

This concept was developed long before scientific calculators became generally available and the equations were therefore published in the form of nomographs or line charts. These can be found in the original publications and in the literature but, since calculators are now generally available, only the equations are given here.

For many bitumens, the ASTM softening point (ring and ball), $T_{R\&B}$, is the same as the temperature required to give a penetration of 800, or $T_{800\ pen}$, and so the equation can be rewritten as:

$$\frac{20 - PI}{10 + PI} \times \frac{1}{50} = \frac{\log_{10} 800 - \log_{10}(\text{pen at } T^\circ C)}{T_{R\&B} - T}$$

If penetration is measured at 25°C, the equation can be rewritten as:

$$PI = \frac{1951.5 - 500 \log_{10} \text{pen}_{25} - 20\ T_{R\&B}}{50 \log_{10} \text{pen}_{25} - T_{R\&B} - 120.15}$$

This method of determining penetration index has been used for many years. It is now known, however, that there can be serious error in assuming that the ASTM softening point occurs at 800 pen, because in some bitumens the paraffin wax component causes errors in the measurement of the softening point. The original definition is now recommended, as discussed by Heukelom (1973). It is, however, for the reasons discussed in sections 18.3 and 18.4, necessary to know the temperature required to give a penetration of 800, $T_{800\ pen}$, and this can be obtained from:

$$T_{800\ pen} = T + \frac{50(\log_{10} 800 - \log_{10} \text{pen } T)(10 + PI)}{(20 - PI)}$$

Viscosity

Although bitumens exhibit visco-elastic behaviour during their service life, their behaviour when heated for application is essentially fluid or viscous. Viscosity is defined as follows: if a sample is sheared by a constant shear stress, τ, N m^{-2} (or Pa), it will deform at a constant rate of shear strain, R, s^{-1}. The dynamic viscosity or the 'coefficient of viscosity', η, is defined as the ratio of shear stress to the rate of strain, $\eta = \tau/R$, Ns m^{-2}.

This is illustrated in figure 18.3. Prior to the adoption of the SI system of units, the unit of viscosity was the Poise, P, and this is still widely used. 1 Poise = 10^{-1} Ns m^{-2}. Also widely used is the term 'kinematic viscosity', which is the Poise divided by the density. The unit is the Stoke, St, m^2 s^{-1} and 1 centiStoke (cSt) = 1 mm^2 s^{-1}.

The range of viscosities is large, as indicated in table 18.1, and so a range of different viscometers is required to measure all viscosities.

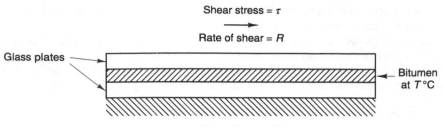

Figure 18.3 *Viscosity*

TABLE 18.1

Range of viscosities

Material	Viscosity at 20°C (Poise)
Water	10^{-2}
Diesel	10^{-1}
Engine oil	10
Cutback	10^3
300 pen bitumen	10^5
100 pen bitumen	10^6
15 pen bitumen	10^8

'Dynamic viscosity' can be measured using a 'sliding plate viscometer' in which the force or shear stress and displacement or shear strain of a sample of bitumen, held between parallel plates in a controlled-temperature bath is measured, as illustrated in figure 18.3. This covers the viscosity range of about 10^2 to 10^{11} Poise. Viscosity can also be measured using rotation or concentric cylinder viscometers, which work on the same principle.

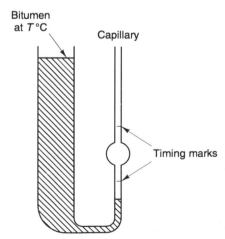

Figure 18.4 *Capillary viscometer*

For measurements at high temperatures, associated, for example, with mixing, it is usually more convenient to use a 'capillary viscometer', the principle of which is illustrated in figure 18.4. A sample of the bitumen is placed in the glass viscometer which is then placed in a constant-temperature bath. The bitumen flows through the capillary and the time for flow between two marks, indicating a fixed volume, is measured. This time is used to calculate the 'kinematic viscosity'. Gravity-flow viscometers of this type have to be calibrated against standard viscosity liquids. Standard methods are IP 70, ASTM D 2170 and AASHTO T 201, and these are suitable for a range of 0.001–1000 P. For higher viscosities, a vacuum can be used instead of gravity as the driving force. Standard methods are IP 222, ASTM D 2171 and AASHTO T 201, and these cover the range $10-10^5$ P.

For routine measurements of viscosity in empirical units, 'efflux viscometers', the principle of which is illustrated in figure 18.5, are normally used. In this type, the time of flow for a given volume of bitumen at a fixed temperature, through a standard sized orifice, is measured. The range is $10^{-1}-10^{-3}$ P. The empirical viscosity units, time in seconds, may be converted approximately to fundamental units using conversion tables. Standard methods are BS 2000–72, IP 72 for the Standard Tar Viscometer, ASTM D 88 and AASHTO T 72 for the Saybolt Furol viscometer. This type of instrument is normally only used for cutbacks or emulsions.

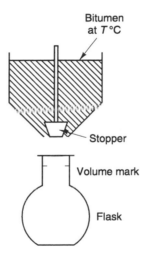

Figure 18.5 *Efflux viscometer*

Fraass breaking point

Some indications about the low-temperature characteristics can be obtained from the 'Fraass breaking point' test, IP 80, although the test is not widely used. A thin steel plate coated with the bitumen is flexed continuously as the temperature is reduced, to determine the temperature at which fracture of the bitumen occurs: it has been shown that the Fraass breaking point occurs when the stiffness modulus (see section 18.4) is 2.1×10^9 Pa.

Grades available

Petroleum bitumens were originally produced by the vacuum distillation at a temperature in the region of 400°C of selected crude oils, from which the light fractions had been removed previously by simple distillation at atmospheric pressure. Vacuum distillation removes the more volatile components in the feedstock, such as gas oil, and is continued until the desired consistency is obtained. This consistency is usually measured by the penetration test, and the softening point (ring and ball) and the grade of product (bitumen grade) are designated by the penetration at 25°C. Penetration grades available commercially are usually described, as in BS 3690, by the penetration limits, such as 300 ± 20, 100 ± 20 and so on.

Although grading by penetration has been adopted in most countries, some North American organisations use viscosity in Poise at 60°C. In some of these specifications, the viscosity refers to the original bitumen, for example, AC-2.5, AC-5, AC-10, AC-20, AC-40, but in others to the residue after the rolling thin film oven test, a test for durability (ASTM D 2872 or AASHTO T 240 – see section 18.5 below), for example, AR-1000, AR-2000, AR-4000, AR-8000 and AR-16000.

Bitumens with a relatively high softening point (ring and ball) for a given penetration can be produced by blowing air through the hot bitumen. This oxidation or blowing process produces bitumens whose consistency is less affected by temperature changes and which are suitable for industrial applications such as roofing and pipe coating. These grades (blown bitumen or oxidised bitumen) are normally described by the mid-points of the softening point (ring and ball) and penetration limits with a prefix 'R', such as 'R85/40'.

In the original method of production, there was no control over the softening point (ring and ball) of the bitumen at any particular penetration, except that provided by crude selection. The terms 'residual' and 'straight-run' refer to this type of bitumen. This process is still sometimes used, but rarely.

Manufacturing techniques are now more versatile than those used originally and permit bitumens of the desired characteristics to be produced from crudes that were unsuitable for processing by the early methods. There are, however, some restrictions to the possibilities that exist to provide bitumens with unusual properties. Although in most cases it is possible to meet normal specification requirements, such specifications do not always ensure that the bitumen will be satisfactory in performance. It is therefore necessary to evaluate new crudes and manufacturing methods by extensive laboratory and field tests before they can be marketed with confidence.

The number of grading systems used for cutbacks and emulsions are many and varied, and depend on the region or country in which they are used. The systems, types and compositions used are usually based on the local conditions, including such factors as application, climate and aggregate type. In the UK, cutbacks are included in BS 3690 and emulsions in BS 434.

18.3 Bitumen test data chart

The routine empirical tests, referred to above, are well known and widely used. They are useful for normal penetration grade bitumens, but there are some types of bitumen, such as those containing significant quantities of paraffin wax, for which these tests give misleading results.

To assist in the interpretation of the routine tests, Heukelom (1973) developed the 'bitumen test data chart' (BTDC) illustrated in figure 18.6. The construction of the chart is such that the test data for normal penetration grade bitumens are represented by straight lines (Class S bitumens – the S referring to the straightness of the line). The chart is based on the routine tests described above, namely 'penetration', 'softening point (ring and ball)' and 'viscosity'. By plotting the routine test results on the chart, it will be seen whether the bitumen in question belongs to Class S, Class W (waxy bitumens) or Class B (blown or industrial grades). Typical bitumens of the three classes are illustrated in figure 18.6.

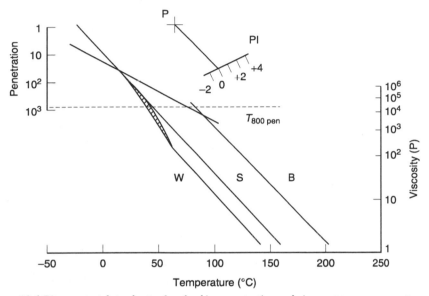

Figure 18.6 *Bitumen test data chart – for checking penetration and viscometer measurements*

Class S bitumens are represented by straight lines. If a line parallel to the plotted points is drawn through the point P, to intersect the PI scale, the PI can be read directly. If the plotted points do not lie on a straight line, the tests should be checked or rejected. The $T_{800\text{ pen}}$ can be read from the intersection of the plotted line with the $T_{800\text{ pen}}$ line.

Class W bitumens are represented by two, normally nearly parallel straight lines, with a transition zone between them, owing to the effects of wax crystallisation interfering with the consistency tests. PI is determined as for Class S bitumens. $T_{800\text{ pen}}$ has two values, depending on whether interest lies in the high or low temperature regions.

Class B bitumens are represented by two straight lines, each having a different slope. There are two effective PIs and two values of $T_{800\text{ pen}}$, depending on whether interest lies in the high or low temperature areas. In practice the PI in the high temperature range is of little practical interest.

It should be noted, as discussed in sections 18.4 and 18.5, that during application and service, all bitumens harden and increase slightly in PI, and that allowance should therefore be made for these factors when using the BTDC.

18.4 Engineering properties

The engineering properties of bitumen are dependent on both temperature, T, and the duration or time of loading, t, of the stress applied, σ. In contrast to many traditional structural materials, such as steel, whose properties are for practical purposes constant, the behaviour under stress varies from elastic to viscous, according to the conditions of stressing. Under stresses of short duration and at low temperatures, bitumen behaves purely elastically, whereas, under conditions when the stress is applied for long periods and/or at high temperatures, the behaviour is purely viscous. Between these two extremes, and this is the range of practical interest for most applications, the behaviour is intermediate or 'visco-elastic'.

Stiffness modulus

To describe the engineering properties of bitumen, a modification and simplification of the approach used for thermoplastic polymers was developed by van der Poel (1954) and subsequently modified by Heukelom (1973). The approach is based on the concept of a 'stiffness modulus', $S_E = S_{bit}$, which is a generalisation of Young's modulus, E, the ratio of stress, σ, to strain, ε, where $E = \sigma/\varepsilon$, as used for conventional structural materials. The magnitude of this 'stiffness modulus' depends on the temperature, T, and the duration or time of loading, t, and it is assumed that the strain, ε, is proportional to the applied stress, σ, or expressed scientifically, $S_{bit(t,T)} = \sigma/\varepsilon_{(t,T)}$. This assumption is acceptable, provided that the strain is less than about 1 per cent. As one of the main objectives of most applications is to restrict or prevent excessive deformation, this restriction to the validity of the concept is acceptable.

Other, more complicated approaches to the description of visco-elastic materials have been developed, usually described as the 'complex modulus' approach, but these are considered to afford little advantage for the majority of practical situations. The main difference from the 'stiffness approach' is that the phase lag, ϕ, between the application of stress and the resulting strain is taken into account, whereas, in the simplification, this is neglected. For most bitumens, the main interest is in the maximum values of stresses and strains, and not in the time interval between them.

The variation of stiffness modulus for a typical bitumen as a function of time of loading for various temperatures is illustrated in figure 18.7, where elastic behaviour is represented by the horizontal portion of the curves and viscous behaviour by lines at a slope of 45°. It will be noted that over a large range, the behaviour is neither purely elastic nor purely viscous: it is visco-elastic.

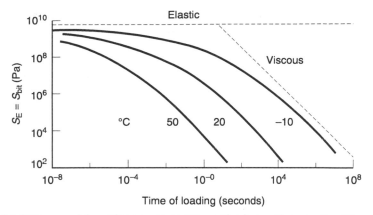

Figure 18.7 *Stiffness modulus of bitumen against time of loading, with curves for different temperatures*

van der Poel Nomograph

Measurement of 'stiffness', $S_E = S_{bit}$, is only possible over the whole range of loading conditions with specialised equipment in a well-equipped laboratory. The results of very many such measurements have, however, been put together in a form of line chart or 'Stiffness Nomograph', by van der Poel (1954) and modified by Heukelom (1973), so that the 'stiffness modulus', of any bitumen, under any specific conditions of time of loading and temperature may be obtained, provided that the hardness of grade, $T_{800\ pen}$, and the type or penetration index are known.

The van der Poel Stiffness Nomograph makes use of the knowledge that the stiffness of bitumen depends on the time of loading, t, the rheological type, PI, the

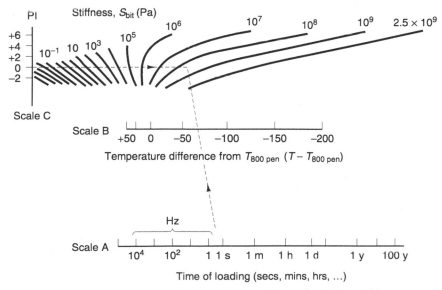

Figure 18.8 *Diagram of van der Poel Nomograph – for predicting the stiffness of bitumen*

temperature, T, and the grade or hardness of the bitumen, $T_{800\text{ pen}}$, which are combined as one parameter, $T - T_{800\text{ pen}}$. This is the reason, referred to in section 18.2, for using $T_{800\text{ pen}}$ for grading rather than the penetration, so that grade can be expressed as a temperature, thus reducing the number of parameters. There are therefore three scales in the van der Poel Nomograph, each representing one of these parameters.

To use the van der Poel Stiffness Nomograph, shown diagrammatically in figure 18.8, the appropriate time of loading, t, is selected on Scale A, and a straight line is drawn from this point to the appropriate point on Scale B, the temperature difference scale. This straight line is extended until it crosses the appropriate PI line, Scale C. At the point of intersection of the two lines, the Stiffness Modulus is read by interpolation between the curved lines, representing fixed values of stiffness modulus.

The van der Poel Nomograph permits the estimation or prediction of the stiffness modulus at any temperature and any time of loading within an accuracy of a factor of 2, which is as good as can be expected from the accuracy of the empirical tests used and with the estimation of temperature in practice.

For the conditions appropriate to most paving situations the stiffness modulus of the bitumen, S_{bit}, can be approximated by the following equation:

$$S_{\text{bit}} = 0.1157 \times t^{-0.368} \times 2.718^{-\text{PI}} (T_{\text{R\&B}} - T)^5$$

where the applicability limits are:

$$t = 0.01 \text{ to } 0.1 \text{ s}; \quad \text{PI} = -1.0 \text{ to } +1.0; \quad T_{\text{R\&B}} - T = 20 \text{ to } 60°\text{C}$$

It should be noted that when determining the stiffness modulus to investigate the service behaviour of bitumen, the results of tests on the bitumen, or estimation, representing the *in situ* conditions should be used. In other words, account must be taken of any changes that have taken place as a result of handling, mixing or spraying and exposure. For example, if used in a mix, the bitumen will have hardened by loss of volatiles and by oxidation during mixing, placing and perhaps during service. It is normal for penetration grades to harden to about 65 per cent of the original penetration or by about one grade during construction, that is 80/100 pen_{25} will have become 60/70 pen_{25}, and there will have been a slight increase in the PI, by the time it is in place.

Range of stiffness values

A pavement structure is subjected to seasonal variations in temperature and to different times of loading, which cover a very wide range, from relatively long periods under standing loads to very short periods under fast moving traffic. As a result, various defects may occur and it is of interest to consider the changes in stiffness modulus of a typical penetration grade bitumen, under the variations in temperature and time of loading. The figures given in table 18.2 illustrate the enormous range that has to be considered in relation to practical performance for conditions that are by no means extreme. The range appropriate to industrial applications tends to be in the more viscous range.

TABLE 18.2

Effects of practical stress conditions on the relative stiffness of 100 pen$_{25}$ bitumen

Time of loading	Condition in practice	Possible defects	Relative stiffness at		
			$-10°C$	$25°C$	$50°C$
1 month	Settlement	Cracking	7×10^{-3}	4×10^{-7}	3×10^{-9}
1 day	Standing load	Deformation	2×10^{-1}	1×10^{-5}	1×10^{-7}
1 hour	Standing load	Deformation	7×10^{-1}	3×10^{-4}	2×10^{-6}
0.4 s	Penetration		8×10^{2}	1	2×10^{-2}
10^{-1} s	Slow traffic	Fatting up & deformation	1×10^{3}	3	4×10^{-2}
10^{-2} s	Fast traffic	Fretting & cracking	3×10^{3}	2×10	3×10^{-1}
10^{-3} s	Fast traffic	Fretting	4×10^{3}	1×10^{2}	3

The figures are relative to the penetration at 25°C, which is often the only measure used in practice. Although the standard penetration test lasts 5 seconds, the effective time has been shown to be (van der Poel, 1954) 0.4 second: since the rate of penetration reduces during the test. The time of loading for non-paving applications, such as revetments for water-retaining structures, can, of course, be many years and this is reflected in figure 18.8. The penetration test alone cannot be expected to provide reliable data about stiffness values of practical interest. However, when used in conjunction with the $T_{800\,pen}$ to obtain PI, it can be used to predict the stiffness modulus under any conditions of interest. The term 'fretting' used in table 18.2, refers to loss of surface particles resulting from brittle fracture of the bitumen.

Permissible strain

Although the data on stiffness modulus can be used in connection with the deformation or flow behaviour of penetration grade bitumens, failure may occur by fracture under critical conditions. It has been shown by Heukelom (1966) that the strain at break depends solely on the stiffness modulus of the bitumen measured at the appropriate time of loading and temperature. From this unique relation between breaking strain and stiffness modulus, the breaking strain of any bitumen may be estimated for any particular conditions.

Heukelom (1966) has shown that the fatigue strain for any particular number of repetitions of loading is determined solely by the value of the stiffness modulus of the bitumen.

Application of the stiffness concept

The stiffness concept, including permissible strain, provides an adequate description of the engineering properties of penetration grade bitumens. In practice, of course, bitumen is usually used in a mixture with mineral aggregates. This subject is dealt with in Chapter 19.

An example of one application of pure bitumen that can be analysed using the stiffness concept is the provision of a slip layer. In the case of soil settlement, which takes place over months or years, the transmission of settlement forces to bearing piles (negative friction) can be reduced, because bitumen behaves viscously under such conditions. The piles must first be coated, stored prior to use and driven by hammering. The stiffness concept can be used to ensure the use of bitumen with suitable properties to resist flow during storage, to resist fracture during driving and to reduce the negative friction.

The equations or nomographs that have been developed for obtaining the stiffness modulus are generally applicable to pure bitumens, which are classed as thermo-rheologically simple materials, but not generally to bitumen containing polymer additives, unless it can be shown that the bitumen plus additive is still a thermo-rheologically simple material. In general, direct testing is recommended for bitumens containing polymer additives. It should be noted that bitumens containing polymer additives should be regarded as materials under development, as factors such as their long-term durability have not yet been proven.

18.5 Non-engineering properties

Although the engineering properties of penetration grade bitumens are of prime importance, the service behaviour will depend on the extent to which these properties are maintained. Two factors which may interfere with the expected behaviour are changes in the engineering properties due to hardening, and possible changes due to the action of water. The latter is to a great extent controlled by the minerals in contact with the bitumen, and this aspect is dealt with in Chapter 19.

Durability

Penetration grade bitumens are relatively inert, but slow surface hardening occurs on exposure to air. The effect is accelerated at elevated temperatures. Hardening can be of two kinds: physical and chemical.

Cooling from the application temperature to the service temperature causes molecular reorientation and hence an increase in viscosity, usually by a factor of less than 3, over a period of several months. The mechanism, a type of age hardening, is reversible on reheating or if the bitumen is recovered by solvents for testing. The changes can therefore only be detected indirectly by the measurement of mix properties, which is discussed in Chapter 19.

A similar phenomenon occurs in the case of paraffin wax, as a result of crystallisation. The process is faster than that referred to above and is also reversible.

A third cause of physical hardening is evaporation of volatile fractions: this is non-reversible. The rate of evaporation depends on the nature and quantity of volatile components and the conditions of exposure.

As with most organic substances, bitumen is slowly oxidised when in contact with atmospheric oxygen, and this causes an increase in viscosity and/or stiffness modulus and an increase in Penetration Index.

Experiments with various layer thicknesses have shown that oxidation is limited to a depth of about 4 microns. In practice, ageing affects the bitumen to a greater depth, as the hard skin that is formed may crack and disintegrate to expose a fresh surface.

The hardening that occurs, by the mechanisms referred to above, is accompanied by a change in the type of bitumen: an increase in PI occurs. The magnitude of both hardening and PI changes is dependent on the exposure conditions that exist during application and during the service life of the bitumen: in other words, on the temperature and time of exposure during application and on the extent to which the bitumen is exposed to the atmosphere during service. Since this is mostly dependent on the type and the porosity of the application, the subject is better dealt with under the relevant application.

Laboratory tests

It should be noted that although bitumens vary in their resistance to ageing, there is no additive that can be used to control the hardening. A variety of laboratory tests is available for assessment, of which the following are typical.

The 'loss on heating test', LOH (BS 2000-45, IP 45, ASTM D 6, AASHTO T 47), has been in use for many years to assess the liability of bitumens to excessive hardening. In this test, the bitumen is stored in a tin, as standardised for the penetration test, on a rotating shelf in an oven for 5 hours at 163°C, and the drop in penetration and loss in mass are measured. The surface area–volume ratio of the sample is, however, not representative of the practical exposure conditions, and the presence of a hardened skin hampers further hardening in the bulk of the sample. The test is mainly useful in detecting the presence of volatiles, and this can be done more easily by the 'flash point'. The test can, however, be useful for obtaining an indication of the extent of hardening during hot storage in a tank, where the surface area–volume ratio is similar to that in the test.

The 'rolling thin film oven test', RTFOT (ASTM D 2872, AASHTO T 240), gives a fairly reliable indication of the effect on bitumen of mixing with aggregate. Cylindrical glass bottles containing the bitumen are fixed in a vertically rotating shelf or wheel so that the bitumen flows along the surface of each container. The assembly is mounted in an oven at 163°C and heated air is periodically blown into the samples. The time test is 75 minutes. As in the loss on heating test for bitumen, the properties are then measured.

18.6 Applications of pure bitumen

In addition to the manufacture of bituminous paints, primers and other surface treatments, which may be regarded as providing relatively low-durability protection, many grades of bitumen, particularly the industrial, blown or 'R' grades, are used mainly for the manufacture of proprietary products for protection and waterproofing. Joint filling compounds, spray applications for flat roofs and pipe coating and lining are examples.

Roofing felts and shingles consist essentially of natural or synthetic fabric impregnated with penetration grade bitumen and then coated with a 'blown grade' of bitumen. Felts are often used for built-up flat roofs where successive layers of felt are bonded using bitumen adhesive. Shingles, which are normally squares cut from rolls of felt, are used as an alternative to tiles on pitched roofs. Flat roofs may also be waterproofed by repeated sprays of bitumen, often incorporating fibre reinforcement.

Repeated sprays, usually blown bitumens or prefabricated mats, are used for membranes where they are protected from mechanical damage by a substantial thickness of soil. They can be used for applications such as small water retaining structures and refuse disposal sites to prevent the spread of pollution.

18.7 Health and safety

Any relevant local laws, regulations, specifications and advice related to Health and Safety aspects must be followed, but it is relevant to review some general principles. In the UK, the Fire and the Health and Safety regulations, such as those of the Institute of Petroleum Model Code of Safe Practice and the Control of Substances Hazardous to Health (COSHH) must be complied with.

Bitumens are handled and applied at high temperatures, which can be above the flash point, and thus present a potential hazard from skin burns and inhalation of substances such as sulphur compounds from hot storage tanks. Skin burns should be plunged into cold water and, once cooled, no further harm will be done as the bitumen provides a sterile covering until it becomes detached.

Bituminous binders contain compounds which are potentially carcinogenic, but the concentration in bitumen is several orders of magnitude lower than the concentration of similar compounds in coal tars and thus does not necessarily constitute a health risk in practice, although appropriate precautions should be taken.

19

Properties of Bitumen–Aggregate Mixes

19.1 General

Bitumen–aggregate mixtures consist of three components: bitumen, mineral aggregates and air. They are mostly used for road and aircraft pavements but are, depending on the properties, also used for water retaining and protection in connection with buildings, dams, canals, sea defences and so on. Their properties are very dependent on volumetric composition as well as the characteristics of the bitumen and aggregate. The need to consider volumetric composition is very important as it takes into account the state of compaction *in situ*, which is not covered when composition by mass is considered. The terminology and symbols used here are those used by The Asphalt Institute, TAI (American association of bitumen producers). Different terminology is used by other authorities, such as the British Standards Institution, Nottingham University and others, but that used here is considered the most suitable, and is the most widely used throughout the world.

Calculation of volumetric composition

Figure 19.1 illustrates the three components together with the terminology used for the composition by mass and volume. The method recommended for calculating the volumetric composition is as follows:

Mass of specimen in air $= M_m$, mass in water $= M_w$, volume $= V_{mb} = M_m - M_w$

Bulk specific gravity of compacted mix	$= G_{mb}$	$= M_m/V_{mb}$
Bulk specific gravity of compacted aggregate	$= G_{sc}$	$= G_{mb}P_s/100$
Theoretical maximum specific gravity of the mix	$= G_{mm\,(th)}$	
	$= 100/(P_s/G_{sb} + P_b/G_b)$	

where G_{sb} = bulk specific gravity of aggregate

and G_b = specific gravity of bitumen

Voids in mix = $V_{a(th)} = ((G_{mm(th)} - G_{mb})/G_{mm(th)}) \times 100$

Voids in mineral aggregate = $V_{ma} = ((G_{sb} - G_{sc})/G_{sb}) \times 100$

A correction may then be required for the amount of bitumen absorbed by the aggregate (ASTM D 2041, AASHTO T 209):

$G_{mm(actual)}$ = (mass of coated mix in air)/(mass of coated mix in air − mass of coated mix in water) = $M_m/(M_m - M_w)$

Volume of bitumen absorbed = $V_{ba} = G_{mb} \times (1/G_{mm(th)} - 1/G_{mm\,(actual)}) \times 100$

Actual voids in mix, allowing for absorption = $V_{a(actual)} = V_{a(th)} + V_{ba}$

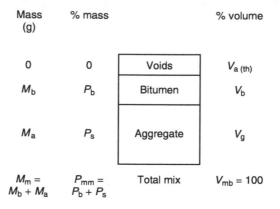

Figure 19.1 *Volume mass relationships for bitumen–aggregate mixtures*

Typical compositions

There is no general agreement on the terminology applied to mixtures of bitumen and aggregate; similar compositions are given different names in different countries and similar names refer to different compositions. For example, in the UK, mixes made with continuously graded aggregates are normally termed 'coated macadams' (BS 4987) and gap-graded mixes are termed 'asphalts' (BS 594). In the UK these specifications are referred to in the *Specification for Highway Works* and the *Design Manual for Roads and Bridges*, which are the main documents required for works for the Department of Transport. The origin of this terminology is largely historical and commercial, but in North America all such mixes are termed 'asphalt concrete'. In this review, the range of mix compositions that are used will be examined, and then consideration will be given to the properties of the various compositions.

There are no *general* requirements concerning the quality of aggregates, except that they must not contain clay, or other water-susceptible components, and that they must be resistant to normal weathering on exposure to the atmosphere, unless protected. An exception is in the case of the wearing course of trafficked surfaces, when skidding resistance must be considered; this is discussed in section 19.6.

TABLE 19.1
Average compositions of typical mixes

	Coated stone	Continuously graded	Gap-graded base course	Gap-graded wearing course	Mastic asphalt
% by volume					
Stone	64.5	44.1	52.0	25.7	27.5
Sand	5.1	32.2	25.7	46.0	18.9
Filler	2.1	4.2	3.5	7.8	27.0
Minerals, V_g	71.7	80.5	81.2	79.5	73.4
Bitumen, V_b	8.3	11.5	13.8	17.5	26.6
Voids, V_a	20.0	8.0	5.0	3.0	0.0
$V_{ma} = V_b + V_a$	28.3	19.5	18.8	20.5	26.6
$V_{fb} = V_b/V_{ma}$ (%)	29.3	59.0	73.4	85.4	100.0
% by mass					
Stone	86.0	52.0	61.0	30.0	30.0
Sand	7.0	38.0	29.0	53.0	26.0
Filler	3.0	5.0	4.0	9.0	32.0
Minerals, P_s	96.0	95.0	94.0	92.0	88.0
Bitumen, P_b	4.0	5.0	6.0	8.0	12.0
Grade (pen$_{25}$) of bitumen	100–300	100–200	40–70	40–70	20–30

Another exception occurs in the case of non-paving mixes required to provide protection against chemical attack, where the aggregate used must usually be resistant to the chemical in question.

In general, the grading, particle shape and packing characteristics required depend on the type of mix and properties required, as discussed below.

Typical compositions of a range of mix types are shown in table 19.1 and the aggregate grading curves in figure 19.2.

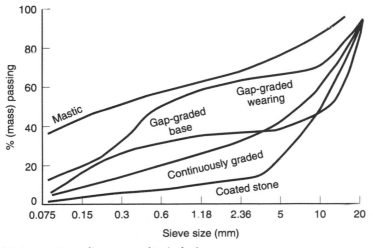

Figure 19.2 *Aggregate grading curves of typical mixes*

A mix used very widely throughout the world has a continuously graded aggregate and is normally called 'asphalt concrete' or 'asphaltic concrete', for example, ASTM D 3515 for 'Hot-Mixed, Hot-Laid Asphalt Paving Mixtures'. In comparison with coated stone, this has rather less coarse aggregate, a significant proportion of sand, filler and rather more of a slightly harder bitumen. The composition is designed to give a low void content after compaction by heavy rolling. This type of mix is very widely used throughout the world for all types of application, particularly for water-retaining structures, not just paving. In the UK, where it is often called 'Marshall asphalt', it is used for aircraft pavements and water-retaining structures, but not for roads. The term 'Marshall asphalt' originates from the test developed by Bruce Marshall that is frequently used for design and control. The test has been standardised as ASTM D 1559 and AASHTO T 245 and in modified form as BS 598: Part 107. However, this type of mix was widely used long before the test was developed and the use of the test is not essential; the term 'Marshall asphalt' is therefore considered inappropriate.

A 'gap-graded' mix, or a type of 'stone-filled sandsheet' or 'Topeka', is widely used for paving in some countries, particularly the UK, where it is called 'hot rolled asphalt' when it conforms to BS 594. This contains much more sand, more filler and more of a harder bitumen, but less stone. As with asphalt concrete, it is applied hot but is usually easier to compact, owing to the rounded sand used and the high bitumen content.

At, or just beyond, the extremes of the compositions of these mixes are 'coated stone' and 'mastic'. 'Coated stone' mixes are almost single-sized aggregate particles coated with a small percentage of very soft bitumen, cutback or emulsion. The void content is high.

'Mastic', which can be defined as a sand–filler–bitumen mortar, having a continuous phase of bitumen, in which coarser mineral particles are dispersed, is made from a large percentage of very hard bitumen, a large percentage of filler, together with a relatively small percentage of aggregate. The void content is zero and the mix is applied at high temperature and screeded to levels: it does not require compaction. Mastic is generally not used for paving, except for the steel decks of long span bridges, but for waterproofing buildings, often in the form of a sandwich construction.

From table 19.1, it can be noted that in general, as the mix changes from the coated stone type to the mastic type, the bitumen content and hardness of the bitumen increase, the filler and sand contents increase, and the aggregate and void contents decrease. Cost increases mainly owing to greater heating costs for harder bitumens and the use of a higher percentage of bitumen.

Since the void content, V_a, gives an indication of the porosity of the mix, it provides a guide to the resistance of the bitumen to hardening and to the possibility of loss of adhesion in the presence of water, two aspects of durability which are discussed in section 19.6. Porosity is, of course, necessary if cutbacks or emulsions are used, since evaporation is required to permit setting or curing. Impermeability is, however, necessary for long life.

The mixes used in practice are quite similar throughout the world and have not been developed specifically to meet any particular climatic conditions: some areas are subjected to large seasonal variations, ranging from tropical to arctic, for example, mid-continent areas. In fact, most types are used quite widely and with very little variation in the grade of bitumen. However, the bitumen grade used for the different types of mix varies from quite hard to really soft or even in some cases, to cutbacks or emulsions.

It is suggested that the bitumen content and grade are changed to offset, as far as possible, some disadvantages arising from the use of less satisfactory aggregate or grading of the minerals used, as discussed in section 19.2 below.

19.2 Load-carrying mechanisms

All mixes used for paving are subjected to similar traffic stresses, although it is often implied that major roads carry heavier traffic: in practice even the heaviest vehicles are found on quite minor local roads and the dominant factor is the number of heavy axle loads. It is, therefore, useful to consider how traffic stresses are distributed or dispersed by, or in, the different types of mix. Non-paving mixes, such as are used, for example, for waterproofing reservoirs, dams and canals, are, of course, subjected to different types of applied stress, such as wave attack, uplift pressures and so on, but the principles are similar.

In the case of 'coated stone', shown diagrammatically in figure 19.3, it can be noted that any stresses applied on the surface will be distributed or dispersed by 'stone to stone' contact and by friction and interlock between the stones. For this to occur, without alteration to the structure of the mix, it is necessary to use aggregate with a high crushing strength, to prevent breakdown at the points of contact. The thin coating of bitumen round each stone serves merely to hold them together and the properties are relatively unimportant. Most open-graded bitumen macadam mixes are of this type.

'Mastic', also illustrated in figure 19.3, on the other hand, distributes the applied stresses within the mortar – quite a different mechanism from that operating in the 'coated stone' type of mix. To resist deformation under the stresses imposed, the mortar must have a high stiffness modulus, which is obtained by the use of a hard bitumen, such as 10–30 penetration at 25°C, and a high filler content which stiffens the bitumen.

In most mixes the stresses are distributed by one or both of these mechanisms. Sometimes the 'mortar mechanism' predominates, in which case the properties of the minerals are relatively unimportant. In others, the 'stone contact mechanism'

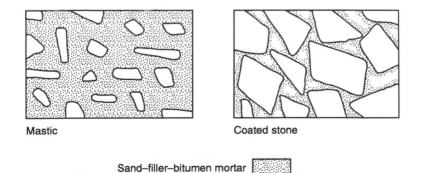

Figure 19.3 *Diagrammatic representation of load-carrying mechanisms for coated stone and mastic*

predominates, in which case the crushing strength and other properties of the stone are very important.

A continuously graded 'asphalt concrete', illustrated in figure 19.4, is a type of mix in which both mechanisms operate. A considerable proportion of the stresses are distributed by the 'stone contact mechanism' but between the stones there is mortar, in which some of the stresses are distributed by the 'mortar mechanism'. 'Dense bitumen macadam', a continuously graded mix, as specified for road bases in the UK (BS 4987), can be of this type but since the compositional limits allowed in the specification are rather wide and since there is no control over the void content, which is normally considered essential for asphalt concrete, such compositions can also behave predominantly as a coated stone type of mix.

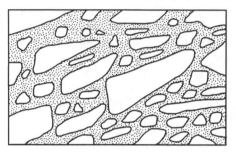

Asphalt concrete

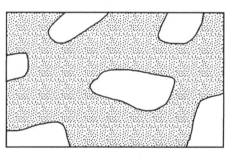

Hot rolled asphalt

Sand–filler–bitumen mortar and voids

Figure 19.4 *Diagrammatic representation of load-carrying mechanisms for asphalt concrete and hot rolled asphalt mixes*

Gap-graded mixes or 'stone filled sandsheet', such as UK 'hot rolled asphalt', also illustrated in figure 19.4, make use of the 'mortar mechanism' to distribute stresses. At first sight this would seem to imply that the larger aggregate particles do not play any significant part in stress distribution. However, there are indications that the resistance to permanent deformation of this type of mix increases with increasing stone content until, at high stone contents, the contact between stones and the build-up of a type of stone skeleton prevents full compaction of the mortar; this in turn reduces the stress distribution ability of the mortar.

This overall comparison of mix compositions and the concept of the mechanisms by which mixes distribute applied stresses can be useful in explaining mix behaviour, particularly when designing and developing economical mixes from locally available materials.

19.3 Types of pavement

Non-paving applications vary significantly in the properties required of the mixes used and each must be considered separately, but it is useful to consider the types of mix used for pavements. Traditionally, road pavements have comprised:

(1) A non-bituminous sub-base, which acts mainly as a working platform to permit construction equipment to operate without damaging the soil or subgrade.
(2) A main structural layer (known in the UK as 'roadbase') to protect the underlying layers and provide support for the subsequent layers. This can be of gravel, crushed stone, cement-bound granular material or a bituminous mixture, depending on the traffic and economic factors.
(3) A bituminous surfacing to protect the underlying layers and to resist traffic stresses and the effects of climate. The lower part of the surfacing is usually called the base course or binder course, and the top part the wearing course. When the roadbase is made with bituminous materials, the base course may be omitted, since one of the main purposes (historically) was to level the underlying unbound materials prior to placing the wearing course.

This type of construction is illustrated in figure 19.5(a). Research including full-scale trials and measurements has provided a better understanding of the properties and performance of pavements and materials, and is resulting in different types of construction. One example of the latter, in the UK, is illustrated in figure 19.5(b). This comprises:

(1) a non-bituminous working platform or sub-base as before;
(2) a lower bituminous roadbase with high resistance to tensile cracking, since it has been shown that the tensile forces are greatest near the bottom of the combined bituminous layers;
(3) the main bituminous structural layer or roadbase;
(4) the wearing course, to provide resistance against the abrasive action of traffic, safety (see section 19.6) and climatic factors.

Since this book deals essentially with the properties of materials and not with the structural design of pavements, the above is a very brief summary of some of the results of pavement design studies, the principles of which can be applied to non-paving applications. The traditional type of pavement, however, has formed the basis of the types of bitumen–aggregate mixtures that are currently used. For example, the more porous and less durable mixes were developed for the lower layers of pavements and for the surfacing of minor roads, where it was thought that less costly materials were acceptable. Later developments have indicated that many of these ideas are no longer valid, and denser and more durable mixes tend

(3) Surfacing

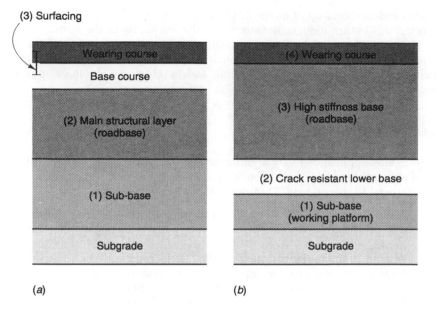

Figure 19.5 *Typical pavement types*

to be more widely used, except for temporary maintenance, lightly trafficked roads and special purposes.

19.4 Mix requirements

Mixes are used in practice to withstand the destructive effects of imposed stresses and climate. Some paving mixes are used for wearing courses and are thus exposed to weather and traffic directly, but others are used mainly to provide structural strength. Mixes can also be used for many other purposes, including waterproof revetments and cores for water-retaining structures such as dams and canals. For these many and varied applications, similar principles, of analysing the stresses and selecting appropriate mix compositions, apply.

In general, to ensure good performance, it is desirable that the mix should be designed initially to have certain properties and that these should be maintained during the service life. These requirements will be considered together with methods by which they can be obtained, including, where appropriate, some comparisons between the different types of mix. These requirements are not dealt with in their order of importance, which will depend, in many cases, on local factors and more particularly on the type of mix that has normally been used for a particular application.

19.5 Engineering properties

Most bitumen–aggregate mixes are used for pavements to carry road vehicles and aircraft. The design of such pavement structures is increasingly being based on structural analysis. This requires the use of the engineering properties, including modulus and strength, of the component layers for analysis of the stress and strain distribution by and within the pavement. Similar principles apply to non-paving applications, including, for example, water-retaining (dams, reservoirs, canals, underground structures) and protecting (sea defences) structures.

Dynamic conditions

As in the case of bitumens, bitumen–mineral aggregate mixtures are visco-elastic under the conditions existing during their service life. Under the conditions appropriate to the distribution of stress within a structure and overall behaviour, the dynamic stiffness modulus of the mix, $S_{mix, dyn}$, when the stiffness modulus of the bitumen S_{bit} is greater than about 10^7 Pa, is the property of major importance. The dynamic stiffness modulus (the terms, elastic stiffness and resilient modulus are also used) is that associated with moving traffic or, for example, wave action on sea defences or the revetments of dams. This dynamic stiffness modulus is dependent only on the stiffness modulus of the bitumen from which the mix is made and the volumetric composition of the mix *in situ*: the volume of aggregate, V_g, the volume of bitumen, V_b, and the volume of air voids, V_a. These parameters take into account the state of compaction of the mix: this is not so if mix composition by mass is considered. If a mix is poorly compacted, the air voids, V_a, will be relatively high and the volumes of aggregate and bitumen, V_g and V_b, correspondingly lower. Under dynamic loading conditions, the type of aggregate and grading are of minor importance; they influence the packing of the aggregate. The mix stiffness modulus is higher as the volume of aggregate is increased. Measurement of the stiffness modulus of the mix under dynamic conditions is complex, time consuming and expensive, since sophisticated equipment is required, but the stiffness modulus can be predicted from the volumetric composition as developed by Bonnaure *et al.* (1977), using the equations below.

The equations are

$$A = 10.82 - 1.342 \, (100 - V_g)/(V_g + V_b)$$
$$B = 8.0 + (5.68 \times 10^{-3} \times V_g) + (2.135 \times 10^{-4} \times V_g^2)$$
$$C = 0.6 \times \log_{10} ((1.37 \times V_g^2 - 1)/(1.33 \times V_b - 1))$$
$$D = 1.12 \times ((A - B)/\log_{10} 30)$$

when S_{bit} is between 5×10^6 and 1×10^9 Pa:

$$\log_{10} S_{mix, dyn} = ((D + C/2 \times (\log_{10} S_{bit} - 8))$$
$$= ((D + C)/2 \times (\log_{10} S_{bit} - 8)) + B$$

when S_{bit} is between 1×10^9 and 3×10^9 Pa:

$$\log_{10} S_{mix, dyn} = B + D + (A - B - D) \times (\log_{10} S_{bit} - 9)/\log_{10} 3$$

These procedures are accurate enough for most practical purposes, although direct measurement may be necessary for some special investigations or if bitumens containing additives, particularly polymers, are used.

An impression of the stiffness modulus of the mix, $S_{mix, dyn}$, under these dynamic conditions can be obtained from the right-hand portion of figure 19.6, where $S_{mix, dyn}$ is plotted against S_{bit}. This method of plotting is simple and convenient, since a number of variables are combined in S_{bit}: these are grade, $T_{800 pen}$, and type, PI, of bitumen, together with the temperature, T, and time of loading, t. It will be noted that, as S_{bit} decreases, either as a result of the use of a softer grade or increases in temperature or time of loading, the value of $S_{mix, dyn}$ decreases, at a rate depending on the volumetric composition of the mix. The implication is that, under dynamic loading conditions, closely packed aggregates, of virtually any particle shape and grading, together with hard bitumen, are required to obtain a high dynamic stiffness modulus.

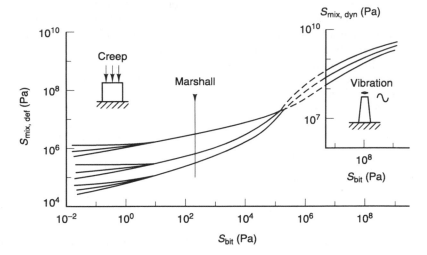

Figure 19.6 *Stiffness of mix against stiffness of bitumen for different mixes*

Quasi-static conditions

Under the conditions related to permanent deformation, when the bitumen is tending towards viscous behaviour, the relation between composition and properties is much more complex. If a mix has poor resistance to permanent deformation, the weakness can manifest itself in a variety of types of poor performance, such as rutting in the wheel tracks, corrugation where shear stresses are high and pushing over kerbs in the case of paving. An analogous situation can occur on canals and dams with steeply sloping revetments. The stiffness modulus of the mix under these quasi-static loading conditions (the term viscous stiffness is sometimes used), $S_{mix, def}$, which are associated with creep speeds, depends on the same factors as for dynamic stiffness modulus, *and* on many additional factors, including the aggregate type, grading, surface texture and interlock.

The stiffness modulus of the mix, under these quasi-static loading conditions, is developed by viscous resistance, or the resistance to sliding between adjacent mineral particles separated by films of bitumen, and this is increased if the film thickness is decreased by close packing of the minerals. Stiffness modulus under these conditions is also developed by frictional resistance between the minerals and by the number and nature of dry contact points between the minerals: these are also maximised by close aggregate packing.

The left-hand part of figure 19.6 provides an illustration of the general relationships. It should be noted that single lines, indicating a simple relationship, $S_{mix, dyn}$ = $f(S_{bit}, V_b, V_g)$ in the right-hand part of the figure, branch into multiple lines (three only are illustrated for simplicity) as S_{bit} reduces, indicating that $S_{mix, def}$ depends on additional factors, $S_{mix, def}$ = $f(S_{bit}, V_b, V_g$, aggregate factors).

The basic principles were established by research using both static and repeated load tests. To keep testing simple and the costs low, static creep tests (uniaxial unconfined constant-stress compression tests) developed by Hills *et al.* (1974) and de Hilster and van de Loo (1977) were recommended, but they recognised that in practice correction factors would have to be applied to allow for the effects of dynamic loading and the later developments are reviewed in Chapter 20. The principles of the static creep test are illustrated in figure 19.7.

These quasi-static properties or creep characteristics, $S_{mix, def}$, control the extent to which mixes will resist permanent deformation. Mixes whose properties are characterised by a relatively flat curve at a high level will, other factors being equal, deform less than mixes whose properties are characterised by a curve with a high slope at a low level. The quality of a mix may be judged by the slope and position of the creep curve. The $S_{mix, def}$ value gives the level and the S_{bit} value reflects the bitumen properties, the time of loading and the temperature. The curve thus represents a material property expressed in rational quantities and is independent of any arbitrarily chosen test conditions such as temperature: the properties at any temperature can be obtained from tests at another temperature.

This is illustrated in figure 19.8. Under the static load applied during the simple creep test, strain, ε_{mix}, is measured as a function of time, t, in seconds, and thus different curves will be obtained at different temperatures, or if different grades of bitumen are used, as indicated in the left-hand portion of the figure. If the results are replotted, as suggested by Hills *et al.* (1974), as shown in the right-hand part of the figure, the curves come together, to form a master curve for the specific mix composition by volume.

Whereas the curves in the right-hand portion of figure 19.6 can be predicted from a knowledge of the volumetric composition of the mix, the left-hand portion can only be determined by direct testing, using a simple creep test such as that developed by de Hilster and van de Loo (1977).

The effects of compositional changes on the creep properties of mixes are illustrated in figure 19.9. Figure 19.9(i) shows the effect of changing the bitumen content of a bitumen–sand mix from 4.8 to 10.5 per cent by mass. Figure 19.9(ii) shows the effect of changing from a rounded sand to a crushed sand of the same grading, while figure 19.9(iii) shows the effect on the properties of different mix types.

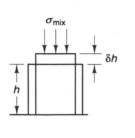

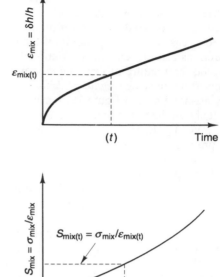

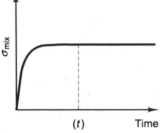

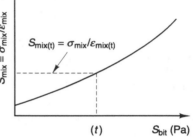

Figure 19.7 *The static creep test*

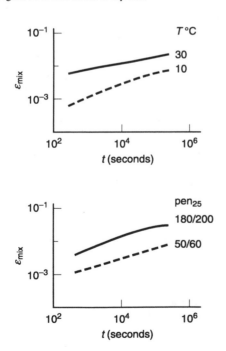

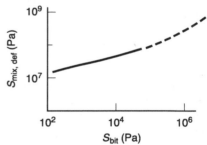

Figure 19.8 *Plotting of creep test results*

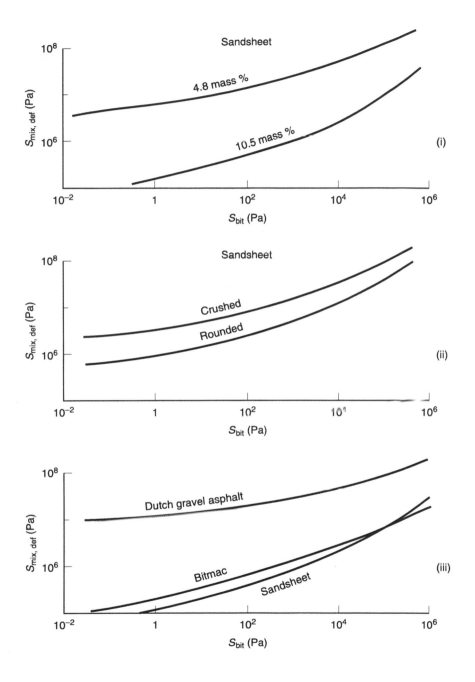

Figure 19.9 *Creep curves for different mixes*

It should be emphasised that a mix with good dynamic properties will not necessarily have good resistance to permanent deformation, as the lines representing S_{mix} sometimes cross one another: the reverse can also occur.

It is useful to consider to what extent the Marshall test (ASTM D 1559, AASHTO T 245), which is used in many areas for the design of paving mixes, provides relevant information on the engineering properties, or stiffness modulus under dynamic and quasi-static loading conditions, which are the properties of major importance to an engineer. The Marshall test is a semi-confined, constant rate of strain compression test, in which the load–deformation relationship is determined; see Chapter 20, section 20.3.

The Marshall test, in effect, provides one point on one of the curves, shown in figure 19.6 and such tests are thus poor measures of the real deformation characteristics of a mix. The test conditions in the Marshall test are between those under which the dynamic and quasi-static properties of interest occur, as illustrated in figure 19.6. The Marshall test is, nevertheless, useful as a guide to the relative behaviour or ranking of mixes within a given mix type and is, therefore, useful in initial mix design, although the situation may change as discussed in Chapter 20.

Fracture and fatigue

The other engineering property required in conjunction with modulus for engineering analysis is the permissible strain, and this has been studied by van Dijk and Visser (1977). The possibility of fracture can then be assessed. In the case of bitumen–aggregate mixes, the permissible strain under conditions of repeated loading, ε_{fat}, smaller than the strain at break for one application, is considered to be the dominant criterion. The permissible strain, ε_{fat}, is dependent on the volume of bitumen, V_b, the type or penetration index of the bitumen, PI, the dynamic stiffness modulus of the mix, $S_{mix, dyn}$, and the number of strain repetitions, N_{fat}. As with the stiffness modulus of the bitumen and mix, the permissible strain can be estimated or predicted, using the procedure developed by Bonnaure et al. (1980), using the following equations:

For controlled strain conditions (with $S_{mix, dyn}$ in Pa)

$$\varepsilon_{fat} = (4.102 \times PI - 0.205 \times PI \times V_b + 1.094 \times V_b - 2.707) \times S_{mix}^{-0.36} \times N^{-0.2}$$

For controlled stress conditions (with $S_{mix, dyn}$ in Pa)

$$\varepsilon_{fat} = (0.300 \times PI - 0.015 \times PI \times V_b + 0.080 \times V_b - 0.198) \times S_{mix}^{-0.28} \times N^{-0.2}$$

Fatigue testing can be carried out using either controlled strain, in effect a fixed amplitude of deformation, in the repeated load test, or controlled stress, in which the stress amplitude is essentially constant. Some authorities, when dealing with the structural design of pavements, associate controlled strain testing with the performance of thin layers and controlled stress with thick layers, but there is no general agreement.

This method of predicting the fatigue characteristics of mixes is believed to be the most widely applicable. Other methods are generally more restricted and it is doubtful whether they would be applicable in all climatic conditions.

In general, improved resistance to fatigue cracking is obtained by using high bitumen contents and densely packed aggregate.

Whatever method of assessing fatigue life is used, any laboratory test result must be modified to allow for the differences between laboratory and field conditions. The main factors involved are that most laboratory tests involve continuous repetition of constant stress or strain, whereas in practice the applied stresses or strains are intermittent and are not of constant magnitude.

The essential or key engineering properties governing the behaviour and performance of mixes are dynamic stiffness modulus and fatigue strain, and deformation resistance together with an acceptable value for total permanent deformation.

19.6 Other mix properties

Durability

The mix must have an acceptable resistance to changes brought about by the influences of water and air.

Adhesion

Most mixes are made from dry heated aggregates and thus initial coating of the aggregate can be assumed. There are also a number of processes which permit the coating of wet aggregates. These are usually based on speciality or proprietary products that are mostly used for lightly trafficked roads and maintenance. Even with a mix having a good initial bitumen coating, the entry of water may, in certain circumstances, result in loss of adhesion between the bitumen and the mineral aggregates. Such loss of adhesion is more likely to occur in mixes having a high void content, such as open-graded bitumen macadam, made with soft bitumens or cutbacks. In fact, failures due to such a loss are rare and, in general, are connected with severe loading and inadequate structural strength, permitting high transient deflection.

The nature of the mineral aggregate and particularly the filler fraction is dominant in respect to adhesion, together with the permeability of the mix. It is generally considered that the use of alkaline minerals, such as limestone, causes fewer adhesion problems than the use of acidic minerals, such as granites. However, natural aggregates are rarely pure and no reliable recommendations, based on aggregate type, can be made. If low void content mixes and relatively hard bitumens are used, adhesion problems are rare.

For dense mixes, a particular type of problem can sometimes occur and this can most easily be explained by means of an example. A dense asphalt concrete mix was designed and placed using crushed granite from a nearby quarry. After the mix had been placed, it was observed, when taking samples a few weeks later, that a significant proportion of the aggregate was uncoated, even though the initial coating had been good. The mix design procedure and functioning of the equipment were checked and did not reveal any obvious cause, but more detailed investigations indicated that the fines and filler fractions, which had been produced from the same granite as all the aggregate, contained a small percentage of clay. This had absorbed moisture from the heavy rainfall that had occurred. The volume of water absorbed was greater than the original air voids, indicating that swelling had occurred. This swelling disrupted the mix and allowed further water absorption to occur.

The basis of the investigation was the percentage of the Marshall stability retained after prolonged soaking, for several days, of the specimens in water at 60°C, instead of the normal time required to bring the samples up to the test temperature. In this particular example, it was found that replacement of the water-susceptible fines by an alkaline filler overcame the problem. In another similar case, the amount of contamination was too great to be cured in this way and material from another quarry had to be used.

Hardening

After initial hardening of the bitumen during the mixing operation, which causes significant hardening owing to the combination of high temperature and exposure in thin films, the entry of air into a mix during the service life causes further ageing of the bitumen. This hardening, principally by oxidation, is accompanied by minor changes in the rheological type of the bitumen (PI increases), but in severe conditions the hardening that occurs can result in embrittlement – the viscous component of the stiffness modulus being insufficient to permit relaxation of stress, resulting in fracture.

Field experiments, in many parts of the world, have demonstrated that the major factor controlling effective service life is the permeability of the mix. If the system is impermeable, long life can be expected, but if the mix is porous, life can be very significantly reduced, by an amount that is very dependent on the climate. In practice, permeability measurement in the field is very difficult, since surface effects, such as a surface skin of bitumen or detritus, can mask the true permeability. It should be pointed out that access of air is from the sides and bottom of a mix as well as from the exposed surface. As a result, permeability must normally be inferred from the void content of the mix.

For continuously graded mixes, of the asphalt concrete type, an acceptable void content of about 5 per cent by volume is often considered as the critical point, below which satisfactory service life will be obtained. There is evidence that at higher void contents, a very reduced life will be obtained but the magnitude is very dependent on the exposure conditions, particularly temperature. The interconnection of the voids is particularly important and the actual void content that is acceptable depends therefore on the particle shape and grading of the aggregate. For example, the interconnection of the voids in a continuously graded mix, made with crushed aggregate, is significantly greater than in the case of a gap-graded mix, made with a rounded sand. Whereas 5 per cent is often considered to be the limit for continuously graded mixes, 7–8 per cent is usually quite acceptable for a gap-graded mix.

Workability

The mix must have good workability so that it can be spread and compacted to the required density without difficulty. If the mix is harsh, as a result of composition or placing at the wrong temperature, it will be difficult to spread without tearing under the screed of the paver, and it will not be possible to compact it adequately in the field. If the mix is not fully compacted, it will not develop *in situ* the desirable engineering properties and durability, referred to above, of which it is inherently capable.

A mix having inherently good stability or resistance to deformation will, if insufficiently compacted, behave as if it were a low stability mix. There may then be a temptation to use a mix of even higher stability, which usually means one of even poorer workability, with the result that the *in situ* stability may well be even lower. This is illustrated in figure 19.10, which was prepared as follows. A mix was made using a standard compaction method and the Marshall stability was measured. Further mixes were then made using reduced compactive effort and the Marshall stability was again measured. It will be noted that the stability reduces significantly as the compactive effort is reduced. A similar dependence on compaction is found for the other engineering properties discussed above.

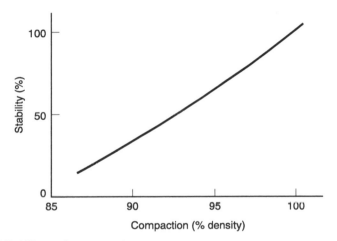

Figure 19.10 *Stability against compaction*

Workability is related to the viscosity of the mix at mixing and compaction temperatures. The low viscosities of the mix that are sometimes desirable to facilitate compaction can be obtained by the use of higher bitumen contents, softer bitumens and by the use of more rounded smooth textured mineral particles, such as single-sized sands. Increase in temperature is not acceptable. In order to obtain good coating of the aggregate particles, mixing must be carried out with the bitumen at a fixed viscosity – in other words, the correct temperature. If the temperature is increased, additional hardening of the bitumen will occur and this will cause the workability to decrease.

Safety

The principal engineering properties together with durability and workability, so far discussed, are applicable to the types of mix used for any application.

When used on a pavement, as a wearing course, as opposed to a base course, base or other component of a pavement structure, the mix must also provide adequate skid resistance or acceptably short braking distances, and must not disintegrate by fretting or ravelling, which would cause danger to traffic from loose stones on the surface.

To provide for this requirement, particularly in wet conditions, the surface should have sufficient texture depth to provide drainage channels, through which the water can be expelled at the tyre/road interface, to provide dry contact. The stones used in the surface should also be sufficiently resistant to polishing and abrasion for the severity of the site in regard to skidding accidents.

Skid resistance at low speeds, below about 50 km h^{-1}, depends on good frictional resistance between the tyre rubber and the surface. The highest resistance is obtainable with smooth treaded tyres and a 'sandpaper' or fine-grained surface texture, known as micro-texture, as illustrated in figure 19.11, in which the large number of sharp points provide high friction and disperse energy.

Micro-texture – low-speed skid resistance

Macro-texture – high-speed skid resistance

Macro- and micro-texture – all-speed skid resistance

Figure 19.11 *Micro- and macro-texture*

In wet conditions the fine projections penetrate through the surface water film to provide dry contact points, although this resistance can be increased by patterned tyre treads. Under these conditions, skid resistance depends primarily on the maintenance of the surface roughness or the micro-texture of the aggregate particles. The micro-texture varies seasonally, becoming less during dry periods, owing to polishing under the action of fine dust particles. In wet periods there is a tendency for the micro-texture to increase owing to abrasion by larger sized particles. The properties of aggregates can be assessed in relative terms by the Polished

Stone Value test (PSV) and the Aggregate Abrasion Value test (AAV), both of which are dealt with in BS 812: Part 3. Both are dependent on traffic volumes, although the effect of traffic is not cumulative and the concepts of fatigue, which must be considered in connection with the engineering properties, do not apply to skid resistance. No general conclusions can be reached regarding the most suitable aggregates, as some with a good PSV have a poor AAV.

At speeds higher than about 50 km h^{-1}, skid resistance depends on an additional factor, the surface texture depth or macro-texture, also illustrated in figure 19.11. This is required in order to provide drainage channels for the surface water to be dispersed, so that there is dry contact between the tyre and the surface above that which can be obtained by patterned tyre treads. The measurement of surface texture depth, using the sand patch method, has been standardised in BS 598: Part 107, but laser measurement is now often preferred.

The role of the bitumen in providing skid resistance is relatively minor in comparison with the aggregate properties and texture depth, although, depending on the type of mix, the durability of the bitumen may affect the exposure of fresh unpolished aggregate particles in the surface. To cover this particular durability requirement for hot rolled asphalt wearing courses in the UK, the permittivity test (a measure of the dielectric constant), BS 2000: Part 357, is included in some BS and DoT standards in order to exclude bitumens from some specific sources. This approach has taken the place of coal tar pitch/bitumen blends and Trinidad Lake asphalt blends, which were used prior to the introduction of permittivity.

Assessment of skid resistance by direct measurement *in situ* is very difficult, owing to the continuous variation caused by daily and seasonal temperature changes, variations in traffic, and the state of polish and abrasion of the aggregates. Regular testing to monitor changes and statistical interpretation of the data is necessary to obtain meaningful results. It is therefore often better to rely on accident statistics and general background information in assessing skid resistance.

Three principal factors must be considered: PSV, AAV and texture depth. The requirements for each must be decided in relation to the volume of traffic, the type of surfacing and the category of the site in regard to the seriousness of skidding. Guidance is provided for UK conditions by the current recommendations of the Department of Transport and the Transport Research Laboratory.

To provide acceptable skid resistance and to reduce the amount of splash and spray on surfaces used for high-speed traffic, another concept is sometimes used. This was first introduced on aircraft runways, where it is known as 'friction course'. This is a highly porous mix of the coated stone type that is placed on a dense impermeable asphalt concrete. Surface water goes into the mix and drains through the mix and at the interface with the asphalt concrete. Hydrated lime is used as part of the filler to improve the adhesion characteristics.

Roads are much more frequently trafficked and cannot be kept as free from detritus as can runways. The more frequent traffic causes densification and thus reduction in porosity and leads to a reduced service life. Modifications to the thickness and grading have been made and the material is called pervious macadam or porous or drainage asphalt. It also provides a surface with reduced noise at the tyre/surface interface. It is not always free draining although it provides a relatively dry surface, but must still be considered as under development.

20

Design of Bitumen–Aggregate Mixes

The objective of mix design is to obtain an economical balance between the sometimes conflicting property requirements. The review of the properties required, see Chapter 19, indicates that, for example, the compositional requirements to obtain high resistance to permanent deformation, for which good aggregate interlock and thin bitumen films are required, are in conflict with those for good fatigue characteristics, for which high bitumen contents are desirable. Workability can be improved by the use of soft bitumen and rounded aggregate particles, but this might also result in poor resistance to permanent deformation. It can thus be seen that most mix design is a compromise.

Because there are severe limitations to the use of open-graded or high void mixes, in terms of life or durability, and hence economics, they are now little used for major works, except for friction courses or porous asphalt. These are special surfacings, discussed in Chapter 19, section 19.6, of limited life, for providing improved surface drainage to improve skid resistance and to reduce splash and spray. Porous mixes are, of course, not used for waterproofing. Their design is purely empirical.

20.1 Types of specification

An acceptable balance, between the sometimes conflicting requirements, has traditionally been achieved on the basis of experience. The type of minerals used and the composition by mass of each of the components, together with the method of production, placing and compaction, have been specified. This empirical or recipe approach is still the basis of many specifications and still dominates in the UK, but changes are under discussion and new standards are expected. Examples of recipe specifications are BS 4987 for coated macadams and BS 594 for hot rolled asphalt.

Such recipe specifications are subject to a number of limitations. For example, the conditions of applied stress, climate and so on, to which the mix is to be subjected, may not be the same as those existing when the experience on which the specification was based was obtained, and there may be differences in the properties of the components. Mix properties and performance, in practice, are dependent on workmanship, as well as on composition, and this aspect is very difficult to specify.

Checking for compliance with the specification is, in principle, relatively simple but there is no means of assessing the seriousness or the practical effects of minor failure to comply. In theory, any material that in any way fails to comply is unacceptable and should be removed. In practice, material that is marginally out of specification is allowed to remain but it is known that, in many cases, the performance will be substandard or the useful life will be shorter than intended or required.

Recipe specifications are restrictive towards new developments, since a new material cannot be used commercially until experience has been sufficient to allow a specification to be written: this may take many years.

The basis of recipe specifications is the aggregate grading curve, and since such curves are based on constant specific gravity of all the component sizes and cannot, except in the most general way, take into account the aggregate properties, shape and packing characteristics, they are not very reliable, particularly when mixtures of different aggregate types are used. Comparison of specified gradings for mixes, that are essentially the same, shows that there are wide variations to take into account the different aggregate characteristics available in different locations from different aggregate sources.

There is a need for a system which takes more fully into account the required engineering properties. These, as discussed, are dependent on the volumetric composition and thus on the specific gravity of each component and on the packing characteristics of the minerals. Aggregates can vary in specific gravity from less than 2.6 to over 2.9, and this factor alone can cause significant changes in the calculated voids. In addition, the resistance to permanent deformation and workability is dependent on aggregate particle shape, surface texture and grading and these are not covered by grading curves.

The need for specifying engineering properties (performance or property specifications) is being recognised in many countries, including the UK, partly to ensure adequate or acceptable performance and partly for use in analytical or mechanistic design systems, which require the use, as input, of the appropriate engineering properties.

Although the emphasis in discussing properties has been on pavements for road vehicles and aircraft, similar principles apply to any application of bituminous mixes. These all require analysis of the effects of applied stresses and appropriate adjustments to the properties and/or layer thickness required.

20.2 Principles of design

In principle, to assess the true packing characteristics and variations in specific gravity, specimens of the complete mix should be prepared to reproduce, as closely as possible, the material that will be used in the field. To be representative it is desirable to compact the specimens, to the same thickness as will be used *in situ*, by rollers, and to a density similar to that which will be attained under the applied stresses during the service life. Densification during service due to repeated stress applications is a similar mechanism to that of compaction, but occurs only where the stress is applied, and thus surface deformation or distortion occurs.

Compaction by rolling is used to prepare specimens for some basic research investigations: specimens of suitable size and shape are then cut using diamond saws. For routine mix design work, compaction must usually be by hammer but the difference between compaction by roller and by hammer should be noted. Rolling tends to cause elongated mineral particles to lie with their longest dimension horizontal and thus the specimens tend to be more anisotropic than those prepared by hammering. The greater use of compaction by vibration is under consideration and this may prove to be a useful compromise.

Knowing the masses and the specific gravities of the mix components and the density of the compacted mix, the volumetric composition can be calculated, as described in Chapter 19, and thus the dynamic modulus and permissible strain predicted. It is important, for example, that the void relationships are acceptable to ensure that there is sufficient bitumen in the mix to provide durability and workability, without overfilling the voids. The latter situation amounts to having the mineral particles floating in bitumen, and thus low stiffness approaching viscous behaviour.

The samples should then be tested to assess their relative resistance to permanent deformation: stiffness modulus under quasi-static conditions. The simple creep test has been suggested for routine testing but it is now more likely that a repeated load creep test, which is a better predictor of performance, will be adopted, as discussed in section 20.5.

20.3 Continuously graded mixes

The main elements of these principles are used in most countries for the design of continuously graded mixes, such as asphalt concrete, for which the Marshall test has normally been employed to assess resistance to permanent deformation. The Marshall test (see Chapter 19, section 19.5) is a semi-confined, constant rate of loading compression test carried out on cylindrical specimens prepared by hammering. The sample deforms as the load is increased, and stability is defined as the maximum load developed before the sample collapses: the deformation at this

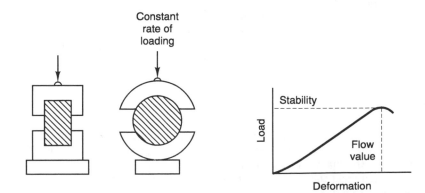

Figure 20.1 *The Marshall test*

point is called the flow value. This is illustrated in figure 20.1. It is appropriate to review the most widely used method before referring to the UK situation and developments that are expected.

Design of asphalt concrete

For the design of continuously graded asphalt mixes, the most widely used procedure worldwide is that published by The Asphalt Institute (TAI) *Mix design methods for asphalt concrete and other hot-mix types*, Manual Series No. 2 (MS-2). This deals with the subject generally and with testing by the Marshall test. It should be noted that 'bitumen' is termed 'asphalt cement' or 'asphalt' in North American terminology.

The aim is to control the void relationships so that there is sufficient bitumen to ensure flexibility (fatigue characteristics), durability and workability, without sacrificing too much resistance to permanent deformation, as measured by the Marshall stability. This latter property is not normally a problem with correctly designed and compacted continuously graded mixes, in which there is good aggregate interlock.

Chapter II of the Manual emphasises the need for attention to detail in testing and the need to ensure that the *in situ* air voids are sufficient to prevent loss of resistance to permanent deformation, by them becoming overfilled: if this occurs the resistance to deformation will be that of the bitumen. In this connection it is emphasised that aggregate gradings close to the maximum density curve (Fuller curve) must be adjusted to increase the voids in mineral aggregates, V_{ma}.

Chapter III of the Manual deals specifically with the Marshall method of mix design. The Marshall test itself has been standardised (ASTM D 1559 and AASHTO T 245) and this standardisation includes the preparation of test specimens, and standardisation of the equipment and of the loading test.

The sample preparation is preceded by determination of the mixing and compaction temperatures for the bitumen being used, which are defined in terms of equi-viscous temperatures for 170 ± 20 centi-Stokes for mixing and 280 ± 30 centi-Stokes for compaction. The mixing temperature determined using this procedure should also be used for full scale work. Specimens are compacted by hammering to a density representative of that which will be attained under the applied stresses by varying the number of hammer blows. Mixes are made at bitumen contents above and below the expected or estimated optimum.

The optimum bitumen content for the grading selected for the initial testing is determined from that required to give maximum Marshall stability, maximum mix density, and for the intended, selected or specified void content, V_a. The optimum is the numerical average of these three values, but this average must also be checked against the voids in the mineral aggregates, V_{ma}, requirement to ensure that this is sufficient for the mix to contain sufficient bitumen, V_b, without reducing the air voids, V_a, unacceptably. This is illustrated in figure 20.2, in which the bitumen content required for maximum density, G_{mb}, was about 6.6 per cent by mass, that required for maximum stability was about 6.3 per cent by mass, and that required for a void content, V_a = 3.5 per cent by volume, was about 6.1 per cent by mass. The average is thus 6.3 per cent by mass. At a bitumen content of 6.3 per cent by mass, the relative density was about 2.35, the stability was about 9.6 kN and the void content was about 3.3 per cent by volume. The V_{ma} was about 17.5 per cent by volume and the flow was about 2.9 mm. As these values can be considered acceptable, the optimum bitumen content was assessed as 6.3 per cent by mass.

If all the requirements are not complied with, the aggregate grading must be changed and the procedure repeated until all the requirements are satisfied.

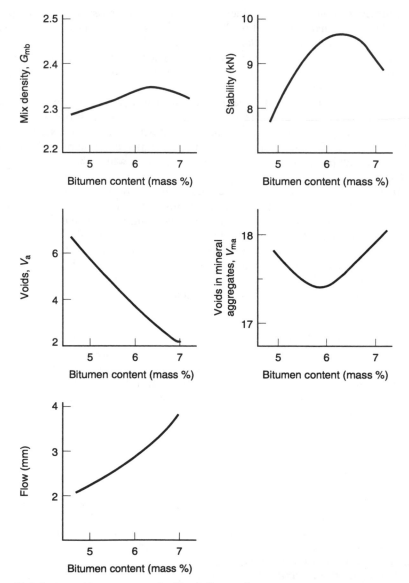

Figure 20.2 *Optimum bitumen content by Marshall procedure*

Once the design of the mix has been completed in the laboratory, plant mix trials and full-scale paving trials must be carried out to check the practicability of the composition. Modifications to the composition may be required at these stages before the Job Standard Mix composition and properties can be fixed. Control of the quality of the production of the mix and paving operations is then required to ensure that the composition and properties of the Job Standard Mix are met, acceptable working tolerances being permitted. Regular monitoring of the mix composition by extraction of the bitumen, grading of the aggregate and the preparation and testing of Marshall specimens is required to check the volumetric composition.

In situ compaction is normally controlled by comparing the density of core samples with corresponding laboratory prepared samples, so that, in effect, the *in situ* voids are checked. Some authorities prefer to compare *in situ* density with the maximum theoretical specific gravity, G_{mm}, of the mix composition to determine the void content, V_a (see section 19.1). This is considered more meaningful than comparison of two four-digit numbers or a percentage of a four-digit number, which is itself a variable. The variability is due to variations in specific gravity and minor variations in composition. The main reason for compaction, in any case, is to obtain the correct void content.

Chapter VI of the Manual deals in detail with the analysis of compacted mixtures and the importance of using the appropriate specific gravity of each mix component for the calculation of the void relationships. This is most important since failure to allow for absorption of bitumen by aggregates can cause serious error in the effective void content.

In general, following these recommendations will provide mixes with acceptable performance, provided that testing is accurate and all the recommendations are taken into account. The procedure is in reasonable conformity with the principles of design discussed in section 20.2.

A similar procedure has been used successfully for the design of asphalt cores for dams and for waterproof revetments on reservoirs, dams and canals. For cores, permeability is a prime requirement and an air voids content of 1–3 per cent by volume is required, although the stability is not critical. For revetments, wave attack must be resisted and good workability is required to facilitate placing and compaction on slopes (pulley systems are usually required to support the placing and compaction equipment), although resistance to flow down the slope is not usually a problem after the mix has been compacted.

UK aircraft pavements

In the UK, continuously graded asphalt concrete mixes are normally only used for aircraft pavements and are designed using a variation of TAI, MS-2. The principal difference is that the voids filled with bitumen, V_{fb}, is used instead of the voids in mineral aggregates, V_{ma}. Since $V_{ma} = V_b + V_a$ and $V_{fb} = (V_b / V_{ma}) \times 100$, it follows that V_{ma} and V_{fb} are both functions of V_b and V_a. This parameter thus covers essentially the same need of ensuring that there is sufficient bitumen for acceptable durability and fatigue resistance. A point sometimes considered important is that the direct use of the volumes of the mix components is preferable to the use of both a volume and a ratio. If two parameters, such as V_{ma} and V_{fb}, covering essentially the same requirement are used there can, if great care is not exercised, be conflict as to which should be dominant.

The procedure regarding the selection and use of the Job Standard Mix composition and the control of compaction in the field are well covered.

UK road pavements

Continuously graded mixes for road pavements in the UK are in accord with BS 4987, Part 1, which is purely empirical. This has been considered satisfactory for the UK where all major aggregate sources are known, but the situation may not be acceptable in the future and there would appear to be a need to control the void relationships. On major road contracts, however, some control over *in situ* compaction has been introduced, in the form of the Refusal Density Test (TRRL, 1987). This is an improvement over the system of comparing core density with laboratory density, since any potential errors due to minor variations in mix composition are removed. Because of the importance of engineering properties, which are to a large extent controlled by the volumetric composition, it is considered that greater emphasis on this aspect is desirable and such developments are under discussion.

20.4 Hot rolled asphalt

For gap-graded mixes in the UK, the traditional empirical specifications, given in BS 594: Part 1, continue to be used. There is, however, an alternative for the design of wearing course mixes, BS 598: Part 107, but not for other mixes, such as base course and roadbase.

This alternative must be considered as a semi-empirical method, since it does not take into account the volumetric composition, which as discussed in Chapter 19 is of major significance in obtaining acceptable properties.

A number of alternative aggregate gradings are specified, which are, in fact, the same as for the recipes, and the test procedure is used only to assess a suitable bitumen content. Specimens are prepared with the aggregate grading selected from the specification using a range of bitumen contents. Marshall stability, flow, mix density and aggregate density are determined and each plotted against bitumen content. The results are plotted in a similar manner to that indicated in figure 20.2. The design bitumen content is the mean value of the requirements for maximum stability, maximum mix density and maximum aggregate density after the addition of an empirical factor, and even this is subject to a specified minimum.

There is no control over *in situ* compaction, and also no provision for changes to the specified aggregate grading. The standard includes details of the calculation of the void relationships, but this is optional and no guidance is given regarding their use. In this connection it should be noted that the BS tests for specific gravity are not so precisely defined as the ASTM and AASHTO specifications.

It should be noted that the Marshall test equipment and procedure specified are not identical with the US standards, for example, the mixing temperature is related to the softening point of the bitumen, not to an equi-viscous temperature, and there are other slight differences. Since hot rolled asphalt has good workability and is thus easy to compact, the temperatures are lower than those considered necessary for continuously graded mixes made with crushed stone, for which the Marshall test was originally intended.

There is no utilisation of the Job Standard Mix concept, and there is little conformity with the principles of design discussed in section 20.2, above. There is also no provision for changing aggregate grading to improve the volumetric composition, nor to change the bitumen–filler ratio which has a significant effect on the properties of gap-graded mixes.

20.5 New developments

Few of the mix design methods currently used conform to the principles of design referred to in section 20.2, and even those which conform in most respects have weaknesses.

Compaction equipment and procedures are a major factor. The size of mould used to prepare specimens should be appropriate to the maximum size of stone used, in order to minimise boundary and edge effects. Compaction by hammering is undesirable as the anisotropic aggregate arrangement obtained by rolling is not achieved. Variation in compactive effort would assist in assessing the workability.

Crushed aggregates frequently have gradings very close to that for maximum density and require modification to increase the V_{ma} value. This is often obtained by increasing the fines content of the mix, but it would often be preferable to decrease the fines, which has a similar effect on V_{ma}. This would increase the stone to stone interlock, reduce cost, since the specific surface area to be coated with bitumen is reduced, provide additional stone in the surface and probably increase the resistance to permanent deformation.

Work by Cooper et al. (1985) at Nottingham concentrated on the grading of continuously graded mixes. The objective was to reduce the guesswork element in deciding what aggregate grading to consider. Use was made of an equation of the form $P = (d/D)^n$, where P = percentage passing a sieve size d, D = maximum stone size, and the grading exponent, n, is a number between 0 and 1. The grading exponent is usually about 0.45 for maximum density and can be increased to give lower density but, for many values of D, the filler contents given by the equation are generally impractical. The equation was therefore modified to

$$P = \frac{(100 - F)(d^n - 0.075^n)}{(D^n - 0.075^n)} + F$$

where F = filler content.

If values for n are in the range 0.5–0.7, the grading will normally produce a mix with an acceptable value of V_{ma}. The results of the equation are illustrated in figure 20.3.

The merit of this equation can be checked by making mixes and compacting them using a kneading action, as discussed by Cooper et al. (1992), with the Refusal Density Test equipment, developed by TRL (1987), using a vibrating hammer. This method of compaction uses 150 mm diameter moulds and can be used for layers of 100 mm thick and is an improvement over a drop hammer. It is

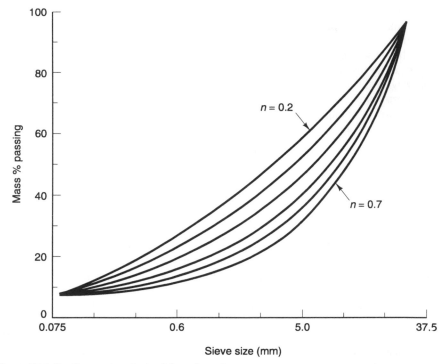

Figure 20.3 *Grading curves obtained from formula*

recommended that several bitumen contents and several levels of compaction be used so that the sensitivity of the mix in this respect can be studied. The volumetric compositions can then be calculated to predict the stiffness modulus and permissible strain.

Studies by Brown *et al.* (1981) concluded that the Marshall test was not a good predictor of resistance to permanent deformation and that, although the creep test was better, the repeated load triaxial test was to be preferred. Since the latter is unsuitable for routine purposes, Cooper and Brown (1989) developed a *R*epeated *L*oad *A*xial (RLA) test.

Parallel to this work Cooper and Brown (1989) developed the American 'resilient modulus' test, which applies a diametral load to cylindrical specimens, to measure dynamic or elastic modulus and fatigue characteristics. The equipment developed, now known as the Nottingham Asphalt Tester (NAT), is illustrated in figure 20.4. The testing frame, shown in (a) can be used for (b) *i*ndirect *t*ensile tests on cylindrical specimens to measure dynamic *s*tiffness *m*odulus (ITSM) and *f*atigue characteristics (ITFT), and (c) for direct static creep or *r*epeated *l*oad *a*xial tests (RLA). It may be used to test either laboratory prepared specimens or cores taken from site. It can be used for mix design, quality control, end product specification, pavement evaluation, failure investigation, the assessment of new products and basic research. Brown *et al.* (1989, 1991) also proposed a procedure, based on the NAT, for mix design.

These test methods and proposals are under active evaluation in the UK 'BITUTEST' project (Gibb and Brown, 1994; Read and Brown, 1994), led by

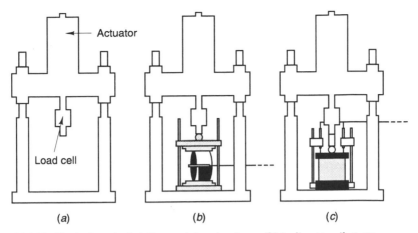

Figure 20.4 *The Nottingham Asphalt Tester: (a) testing frame, (b) in direct tensile tests; (c) repeated load axial tests*

Nottingham University. This is a project involving collaboration between the Department of Transport, the Science and Engineering Research Council, nine companies and fourteen UK highway authorities. It is anticipated that it will lead to a new approach to mix design, which will be incorporated into British Standards and this, in principle, could be further developed for use in other areas. The BITUTEST project is aimed at developing low-cost practical tests. In the USA the Strategic Highway Research Program (SHRP) was much more ambitious and the tests developed were more sophisticated and thus more costly.

The trend towards the specification of engineering properties and the use of performance-related tests are likely to be accelerated by such developments. The relatively simple tests developed for the direct measurement of engineering properties are also necessary to deal with the interest in polymer and other additives for bitumen, for which the systems used for the prediction of properties (see section 19.5) are not generally applicable.

References

Bonnaure, F., Gest, G., Gravois, A. and Ugé, P. (1977) A new method of predicting the stiffness modulus of asphalt paving mixtures, *Proceedings, Association of Asphalt Paving Technologists*, Vol. 46, pp. 64–104.

Bonnaure, F., Gravois, A. and Udron, J. (1980) A new method of predicting the fatigue behaviour of asphalt paving mixtures, *Proceedings, Association of Asphalt Paving Technologists*, Vol. 49, pp. 449–529.

Brown, S.F., Cooper, K.E. and Pooley, G.R. (1981) Mechanical properties of bituminous materials for pavement design, *Proceedings, Eurobitume Symposium*, pp. 143–147.

Brown, S.F., Cooper, K.E., Preston, J.N. and Bell, C.A. (1989) Development of a new procedure for bituminous mix design, *Proceedings, 4th Eurobitume Symposium*, Madrid.

Brown, S.F., Preston, J.N. and Cooper, K.E. (1991) Application of new concepts in asphalt mix design, *Proceedings, Association of Asphalt Paving Technologists*, 1991, Vol. 60, pp. 264–286.

Cooper, K.E. and Brown, S.F. (1989) Development of simple apparatus for the measurement of the mechanical properties of asphalt mixes, *Proceedings, 4th Eurobitume Symposium*, Madrid, pp. 494–498.

Cooper, K.E., Brown, S.F. and Pooley, G.R. (1985) The design of aggregate gradings for asphalt basecourses, *Proceedings, Association of Asphalt Paving Technologists*, Vol. 54, pp. 324–346.

Cooper, K.E., Brown, S.F., Preston, J.N. and Akeroyd, F.M.L. (1992) Development of a practical method for the design of hot mix asphalt, *Transportation Research Record*, 1317, pp. 42–57 (Washington, D.C.).

de Hilster, E. and van de Loo, P.J. (1977) The creep test: influence of test parameters, *International Colloquium on Plastic Deformability of Bituminous Mixes*, Federal Institute of Technology, Zurich, pp. 175–215.

Gibb, J.M. and Brown, S.F. (1994) A repeated load compression test for assessing the resistance of bituminous mixtures to permanent deformation, paper presented to *Symposium on Performance and Durability of Bituminous Mixtures*, Leeds, March.

Heukelom, W. (1966) Observations on the rheology and fracture of bitumens and asphalt mixes, *Proceedings, Association of Asphalt Paving Technologists*, Vol. 35, pp. 358–399.

Heukelom, W. (1973) An improved method of characterising asphaltic bitumens with the aid of their mechanical properties, *Proceedings, Association of Asphalt Paving Technologists*, Vol. 42, pp. 62–98.

Hills, J.F., Brien, D. and van de Loo, P.J. (1974) The correlation of rutting and creep tests on asphalt mixes, *Paper IP-001*, Institute of Petroleum.

Krom, C.J. (1950) Determination of the ring and ball softening point of asphaltic bitumen with and without stirring, *Journal of the Institute of Petroleum*, Vol. 36, p. 36.

Pfeiffer, J.Ph. and van Doormaal, J. (1936) The rheological properties of asphaltic bitumens, *Journal of the Institute of Petroleum*, Vol. 22, pp. 414–440.

Read, J.M. and Brown, S.F. (1994) Fatigue characterisation of bituminous mixtures using a simplified test method, paper presented to *Symposium on Performance and Durability of Bituminous Mixtures*, Leeds, March.

TRRL (1987) Percentage refusal density test, *Contractor Report No. 1*, Department of Transport, London.

van der Poel, C. (1954) A general system describing the visco-elastic properties of bitumens and its relation to routine test data, *Journal of Applied Chemistry*, Vol. 4, Part 5, pp. 221–236.

van Dijk, W. and Visser, W. (1977) The energy approach to fatigue for pavement design, *Proceedings, Association of Asphalt Paving Technologists*, Vol. 46. pp. 1–40.

Further reading

Claessen, A.I.M., Edwards, J.M., Sommer, P. and Ugé, P. (1977) Asphalt pavement design – the Shell method, *Proceedings, Fourth International Conference, The Structural Design of Asphalt Pavements*, University of Michigan, Vol. 1, pp. 39–74.

Construction Materials (1994) Spon, London.

Hunter, R.N. (1994) *Bituminous Materials in Road Construction*, Thomas Telford, London.

Shell (1978) *Shell pavement design manual – asphalt pavements and overlays for road traffic*, Shell International Petroleum Company Ltd.

Shell (1985) *Addendum to the Shell pavement design manual*, Shell International Petroleum Company Ltd.

Strabag (1964-81) *Asphaltic Concrete for Hydraulic Structures*, Work executed during

1964–1981, Booklets 37, 40, 43 and 44, Strabag Bau AG, Cologne, Germany.

Strabag (1990) *Asphaltic Concrete for Hydraulic Structures – Asphaltic Concrete Cores for Earth- and Rockfill Dams*, Booklet 45, Strabag Bau AG, Cologne, Germany.

White, M.J. (1992) *Bituminous Materials and Flexible Pavements – an Introduction*, British Aggregate Construction Materials Industries, 156 Buckingham Palace Road, London.

Whiteoak, D. (1990) *Shell Bitumen Handbook*, Shell Bitumen UK.

V SOILS

Introduction

Every work of construction in civil engineering is built on soil or rock and in many instances these are also the raw materials of construction. The study of soil and rock materials is an important part of a wider area of study often called *geotechnology*. This is an area common to civil engineering and engineering geology and forms a quantitative branch of engineering geology dealing with the mechanics of behaviour of soils and rocks.

The geotechnical engineer has to find answers to difficult questions, for example

Will a given soil provide permanent support for a proposed building?
Will a given soil compress (or swell) on application of the load from a proposed building, and by what amounts, and at what rates?
What will be the margin of safety against failure or excessive or unequal compression of the soil?
Are natural or constructed soil slopes stable or likely to slide?
What force is exerted on a wall when a given soil is packed against it?
At what rate and to what pattern will water flow through a soil?
Can a given soil be improved in any way by treatment or admixture?

All these require an understanding of soils as materials and of the mechanics of materials. Absolute answers are unobtainable and the engineer has to practise the art as well as the science of geotechnology. Construction work proceeds in the state of knowledge existing at the time the questions are asked. This knowledge, both of precedents and discoveries, is continually growing through research and not least in the area of *soil as a material*. Not all researches lead to an immediate or comprehensive solution to engineering problems but a better understanding of the behaviour of soil as an engineering material cannot but help in the amelioration of these problems. The analysis of behaviour of masses of soil when built upon, or cut to new shapes or used as materials of construction falls properly into texts in soil mechanics, for example Berry and Reid (1987) and Barnes (1995). In this Part of the book the emphasis is on the nature of soil itself as a material and within the limitations of space available the author has selected topics of importance to the student engineer; any shortcomings in this selection are those of the author.

There is no clear dividing line between the areas of study of soil as a material and soil mechanics, and both are equally important. There is obviously no justification for using the more sophisticated and rigorous methods of analysis developed in soil mechanics if the measurement of the material characteristics of the soil needed in the analysis is not made in a correspondingly refined and perceptive way.

Soils are naturally occurring, largely inert, materials of a particulate kind derived, as products of weathering, from rocks. The term 'soil' as used in this book therefore excludes the harder, massive or cemented and feebly weathered materials of the planet's surface, called *rocks* by the engineer. Again there is no clear division between soils and rocks, as briefly defined – many aspects of behaviour being common.

350

21

Formation, Exploration and Sampling of Soils

A detailed study of the origin, formation and distribution of soils would be rewarding but is beyond the scope of this book. The student of civil engineering has the complementary subject of geology in his curriculum and in studying this and coupling it with his own observations and experience he will extend his appreciation of the origins of soils and of natural aggregates for concrete and road construction.

The physical characteristics of soil are important factors in most civil engineering constructions and it is the assessment of these physical characteristics that is the principal concern here. Chemical and biological characteristics of soils are of importance in agriculture, horticulture and land management but they generally have less relevance to the subsoils of concern in civil engineering. In civil engineering usage, the term 'soil' describes the uncemented or weakly cemented material overlying the harder rock on the planet's surface. Soils exist in great variety, and are the accumulated result of many separate factors and processes. Their characteristics depend on the parent rocks from which they are derived; on the weathering of these rocks and the weathering of the soil itself at its various stages of formation; on the means of transport bringing the soil to its present location; on the manner of deposition of the soil; on its history of loading, drainage, wetting and drying and on many other processes.

The parent rocks themselves occur in great variety. The rock may have formed by the cooling and hardening of a molten material, this group being termed *igneous rocks*. The igneous rocks include granites, basalts and gabbros. On the other hand the rock may have formed from the accumulation of transported and deposited weathered rock fragments, subsequently consolidated to give relatively hard rocks; water has usually been the transporting agent. These are the *sedimentary rocks* – the sandstones, limestones, shales and conglomerates. Some of these igneous and sedimentary rocks have been altered over the ages by heat or pressure or both. The result is the *metamorphic rocks* – very different from the original materials. Metamorphic rocks include schist, gneiss, slate and marble. The processes of metamorphism and weathering over aeons of time have transformed some of the rock material into the infinite variety of soils seen today. The contin-

351

uing process of pedogenesis, as it is called, is one in which the hard rocks are slowly weathered and broken down and their constituent minerals chemically altered.

If the soil debris of today is derived from weathering of the rocks in place, it is described as a *residual soil*. Soils are commonly formed, however, of *transported material* – the weathered rock fragments having arrived in their present location after periods of transport by glacier, river or wind. The material may be much altered during transport and after deposition it experiences further weathering. Once settled in place the surface of the soil deposit will in time accumulate a growth of organic material, humus. The great variety of the processes of pedogenesis means that the physical and engineering characteristics of soils are likely to differ within short distances. Mapping of these characteristics over even a limited area of ground and in depth below its surface therefore requires a detailed programme of soil sampling and testing.

An outline of the formation processes follows. For a more extended geomorphological description of the shaping of the planet's landforms and materials, see Fookes and Vaughan (1986).

21.1 Residual soils

When rocks are weathered and the products accumulate in place, without transport occurring, the products are described as residual soils. In a sense the humus-bearing top soils and weathered subsoils of any formation are residual soils but the term is usually used for rock products such as laterites. Significant thicknesses of these soils are found in various parts of the world and from their distribution it appears that very warm humid regions are most favourable to the kind of weathering that produces residual soils.

21.2 Transported soils

Glacially transported soils

The superficial geology of much of Scandinavia, Britain and north-west Europe is derived from glacial action. Although the main elements of geographical relief were formed earlier, the land surface was modified during the Ice Age by complex erosion processes followed by grinding of the englacial material and then by the deposition of glacial drift to form many of the scenic features of today. This glacial activity ended some 10000 years ago but was a major means of transport of soils that are encountered now in these areas by the civil engineer. The rock debris carried in or on the glaciers of the times was ultimately deposited in one of a great variety of forms. A study of these forms provides valuable information to the materials engineer in his search for quantities of suitable natural aggregates and fill materials for construction work.

The non-stratified drift is a direct glacial deposit containing a wide range of sizes of soil particles and is commonly termed *glacial till*. The manner of deposition has produced quite heavily consolidated soils so that in their natural state today these tills are rather stiff dense soils and are usually fissured. The shape and orientation of the stone particles in them depend on such factors as the nature of the parent rock, and the distance, manner and direction of transport by glacial actions.

Some of the glacial debris was transported with melt water before final deposition and as a result of this is likely to be sorted into size groups of sands and gravels stratified after the manner of river deposits.

Water-transported soils

Soils formed of waterborne material tend to vary erratically in physical properties from place to place and depth to depth, a consequence of the continual change occurring in the streams and rivers which formed them. Fast-flowing steep streams in mountain areas transport all but the largest rock fragments downstream and the fragments become progressively broken up, worn and deposited. As the steepness of the streams diminishes the larger particles being transported are left behind. On reaching the plains and estuaries only the finer sizes remain in transport. Seasonal extremes of flood and drought, meandering and braiding of stream channels all lead to wide variations in the erosion and deposition patterns of waterborne deposits and hence in the engineering and other characteristics of the soils forming these deposits. These variations can be sudden, both laterally and with depth.

The mechanical process of abrasion and grinding accounts for most of the reduction in particle sizes from the fragments of the parent rocks. Such action may reduce particle sizes to 0.001 mm (1 μm or 1 micron) and in the case of rock flours derived from glacial grinding of rocks to yet smaller sizes. In appearance these soil particles are mainly blocky and angular although perhaps more or less rounded by abrasion during transport. A specimen of sand is shown in figure 21.1. Larger and smaller particle sizes from the gravels through to the silts may have similar

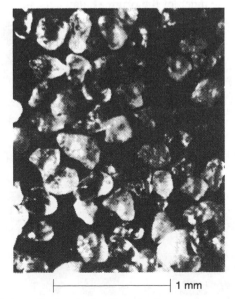

├─────────────────────────┤ 1 mm

Figure 21.1 *Enlarged photograph of a size fraction of a river sand*

appearance and these soils are described as *granular, cohesionless* and *coarse grained.*

Most soil material smaller than about 1 μm is the result of chemical processes of solution and crystallisation and these minute particles form the very significant clay fraction of soils. These clays comprise the *cohesive, fine-grained* fraction. The particles, in addition to being minute in size, are also unlike the coarse-grained soils in appearance, being predominantly plate or rod shaped. The development of electron microscopy with its high magnification has enabled clay particles and to some extent the structural grouping of clay particles to be photographed. In figure 21.2 the appearance of halloysite clay particles is seen and, in figure 21.3(a), (b) particle arrangements in a mainly illitic silty clay. The smallest soil particles transported in rivers are carried in suspension and on reaching lakes or ponded areas settle there to form soft, mainly clay and often organic deposits termed lacustrine clays. This process is a continuing one with many present-day examples in the course of formation.

The minute particles of clay size are less affected by gravity forces during sedimentation than by the interparticle forces of attraction and repulsion and the latter forces are in turn much influenced by the electrolyte content of the water in which the clays are settling during the formation process. Even so-called fresh water in rivers and lakes contains concentrations of ions and there is a grading of possible depositional environments from fresh through brackish water to marine conditions. These differing environments lead to a complex variety of *micro-fabric forms* in clays and these are a subject of continuing study. In the distant past the British Isles were covered by a succession of seas and one of the marine deposits which has received particular study from engineers is London clay. This is an extensive deposit of the Eocene age some 30 million years ago and today exists as a rather uniform stiff-fissured soil in thicknesses of up to 150 m or so. Erosion in late Tertiary and Pleistocene times removed some 200–300 m of overlying material, including much of the original clay deposit.

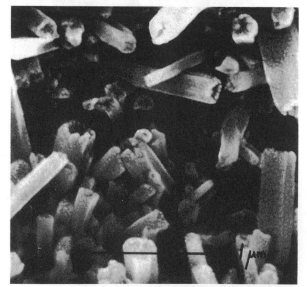

Figure 21.2 *Electron microscope photograph of halloysite from Bedford, Indiana (N.K. Tovey)*

Fine sediments laid down in the late glacial period have certain special features owing to the cyclic nature of their deposition. The lakes were then confined by land formations or by ice damming or both and the sediments reaching the lakes were seasonally released with the melt water from the glaciers. The coarser particle sizes in the sediment settled rapidly to the floor of the lake but the clays settled slowly over the seasonal period, the process being repeated in succeeding seasons. The composite soil as it is encountered today is termed a varved sediment or varved clay or, where the layers are less distinct, a laminated clay. A succession

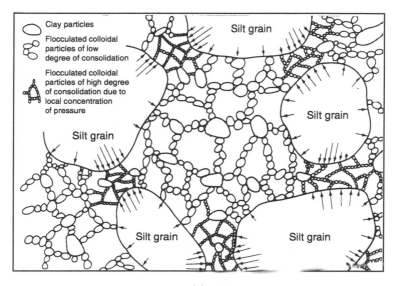

(a)

Figure 21.3 *(a) Hypothetical structure of undisturbed marine clay suggested by A. Casagrande in 1932. (b) Electron microscope photograph of a marine clay from the Drammen valley, Norway. This clay is a very sensitive, illitic, marine-deposited, leached silty clay*

of nearly identical layers or varves is typical of such sediments and, as indicated in figure 21.4, the laminations in a sample are more apparent to the observer if the soil is allowed to dry a little. The engineering properties of such sediments obviously vary with the directions in which the properties are measured. The larger features of clays – the *macro-fabric* – usually have a dominant influence on engineering behaviour. These features include layering and varves, vertical root channels and tension cracks, fissuring and jointing.

Figure 21.4 *Laminated clay partly dried to display the layering*

In certain parts of the world, notably Scandinavia and South-East Asia, sediments were transported in post-glacial times to the sea and deposited there in salt water. The clay particles tended to flocculate, that is, to form rather uniform groups of platelets in a very loose structure. These marine clays are therefore soft and highly compressible, although some may have been subjected to desiccation, giving a crust of stiffer soil.

Part of the subsequent geological change was the gradual raising of the land surface relative to the sea, a change which is still continuing. This upheaval raised many of the marine-deposited sediments above the present sea level. Artesian or other conditions of groundwater flow then resulted in a slow percolation of fresh water through the clays, gradually exchanging the salt water in the pore spaces in the clays with fresh, the process being called leaching. This leaching process altered the electrolyte content of the porewater and in various ways altered the bonds between the minute particles making up the clay structure; the bonds were greatly weakened without the structural arrangement of the particles being appreciably altered. The result was dramatic in engineering terms in that the leached sediment, originally rather soft and compressible, became an unstable sensitive or 'quick' clay. Such clays today have the consistency of a liquid when remoulded and any disturbance of the natural state leads towards that remoulded condition and to a rapid and usually substantial loss of strength. A second result of leaching is that the already compressible sediment becomes more compressible. These leached marine clays are thus exceptionally difficult materials to deal with in engineering construction works. Some Canadian clays have a history not unlike that described above. The water in which the clays were initially deposited was a more variable electrolyte, fresh, brackish or marine.

Wind-transported soils

In some regions of the world, particularly in the Americas and central China, extensive soil formations are the result of the deposition of windborne particles. The principal aeolian or windborne deposit is termed *loess*. This is a rather fine-grained, silty and remarkably uniform soil with a rather loose density. The particles are often cemented and when dry the soils are quite strong. Loess often shows a vertical jointing, attributed to a contraction after deposition, and sloping faces often weather to near vertical faces. The cementing material dissolves on wetting so that the soil is easily eroded by streams and tends to subside or settle quite suddenly when wetted. A 'loessic' component has been noted in soils in East Anglia, UK and in France and Belgium.

Soils transported by man

In the consideration of soils in relation to engineering works, man himself should be included as an agent of transportation. Soils which have been excavated from one place and transported to and placed in another place are described as *fills*. Fills may be materials other than soils as we have described them – they may comprise randomly dumped rock spoil from a tunnel or other construction works, or colliery or other industrial waste products or even municipal refuse. Fills are therefore likely to be variable materials in quality and density and if encountered on the site of a proposed construction they should be presumed to be non-uniform and compressible, and possibly contaminated. On the other hand, soil fills may be placed and densified under controlled conditions giving a uniform firm material of predictable and dependable behaviour. The control and compaction of fills is introduced in Chapter 28.

21.3 Exploration and sampling

To consider soils among the other civil engineering materials it is necessary to distinguish the principal soil groups one from another and, as with the other materials, to establish parameters by which the behaviour of soils can be described quantitatively. The complex formation processes just outlined have, however, produced complex and variable materials. Such engineering characteristics as *permeability, compressibility* and *strength* usually differ in value from point to point in a soil mass and at any one point these characteristics differ with the direction in which they are measured. Soils showing this behaviour are described as *anisotropic*. If the characteristics were practically unvarying in value throughout a soil, that soil would be described as *isotropic*. It is questionable whether any soil is isotropic, although the design engineer will commonly assume isotropic behaviour in soils when he is analysing their projected behaviour in engineering works and will use a simplified sequence of soil strata sufficiently representative of the actual soils to allow design calculations to be made with confidence. This rather sweeping assumption demands an exercise of judgement by the engineer based on knowledge of the actual properties of the soil, measured using the best available methods, together with a good appreciation of the behaviour of soil as a material and an awareness of the susceptibility of his analysis to approximations of soil properties, say, for example, their directional variation. Many case histories exist to support the reliability of this simplified approach when natural strata or fills are sensibly uniform.

It was observed earlier that many soils show a fabric of laminations, varves, fissures, organic matter or root networks. A quantitative description of the engineering characteristics of a particular soil stratum taking account of these features of fabric should apply convincingly to the whole stratum as it exists in nature. The sampling and testing of soils to provide this description is part of an important activity of the civil or geotechnical engineer, namely, site investigation or more particularly *soil exploration*.

Given that all existing information about a site has been assembled, the extent of exploration will depend on the character of the ground and the type of construction work to be undertaken. Geological and geophysical mapping will supplement existing data and enable further exploration to be well planned. This in turn should ensure that all soils, rocks and groundwater conditions are sufficiently described to allow safe and economical project design. The lateral extent and depth of exploration should include all soils affected by the proposed work. Methods of investigation at this stage include trial pits, shafts, boreholes and probings. From samples, of various qualities according to soil type and sampling procedure, a geotechnical description of the soil strata is assembled. For some background and case histories, see Site Investigation in Construction, Part 1 (1993).

Soil profile

The results of a soil exploration are often presented in the form of a *soil profile*. A zone which is distinguishable from its neighbours above and below is called a soil horizon and the sequence of horizons from the ground surface to the limit of depth of the exploration is called the soil profile. The variation, with depth, of any property of the soil – for example, water content or density – may be displayed by a curve as shown in figure 21.5.

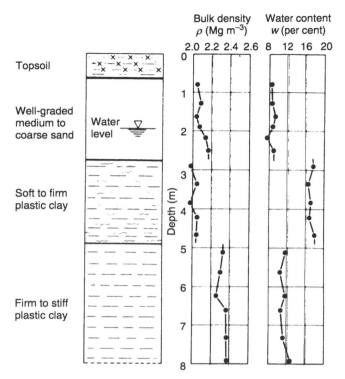

Figure 21.5. *Vertical profile of soil strata showing variations of properties with depth, the properties in this illustration being density and water content*

22

Bulk Properties

As part of a basic description of soil, it is necessary to establish parameters defining some physical properties of the mixtures of solid particles, water and sometimes air or gas which make up a soil. Suppose a sample of natural soil is taken without altering it in any way so that it is entirely typical of the soil stratum from which it came. Let its total volume be denoted by V and its total mass by M. The constituent solid particles, water and air will each have *absolute* volumes and masses which when added together will give V and M respectively. It is convenient to present these constituents diagrammatically in terms of their *absolute* volumes, stacking the absolute volumes one on the other, as in figure 22.1. Thus

$$V = V_s + V_w + V_{air}$$

where V_s is the absolute volume of all solid particles, V_w the volume of porewater and V_{air} the volume of pore air, or gas, in the sample of total volume V. Also

$$V = V_s + V_v$$

where $V_v = V_w + V_{air}$ is the absolute volume of voids or pore spaces in the sample.

The same diagrammatic form is suited to presenting the masses of the constituents of the sample. Thus

$$M = M_s + M_w$$

where M_s is the mass of dry solid particles and M_w the mass of water in the pore spaces in the sample.

The units of volume are usually m^3, l or ml and those of mass Mg, tonne (t), kg or g. The particle density of the solids

$$\rho_s = \frac{M_s}{V_s}$$

and the density of fresh water

$$\rho_w = \frac{M_w}{V_w} = 1000 \text{ kg m}^{-3} = 1 \text{ t m}^{-3} = 1 \text{ g ml}^{-1} \text{ at } 20°C$$

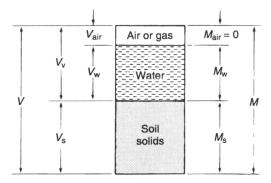

Figure 22.1 *Soil solids, water and gas phases in soil*

22.1 Specific gravity

The symbol G is used for the specific gravity of the solid particles of soil. Specific gravity is defined as the ratio of the mass of the soil particles to the mass of the same (absolute) volume of water. Thus using the notation of figure 22.1

$$G = \frac{M_s}{V_s \rho_w} = \frac{\rho_s}{\rho_w}$$

and is dimensionless. Physicists prefer the term *relative density* for this property of materials. It will be noted later in this chapter however, in section 22.4, that relative density has another meaning when applied to engineering soils.

The value of G is usually determined as an average for a suitably representative sample of soil, although strictly it may vary among the mineral constituents of the soil and from size fraction to size fraction. Clean sands generally have a specific gravity close to 2.65 and clays a somewhat higher value around 2.72 or more. For many purposes it is sufficient simply to use these values. Low values in natural soils would suggest the presence of organic matter.

Tests for the determination of particle density, ρ_s (Mg m^{-3}) of soils are detailed in BS 1377: Part 2. Particle density ρ_s is numerically equal to specific gravity G. The average of two determinations is taken and reported to the nearest 0.01 Mg m^{-3}. The main difficulty in the test is that of removing air from the soil–water mixture during the procedure for measuring the volume of the soil. If air is not completely removed, a low value is obtained. An investigation of the reproducibility of specific gravity results (Sherwood, 1970) has shown that unless the detailed procedures of BS 1377 are closely followed, a large proportion of specific gravity results will be outside the desired tolerance of ±0.05 on the true value required for the proper compaction control of earthworks.

22.2 Water content

The symbol w is used for the water, or moisture, content of soils and the term is defined as

$$w = \frac{M_w}{M_s} \text{ (usually expressed as a percentage)}$$

The value of w is determined by measuring the loss of water on drying suitably representative samples of soil, again following procedures such as those in BS 1377: Part 2. Particularly in fine-grained soils, the drying procedure may influence the results obtained. To avoid overheating a drying temperature of 105–110°C is specified. The period of drying varies with the soil type but is usually about 24 hours.

Water content values up to 10 per cent are reported to two significant figures and above 10 per cent to the nearest whole number. Values of natural water content in soils may range from under 5 per cent in gravels and sands to 50 per cent or more in the fine-grained cohesive soils. The presence of organic matter may increase w considerably.

22.3 Density

The mass of unit total volume of a particulate solid (soil) is described as its density, or bulk density, and is denoted here by the symbol ρ. The dry density ρ_d is the mass of dry soil contained in a unit total volume.

Thus, referring again to figure 22.1:

$$\rho = \frac{M}{V}$$

$$\rho_d = \frac{M_s}{V} = \frac{\rho}{1 + w}$$

The unit in each case has the form Mg m^{-3}, t m^{-3} or kg m^{-3}.

Tests for the determination of density are given in BS 1377: Parts 2 and 4. Densities are expressed to the nearest 0.01 Mg m^{-3}.

22.4 Derived soil properties and interrelationships

Voids ratio and porosity

These are terms describing the proportion of the total volume of a soil which is not occupied by solid constituents. The first term, voids ratio, is the more commonly used. Voids ratio e and porosity n are given by

$$e = \frac{V_v}{V_s}$$

and

$$n = \frac{V_v}{V}$$

Hence

$$e = \frac{n}{1 - n}$$

and

$$n = \frac{e}{1 + e}$$

Voids ratio and porosity are dependent on a determination of G and bulk density. The values of e and n (the latter always less than unity) are given to two decimal places.

Degree of saturation

This is a measure of the proportion of the available voids volume in the soil which is filled with water. The symbol S_r is used for the degree of saturation, where

$$S_r = \frac{V_w}{V_v} \text{ (usually expressed as a percentage)}$$

The determination of S_r is dependent on determinations of G, w and the bulk density ρ of the soil.

Natural soils below the water table or groundwater surface (defined in Chapter 25, section 25.2) will usually be saturated, that is, $S_r = 100$ per cent. The finer-grained soils may also be saturated for some distance above the water table owing to capillarity and adsorption actions. Fill materials can be compacted to a high degree of saturation but it is not possible by mechanical compaction – rolling, kneading, impact and vibration – to achieve 100 per cent saturation.

Air voids content

The compaction of fill materials is sometimes described in terms of air content. Air voids content is given by

$$V_a = \frac{V_v - V_w}{V} \times 100 \text{ per cent}$$

The value will be small in the case of well-compacted fills placed at suitable water contents – usually 5–10 per cent.

Density

The density of a given soil is expressed in a number of different ways; thus, referring again to figure 22.1, natural or bulk density ρ, dry density ρ_d, saturated density ρ_{sat} and submerged or buoyant density ρ' are given by

$$\rho \quad = \frac{M}{V} = \frac{G + S_r e}{1 + e} \rho_w = \frac{(1 + w) G}{1 + e} \rho_w$$

$$\rho_d \quad = \frac{M_s}{V} = \frac{G}{1 + e} \rho_w = \frac{\rho}{1 + w}$$

$$\rho_{sat} \quad = \frac{G + e}{1 + e} \rho_w \text{ for } S_r = 100 \text{ per cent}$$

$$\rho' \quad = \rho_{sat} - \rho_w = \frac{G - 1}{1 + e} \rho_w$$

In the above equations it is assumed that the porewater is fresh water with a specific gravity $G_1 = 1$. Strictly, if the pore liquid is salt water or some other liquid for which $G_1 \neq 1$, the equations become

$$\rho = \frac{G + G_1 S_r e}{1 + e} \rho_w$$

$$\rho_{sat} = \frac{G + G_1 e}{1 + e} \rho_w$$

$$\rho' = \frac{G - G_1}{1 + e} \rho_w$$

Unit weight

The weight of unit total volume of a soil is described as its unit weight and is denoted by the symbol γ. It is obtained by multiplying the density ρ by the acceleration due to gravity g in m s^{-2}.

Thus a soil of density 1.75 Mg m^{-3} will have a unit weight

$$\gamma = \rho g = 1.75 \times 9.81 = 17.16 \text{ kN m}^{-3}$$

and for fresh water

$$\gamma_w = \rho_w g = 9.81 \text{ kN m}^{-3}$$

Unit weights are required when calculating the vertical forces or pressures exerted by soils.

Density index or relative density

The natural or compacted density of a given soil will vary from place to place. The drainage characteristics, compressibility and strength and other aspects of engineering behaviour of a soil are related to its density. The variation in density may be examined by determining the variation in natural or dry density values. In cohesionless soils, however, it has been found convenient and useful in practice to refer the actual density of a soil to the range of densities over which the soil can exist or can be placed.

The actual density is then described using the concept of the density index I_D where

$$I_D = \frac{e_{max} - e}{e_{max} - e_{min}}$$

in which e is the actual voids ratio of the cohesionless soil, and e_{max} and e_{min} are respectively the greatest and least values of voids ratio obtainable with the soil.

Tests for the determination of maximum and minimum dry densities are given in BS 1377: Part 4. The density index may be derived from

$$I_D = \left(\frac{\rho_d - \rho_{dmin}}{\rho_{dmax} - \rho_{dmin}} \right) \left(\frac{\rho_{dmax}}{\rho_d} \right)$$

Bulk properties

These parameters describing the soil bulk are much used and their names are part of the language of geotechnology. They should be studied until the student engineer is wholly familiar with them.

For a sample of a natural or compacted soil the determination of parameters ρ, w and G enables the others to be derived. The following convenient relations between the parameters may also be found useful. If $V = 1$ then

$$V_s = \frac{1}{1+e}$$

and

$$V_v = \frac{e}{1+e}$$

allowing easy derivation of the above densities. Also

$$S_r e = wG$$

For a saturated sand therefore $e = 2.65w$ approximately.

Some typical values of bulk properties of soils are given in table 22.1 to indicate the magnitudes of the various parameters.

TABLE 22.1
Bulk properties of some natural soils

Description	Porosity n (per cent)	Voids ratio e	Water content at $S_r = 1$ w (per cent)	Density (Mg m⁻³) ρ_d	ρ_{sat}
Uniform sand, loose	46	0.85	32	1.44	1.89
Uniform sand, dense	34	0.51	19	1.75	2.08
Mixed grained sand, loose	40	0.67	25	1.59	1.98
Mixed grained sand, dense	30	0.43	16	1.86	2.16
Glacial till, very mixed grained	20	0.25	9	2.11	2.32
Soft glacial clay	55	1.2	45		1.76
Stiff glacial clay	37	0.6	22		2.06
Soft slightly organic clay	66	1.9	70		1.57
Soft very organic clay	75	3.0	110		1.43
Soft bentonite	84	5.2	194		1.28

Consistency of fine-grained soils (per cent)

Soil	w	w_L	w_P	I_P	I_L
Soft clay, Chicago	26	32	18	14	57
Stiff clay, London	24	80	28	52	− 8
Quick silty clay, Oslo	40	28	20	8	250

23

Coarser- and Finer-grained Soils

23.1 Particle size distribution

As already seen in Part III (concrete), coarser-grained soils and aggregates are well described by their *particle size distribution* curves. The most common form of the plot is the cumulative diagram displaying the percentage by weight of the material finer than any given size, the latter being presented on a logarithmic scale. Individual soil particles are not of course spherical but for most coarser-grained soils, except those containing quantities of micaceous material, the particles are nearly equidimensional and when the particles are matched against sieves of known aperture sizes a satisfactory estimate of the amounts of material finer than each size is easily obtained. The preparation of samples for sieving (mechanical analysis) and the sieve test procedures are described in BS 1377: Parts 1 and 2. The lower limit of size analysed by sieving is usually fine sand at 63 μm.

The particle size distribution of the silty fractions of soils is of interest too and these sizes lie below 63 μm. The size distribution over the silt range is indirectly assessed from determinations of the velocities of sedimentation of the particles in water, assuming these to be perfect spheres having the same specific gravity as the soil particles. The silts reach their terminal velocity almost immediately and thereafter fall in water at a constant velocity which, for the silt sizes, is proportional to the square of the particle diameter (Stokes' law). The sedimentation test procedures are also described in BS 1377: Part 2.

In a soil both procedures may be required to define the particle size distribution curve. An example of such a curve is given in figure 23.1 which also shows the particle size boundaries used in BS 1377 to distinguish silts, sands, gravels and cobbles. Soils with similar size distribution curves would in a general sense be expected to show similar engineering characteristics. Predictions of behaviour are made cautiously because the size distribution alone does not convey the arrangement and density of packing of the particles and other factors which influence the behaviour of the soil.

Although the size distribution plot is in the form of a curve whose shape depends on the chosen method of presentation as well as on the soil itself the plot

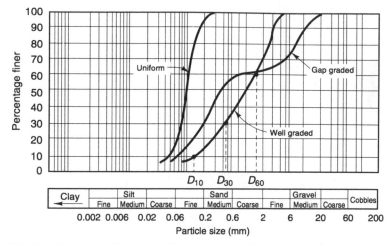

Figure 23.1 *Cumulative particle size distribution curves on the conventional semi-logarithmic plot*

can be used to describe and classify coarse-grained soils. It has long been known by engineers that the fine grains in a soil exercise the largest influence on its behaviour. Accordingly the term *effective size* or *ten per cent size* (D_{10}) is much used in describing soils. The effective size of a soil is the particle diameter at which 10 per cent by weight of the soil is finer in size. This D_{10} value is conveniently read from the particle size distribution curve as shown for the well graded soil in figure 23.1.

The steepness of the curve is given a numerical value in the *uniformity coefficient C_u* where $C_u = D_{60}/D_{10}$ and D_{60} is the particle diameter at which 60 per cent by weight of the soil is finer. If the range of sizes present in a soil is small it is described as a *uniform soil* or *uniformly graded*. Such *poorly graded* soils will have C_u values of 2 or less. A *well-graded* soil is one which can give a dense, rather strong soil. It will have a wide range of particle sizes and the size distribution curve will be smooth and concave upwards, with no deficiencies or excesses of sizes in any size range. A well-graded soil has a C_u value of 5 or more.

The shape of the central portion of the curve is assessed by the *coefficient of curvature C_c* where

$$C_c = \frac{D_{30}^2}{D_{10}D_{60}}$$

For the grain sizes and amounts to be so arranged that a dense packing is possible the distribution curve should be concave upwards giving a C_c value between 1 and 3. Some poorly graded soils have a deficiency of intermediate sizes and are described as *gap graded*. This shows on the size distribution curve as a 'step' and an example is given in figure 23.1.

23.2 Particle shape and surface texture

The coarser-grained soils are commonly equidimensional in particle shape, that is, in any particle the length, width and height are nearly equal. In cases where a more specific description is warranted, reference may be made to BS 812: Parts 102 and 105 for definitions and illustrations of form, angularity and texture of particles. Use of a limited range of descriptive terms – flaky, elongated; rounded, angular; smooth, rough and others – is sufficient for this purpose; see Chapter 12, tables 12.11 and 12.12.

23.3 Atterberg limits

When a soil contains an appreciable quantity, say 20 per cent or more by weight, of material finer than the 63 μm size, a description based on size distribution alone is insufficient. The size distribution of clays and silty clays is of interest but in practical engineering terms too difficult to determine and there are other characteristics which are more simply obtained and of more immediate relevance. The behaviour of clay is related to its mineralogical composition, water content and the micro- and macro-fabric of its particles as outlined in the discussion on sampling and elsewhere. A full study of a clay would require all these and other factors to be accounted for.

The water content is an obviously important and easily measured property of a clay. If the water content of a natural firm clay is allowed to increase, the clay will soften and become more compressible. In a different type of clay the effect will be similar but different in degree. The change in the state of a clay as the water content is changed has been found to be a useful and simple way of distinguishing one clay from another and this is the basis of soil classification applied to clays.

The *Atterberg limits* are empirical and somewhat arbitrary divisions, in terms of water content, between states of a clay. As the water content of a clay element is increased from the dry solid condition it will pass through a semi-solid, friable or crumbly state into a plastic state. In the plastic state the clay can be kneaded without rupture or cracking of the soil. On further wetting the clay has the consistency of a liquid, becoming less viscous as the water content is increased. These states are listed in figure 23.2. The water contents at the boundaries of these states are called the Atterberg limits – they are the *shrinkage limit* w_S, the *plastic limit* w_P and the *liquid limit* w_L. Tests for the liquid limit, the plastic and a shrinkage limit are made on that fraction of soil finer than 425 μm and the procedures are

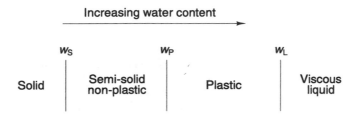

Figure 23.2 *State of clay soil over a range of water contents showing the relative placing of the index limits* w_S, w_P *and* w_L

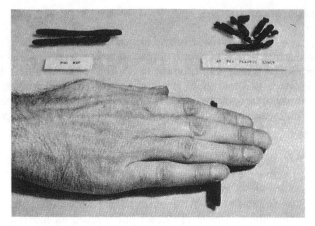

Figure 23.3 *The determination of the plastic limit from BS 1377: 1990 reproduced by permission of the British Standards Institution*

described in BS 1377: Part 2. These tests are empirical in nature and since the results are to be used in the comparison of soils the tests must follow the standardised procedures.

Essentially the *liquid limit* is determined by relating the water content of a clay to the penetration of a standard cone into the wet clay during a period of 5 ± 1 s. The liquid limit w_L is the value of the water content w, expressed to the nearest whole number, corresponding to a cone penetration of 20 mm. The *plastic limit* is determined by rolling threads of the clay on a flat surface. When these threads begin to crumble at just over 3 mm diameter (figure 23.3), the water content is measured as the plastic limit w_P. The *shrinkage limit* is the water content at which an initially dry specimen of the clay is just saturated, without change in total volume. The liquid and plastic limits are the principal limits for classification purposes.

The *plasticity index* is defined as

$$I_P = w_L - w_P$$

and gives a measure of the extent or range of water contents over which a soil is in the plastic state.

The *liquidity index* is defined as

$$I_L = \frac{w - w_P}{w_L - w_P} = \frac{w - w_P}{I_P}$$

where w is the natural water content of a clay having limits w_L and w_P. Thus the liquidity index, which is usually expressed as a percentage and which may be negative, gives a measure of the consistency of the natural soil. At a liquidity index of 100 per cent the natural clay is at its liquid limit and it will experience a loss of strength on disturbance or remoulding. A liquidity index of zero signifies that the natural clay is at its plastic limit and stiff, and will tend to crumble on remoulding. An alternative form, the *consistency index* is sometimes used where consistency index $= 1 - I_L$

It is worth recalling again the emphasis placed on testing undisturbed soils in the assessment of their future behaviour in engineering works. The liquid and plastic limit tests are made on highly *disturbed* soil–water mixtures and cannot be expected to yield information which depends on the fabric and structure of the soil *in place*. Sherwood (1970) has reported on an investigation of the repeatability of the results of liquid and plastic limit tests when carried out by a single experienced operator, by several different operators and by several different testing laboratories. The liquid limit determinations were made with the Casagrande apparatus method, BS 1377: Part 2. The repeatability was found to be disturbingly poor in view of the important engineering decisions which may rest on the test results. The quality of the information should be good however if a generous number of specimens are tested to give average values for the limits and if the procedures of BS 1377 are meticulously followed.

23.4 Activity

The affinity of a clay for water will depend in part on the mineralogical composition of the clay and on the fineness of particle size – composition and fineness being related. Fineness of a particulate material is often quantified as its specific surface, that is, the total surface area of all its particles per unit of mass or absolute volume. Montmorillonite clays have a high specific surface relative to the kaolinite clays. A simple ratio giving a measure, for engineering purposes, of the mineralogical composition of a clay was proposed by Skempton (1953). This ratio is called the 'activity' A of a clay and

$$A = \frac{I_P}{(\text{per cent clay})}$$

where I_P is the plasticity index of the clay, and 'per cent clay' is the mass of clay fraction (finer than 2 µm in size) expressed as a percentage of the dry mass of the soil. For a clay of particular physicochemical characteristics, A has been found to be a constant and a measure of the colloidal activity of the clay. Skempton gave the following classification of activity for clays

inactive $A < 0.75$
normal $0.75 < A < 1.25$
active $A > 1.25$

There appears to be a relationship between the components of shear strength of clays and their activity, and between activity and difficulties in obtaining deep samples of some clays. The largest group of clays has activity in the normal range and for these $A = 1$ approximately. Wilun and Starzewski (1975) quote an approximate classification from Polish standards, based on clays for which $A = 1$, and this is given in table 23.1.

TABLE 23.1

Description of cohesive soils

Description of cohesiveness of soil	Clay fraction (per cent)	I_P (per cent)
Cohesionless	0–2	<1
Slightly cohesive	2–10	1–10
Medium cohesive	10–20	10–20
Cohesive	20–30	20–30
Very cohesive	30–100	>30

In the case of a discrepancy in the description of a soil by the plasticity index and the clay fraction, the colloidal activity should be quoted after the description. For example, a soil containing 25 per cent clay and having I_P = 38 per cent would be described in terms: heavy cohesive silty clay (25 per cent clay fraction, A = 1.5).

23.5 Chemical composition

In chemical composition soils are complex and variable minerals. For engineering purposes, however, tests are usually restricted to the determination of organic matter content, sulphate content of soil and groundwater, and determination of the pH value. Details of test procedures are given in BS 1377: Part 3. The carbonate content may also be of interest since chalk soils are known to be particularly frost susceptible.

The presence of appreciable organic matter indicates a compressible soil and even a small percentage of decomposed organic matter is sufficient to react unfavourably with Portland cement when the latter is used as an admixture in soil stabilisation. Taken together with the pH value, the sulphate content gives a measure of the aggressiveness of the soil chemicals towards Portland cement concretes and other materials of construction. Sulphates occur in a number of clay soils and may also be troublesome in fills of colliery shale, brick rubble, ash and some industrial wastes. Sulphate can only continue to reach concrete by movement of groundwater so that concrete which is wholly and permanently above the water table is unlikely to be seriously attacked.

The reaction of most natural soils in Britain to pH tests varies between about pH 4.0 for very acid fen, heath and peat soils to about pH 8.0 for chalky alkaline soils. A knowledge of the acidity or alkalinity is of interest to the engineer mainly because of its influence on corrosion of concretes or metalwork laid in contact with the ground.

As a result of the increasing development of derelict land for construction, the need for more extended chemical analyses for contaminants in the ground has also grown. The diverse manufacturing industries of the nineteenth century, gas and tar works, and decades of industrial and domestic waste tippings have provided a legacy of potential hazards from toxic substances, gases and leachates. These contaminants have to be assessed and, if necessary, dealt with in advance of construction.

24

Soil Classification

As emphasised in the Introduction, an important aim of the geotechnical engineer is to quantify parameters which describe soil behaviour so that he can communicate these to others or use them in his own design analysis. Individually these parameters may allow a soil to be classified in terms of the degree to which it possesses a particular property – for example density. A soil may be described as very loose, loose, medium dense, and so on. In a more general way engineers have found it useful to allocate soils to groups, each group having relatively consistent behaviour and distinguishable from soils in other groups. This placing in groups is called *soil classification*. By adopting a concise and reasonably systematic method of designating the various types of soils, many difficulties in the communication of descriptions and comparisons of soils are overcome.

Several systems of classification have been proposed but none is universally accepted. A system presented by Casagrande has been widely used over several decades, and the British Soil Classification System for Engineering Purposes (BSCS) is a further development of it. The BSCS is described in BS 5930: Code of practice for site investigations. It remains discretionary in application but is particularly useful where soils are to be used as construction materials, as in earthworks. The BSCS places soils into groups defined by the particle size distribution of their coarser particles and the plasticity of their finer particles, characteristics that play a major role in determining soil engineering properties. Taken together with descriptions of the material in-place, the classifications may enable predictions of soil behaviour to be made and solutions to engineering problems reached. Without further data from tests developed to measure specific parameters, for example compressibility and shear strength, such predictions should however be used cautiously.

24.1 Soil description

Description of a soil should include its mass characteristics, material characteristics, geological description and its soil classification. The mass characteristics are those features of spacing and surface texture of layers, bedding planes, joints, fissures and the like. The relative density of sands and gravels and the estimated

TABLE 24.1
Field identification and description of soils

	Basic soil type	Particle size (mm)	Visual identification	Particle nature and plasticity	Composite soil types (mixtures of basic soil types)		
Very coarse soils	BOULDERS		Only seen complete in pits or exposures	Particle shape:	Scale of secondary constituents with coarse soils		
		— 200					
	COBBLES		Often difficult to recover from boreholes	Angular	Term		% of clay or silt
		— 60		Subangular			
Coarse soils (over 65% sand and gravel sizes)	GRAVELS	coarse	Easily visible to naked eye; particle shape can be described; grading can be described	Subrounded Rounded Flat Elongate	slightly clayey ⎫ GRAVEL or slightly silty ⎭ SAND		under 5
		— 20					
		medium	Well graded: wide range of grain sizes, well distributed. Poorly graded: not well graded. (May be uniform: size of most particles lies between narrow limits; or gap graded: an intermediate size of particle is markedly under-represented)		— clayey ⎫ GRAVEL or — silty ⎭ SAND		5 to 15
		— 6					
		fine			very clayey ⎫ GRAVEL or very silty ⎭ SAND		15 to 35
		— 2		Texture:			
	SANDS	coarse	Visible to naked eye: very little or no cohesion when dry; grading can be described	Rough Smooth Polished	Sandy GRAVEL ⎫ Sand or gravel are important second constituent of the Gravelly SAND ⎭ coarse fraction		
		— 0.6					
		medium	Well graded: wide range of grain sizes, well distributed. Poorly graded: not well graded. (May be uniform: size of most particles lies between narrow limits; or gap graded: an intermediate size of particle is markedly under-represented)		For composite types described as: clayey: fines are plastic, cohesive; silty: fines non-plastic or of low plasticity		
		— 0.2					
		fine					
		— 0.06					
Fine soils (over 35% silt and clay sizes)	SILTS	coarse	Only coarse silt barely visible to naked eye; exhibits little plasticity and marked dilatancy; slightly granular or silky to the touch. Disintegrates in water; lumps dry quickly; possess cohesion but can be powdered easily between fingers	Non-plastic or low plasticity	Scale of secondary constituents with fine soils		
		— 0.02			Term		% of sand or gravel
		medium					
		— 0.006			sandy ⎫ CLAY or gravelly ⎭ SILT		35 to 65
		fine					
		— 0.002					
	CLAYS		Dry lumps can be broken but not powdered between the fingers; they also disintegrate under water but more slowly than silt; smooth to the touch; exhibits plasticity but no dilatancy; sticks to the fingers and dries slowly; shrinks appreciably on drying usually showing cracks. Intermediate and high plasticity clays show these properties to a moderate and high degree, respectively	Intermediate plasticity (Lean clay)	— CLAY SILT		under 35
					Examples of composite types (indicating preferred order for description)		
				High plasticity (Fat clay)	Loose, brown, subangular, very sandy, fine to coarse GRAVEL with small pockets of soft grey clay		
Organic soils	ORGANIC CLAY, SILT or SAND	Varies	Contains substantial amounts of organic vegetable matter		Medium dense, light brown, clayey, fine and medium SAND		
					Stiff, orange brown, fissured sandy CLAY		
					Firm brown, thinly laminated SILT and CLAY		
	PEATS	Varies	Predominantly plant remains usually dark brown or black in colour, often with distinctive smell; low bulk density		Plastic, brown, amorphous PEAT		

Compactness/strength		Structure			Colour
Term	Field test	Term	Field identification	Interval scales	
Loose		Homo-geneous	Deposit consists essentially of one type	Scale of bedding spacing	Red
	By inspection of voids and particle packing				Pink
Dense		Inter-stratified	Alternating layers of varying types or with bands or lenses of other materials	Term / Mean spacing, mm	Yellow / Brown
					Olive
			Interval scale for bedding spacing may be used	Very thickly bedded / over 2000	Green
					Blue
Loose	Can be excavated with a spade; 50 mm wooden peg can be easily driven	Hetero-geneous	A mixture of types	Thickly bedded / 2000 to 600	White
				Medium bedded / 600 to 200	Grey
Dense	Requires pick for excavation; 50 mm wooden peg hard to drive	Weathered	Particles may be weakened and may show concentric layering	Thinly bedded / 200 to 60	Black / etc.
Slightly cemented	Visual examination; pick removes soil in lumps which can be abraded			Very thinly bedded / 60 to 20	
				Thickly laminated / 20 to 6	Supplemented as necessary with:
				Thinly laminated / under 6	Light
					Dark
					Mottled
					etc.
Soft or loose	Easily moulded or crushed in the fingers	Fissured	Breaking into polyhedral fragments along fissures. Interval scale for spacing of discontinuities may be used		and
Firm or dense	Can be moulded or crushed by strong pressure in the fingers				Pinkish
		Intact	No fissures		Reddish
Very soft	Exudes between fingers when squeezed in hand	Homo-geneous	Deposit consists essentially of one type		Yellowish
					Brownish
Soft	Moulded by light finger pressure	Inter-stratified	Alternating layers of varying types. Interval scale for thickness of layers may be used	Scale of spacing of other discontinuities	etc.
Firm	Can be moulded by strong finger pressure	Weathered	Usually has crumb or columnar structure	Term / Mean spacing, mm	
Stiff	Cannot be moulded by fingers. Can be indented by thumb			Very widely spaced / Over 2000	
Very stiff	Can be indented by thumb nail			Widely spaced / 2000 to 600	
				Medium spaced / 600 to 200	
Firm	Fibres already compressed together			Closely spaced / 200 to 60	
Spongy	Very compressible and open structure	Fibrous	Plant remains recognisable and retain some strength	Very closely spaced / 60 to 20	
Plastic	Can be moulded in hand, and smears fingers	Amor-phous	Recognisable plant remains absent	Extremely closely spaced / under 20	

375

strength of clays form part of this description. These characteristics can only be observed in undisturbed samples or in pits and shafts. The material characteristics include soil name, particle shape and composition and colour.

Examples of soil description

Firm, closely fissured yellowish brown CLAY of high plasticity – London Clay. Dense yellow fine SAND with thin lenses of soft grey silty CLAY – Recent Alluvium.

Where an assessment of a soil is required in the field or without laboratory testing, the visual and textural means of identification outlined in table 24.1 give a guide.

Particles of 60 μm size are at about the limit of visibility unaided. At this size a moist sand feels harsh but not gritty when rubbed between the fingers. The presence of coarser sizes makes the soil feel more gritty, and finer sizes make it less harsh.

Observation of *dilatancy* helps to distinguish silts and fine sands from clays. A portion of soil is taken and water mixed in until the consistency is soft but not sticky. Some of this is formed into a pat about 25–30 mm in diameter and 5–8 mm thick in the open palm of one hand. With the hand held horizontal it is struck against the other hand several times. If the soil is a dilatant material, water will flow to the surface of the pat giving it a shiny 'livery' appearance. Further, on squeezing the pat with the fingers the water recedes into the pat and the surface dulls, and the pat stiffens and crumbles. The property of dilatancy belongs to the granular materials, and the silts and very fine sands show marked reactions to the test. If the material is substantially clay, then the surface of the pat will not alter in appearance in this test.

Soils comprising mixtures of the basic types are usual and are appropriately described. Where a layering or fissuring is evident, this too is included in the description, for example – Stiff brown fissured sandy CLAY.

Most natural sediments include macrofabric features, for example laminations or fissures, which play important roles in their engineering behaviour. A number of authors have suggested ways of measuring and quantifying the fabric of layered sediments and clay soils containing discontinuities such as fissures (McGown *et al.*, 1980). Fabric features include spacing, thickness, extent, surface geometry and coating, orientation etc. and these can be classified into groups.

Discontinuities for example are measured either in cavities cut into undisturbed soil or on oriented undisturbed block samples of adequate volume. The measured directional data are commonly presented as stereographic plots which enable any principal sets of discontinuities to be recognised.

From the fabric features, predictions can be made of the directional variations of certain engineering properties of the sediments.

Where the geological formation can be named with confidence, this should be included in the soil description.

24.2 British Soil Classification System for Engineering Purposes

The diversity of soils in place leads to diversity of descriptions, when the size distribution, plasticity, compactness, strength, structure and colour are included. The size distribution and plasticity of a disturbed sample of soil are sufficient, however, for *soil classification* purposes to give a guide to the engineering behaviour of the soil as a construction material in earthworks.

To apply the BSCS the portion of coarse material exceeding 60 mm (i.e. retained on the 63 mm test sieve) is picked out and its proportion of the whole soil recorded.

TABLE 24.2

Names and descriptive letters for grading and plasticity characteristics

		Descriptive name	Letter
Coarse components	Main terms	GRAVEL	G
		SAND	S
	Qualifying terms	Well graded	W
		Poorly graded	P
		Uniform	Pu
		Gap graded	Pg
Fine components	Main terms	FINE SOIL, FINES may be differentiated into M or C	F

		SILT (M-SOIL)*
		    plots below A-line of plasticity chart
		    of figure 24.1 (of restricted plastic range)
			M

		CLAY plots above A-line (fully plastic)	C
	Qualifying terms	Of low plasticity	L
		Of intermediate plasticity	I
		Of high plasticity	H
		Of very high plasticity	V
		Of extremely high plasticity	E
		Of upper plasticity range† incorporating groups I, H, V and E	U
Organic components	Main term	PEAT	Pt
	Qualifying term	Organic may be suffixed to any group	O

* See note 5 following table 24.3.

† This term is a useful guide when it is not possible or not required to designate the range of liquid limit more closely, e.g. during the rapid assessment of soils.

The particle sizes and plasticity characteristics are divided into a number of clearly defined ranges, each being referred to by a descriptive name and letter as shown in table 24.2. Any soil can be placed in one of a number of soil groups shown in table 24.3. The groups are formed from combinations of the ranges of characteristics. In group symbols the dominant size fraction is placed first, for example CS, sandy CLAY; SC, very clayey SAND; S–C (spoken 'S dash C'), clayey SAND. Where field identification is used, only the main soil groups need be identified.

Soils exhibiting plasticity are readily classified from figure 24.1.

Classification systems are widely used for earthworks design in highway and airfield construction. Empirical relationships have been established between the classification groups and factors such as pavement thickness, frost susceptibility, shrinkage and swelling, drainage characteristics and compaction methods. Some of these are illustrated in table 24.4, from BS 6031.

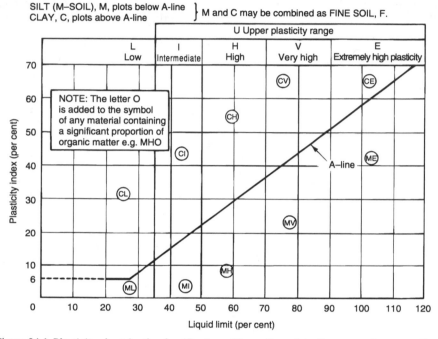

Figure 24.1 *Plasticity chart for the classification of fine soils and the finer part of coarse soils (measurements made on material passing a 425 μm British Standard sieve)*

TABLE 24.3
British Soil Classification System for Engineering Purposes

Soil groups (see note 1) (GRAVEL and SAND may be qualified Sandy GRAVEL and Gravelly SAND, etc. where appropriate)			Sub-groups and laboratory identification				
			Group symbol (see notes 2 and 3)	Sub-group symbol (see note 2)	Fines (% less than 0.06 mm)	Liquid limit (%)	Name
COARSE SOILS less than 35% of the material is finer than than 0.06 mm	GRAVELS More than 50% of coarse material is of gravel size (coarser than 2 mm)	Slightly silty or clayey GRAVEL	G / GW / GP	GW / GPu / GPg	0 to 5		Well-graded GRAVEL / Poorly graded/Uniform/Gap-graded GRAVEL
		Silty GRAVEL	G-M / G-F	GWM / GPM	5 to 15		Well-graded/Poorly graded silty GRAVEL
		Clayey GRAVEL	G-C	GWC / GPC			Well-graded/Poorly graded clayey GRAVEL
		Very silty GRAVEL	GF / GM	GM			Very silty GRAVEL: subdivide as for GC
		Very clayey GRAVEL	GC	GCL / GCI / GCH / GCV	15 to 35		Very clayey GRAVEL (clay of low, intermediate, high, very high, extremely high plasticity)
	SANDS More than 50% of coarse material is of sand size (finer than 2 mm)	Slightly silty or clayey SAND	S / SW / SP	SW / SPu / SPg	0 to 5		Well-graded SAND / Poorly graded/Uniform/Gap-graded SAND
		Silty SAND	S-M / S-F	SWM / SPM	5 to 15		Well-graded/Poorly graded silty SAND
		Clayey SAND	S-C	SWC / SPC			Well-graded/Poorly graded clayey SAND
		Very silty SAND	SF / SM	SM			Very silty SAND; subdivided as for SC
		Very clayey SAND	SC	SCL / SCI / SCH / SCV	15 to 35		Very clayey SAND (clay of low, intermediate, high, very high plasticity)
FINE SOILS more than 35% of the material is finer than 0.06 mm	Gravelly or sandy SILTS and CLAYS 35% to 65% fines	Gravelly SILT	FG / MG	MG			Gravelly SILT; subdivide as for CG
		Gravelly CLAY (see note 4)	CG	CLG / CIG / CHG / CVG		<35 / 35 to 50 / 50 to 70 / 70 to 90	Gravelly CLAY of low plasticity / of intermediate plasticity / of high plasticity / of very high plasticity
		Sandy SILT	FS / MS	MS			Sandy SILT
		Sandy CLAY (see note 4)	CS	CLS, etc.			Sandy CLAY; subdivide as for CG
	SILTS and CLAYS 65% to 100% fines	SILT (M-SOIL)	F / M	M			SILT; subdivide as for C
		CLAY (see notes 5 and 6)	C	CL / CI / CH / CV		<35 / 35 to 50 / 50 to 70 / 70 to 90	CLAY of low plasticity / of intermediate plasticity / of high plasticity / of very high plasticity
ORGANIC SOILS	Descriptive letter 'O' suffixed to any group or sub-group symbol				Organic matter suspected to be a significant constituent. Example MHO Organic SILT of high plasticity		
PEAT	Pt	Peat soils consist predominantly of plant remains which may be fibrous or amorpheus.					

Note 1. The name of the soil group should always be given when describing soils, supplemented, if required, by the group symbol, although for some additional applications (e.g. longitudinal sections) it may by convenient to use the group symbol alone.

Note 2. The group symbol or sub-group symbol should be placed in brackets if laboratory methods have not been used for identification, e.g. (GC).

Note 3. The designation FINE SOIL or FINES, F, may be used in place of SILT, M, or CLAY, C, when it is not possible or not required to distinguish between them.

Note 4. GRAVELLY if more than 50% of coarse material is of gravel size. SANDY if more than 50% of coarse material is of sand size.

Note 5. SILT (M-SOIL), M, is material plotting below the A-line, and has a restricted plastic range in relation to its liquid limit, and relatively low cohesion. Fine soils of this type include silt-sized materials and rock flour, micaceous and diatomaceous soils, pumice and volcanic soils, and soils containing halloysite. The alternative term 'M-soil' avoids confusion with materials of predominantly silt size, which form only a part of the group. Organic soils also usually plot below the A-line on the plasticity chart, when they are designated ORGANIC SILT, MO.

Note 6. CLAY, C is material plotting above the A-line, and is fully plastic in relation to its liquid limit.

TABLE 24.4
Field characteristics of soils and other materials used in earthworks

Material	Major divisions	Sub-groups	BSCS* group symbol	Casagrande group symbol	Drainage characteristic†	Potential frost action	Shrinkage or swelling properties	Value as a road foundation when not subject to frost action	Bulk density before excavation (Mg m⁻³)	
									Dry or moist	Submerged
Coarse soils and other materials	Boulders and cobbles	Boulder gravels	—	—	Good	None to very slight	Almost none	Good to excellent	—	—
	Other materials	Hard: Hard broken rock, hardcore, etc.	—	—	Excellent	None to slight	Almost none	Very good to excellent	—	—
		Soft: Chalk, soft rocks, rubble	—	—	Fair to practically impervious	Medium to high	Almost none to slight	Good to excellent	1.10 to 2.00	0.65 to 1.25
	Gravels and gravelly soils	Well-graded gravel and gravel–sand mixtures, little or no fines	GW	GW	Excellent	None to very slight	Almost none	Excellent	1.90 to 2.10	1.15 to 1.30
		Well-graded gravel sand mixtures with excellent clay binder	GWC	GC	Practically impervious	Medium	Very slight	Excellent	2.00 to 2.25	1.00 to 1.35
		Uniform gravel with little or no fines	GPu	GU	Excellent	None	Almost none	Good	1.60 to 1.80	1.00 to 1.11
		Gap-graded gravel and gravel–sand mixtures, little or no fines	GPg	GP	Excellent	None to very slight	Almost none	Good to excellent	1.60 to 2.00	0.90 to 1.25
		Gravel with fines, silty gravel, clayey gravel, poorly graded gravel–sand–clay mixtures	GM/GC	GF	Fair to practically impervious	Slight to medium	Almost none to slight	Good to excellent	1.80 to 2.10	1.10 to 1.30
	Sands and sandy soils	Well-graded sands and gravelly sands, little or no fines	SW	SW	Excellent	None to very slight	Almost none	Excellent to good	1.80 to 2.10	1.05 to 1.25
		Well-graded sand with excellent clay binder	SWC	SC	Practically impervious	Medium	Very slight	Excellent to good	1.90 to 2.10	1.15 to 1.30
		Uniform sands with little or no fines	SPu	SU	Excellent	None to very slight	Almost none	Fair	1.65 to 1.85	1.00 to 1.15

Fine soils category	Description	(British)*	(Unified)	Compaction/permeability	Drainage	Frost heave	Workability	Dry density	Factor
Fine soils — Soils having low compressibility	Gap-graded sands, little or no fines	SPg	SP	Excellent	None to very slight	Almost none	Fair to good	1.45 to 1.70	0.90 to 1.00
	Sands with fines, silty sands, clayey sands, poorly graded sand–clay mixtures	SM/SC	SF	Fair to practically impervious	Slight to high	Almost none to medium	Fair to good	1.70 to 1.90	1.00 to 1.15
	Silts (inorganic) and very fine sands, rock flour, silty or clayey fine sands with low plasticity	ML/SCL MS/CLS	ML	Fair to poor	Medium to very high	Slight to medium	Fair to poor	1.70 to 1.90	1.00 to 1.15
	Clay of low plasticity	CL	CL	Practically impervious	Medium to high	Medium	Fair to poor	1.60 to 1.80	1.00 to 1.11
	Organic silts of low plasticity	MLO	OL	Poor	Medium to high	Medium to high	Poor	1.45 to 1.70	0.90 to 1.00
Soils having medium compressibility	Silt and sandy clays (inorganic) of intermediate plasticity	CIS	MI	Fair to poor	Medium	Medium to high	Fair to poor	1.55 to 1.80	0.95 to 1.11
	Clays (inorganic) of intermediate plasticity	CI	CI	Fair to practically impervious	Slight	High	Fair to poor	1.60 to 2.00	1.00 to 1.10
	Organic clays of intermediate plasticity	CIO	OI	Fair to practically impervious	Slight	High	Poor	1.50	0.50
Soils having high compressibility	Micaceous or diatomaceous fine sandy and silty soils, elastic silts	—	MH	Poor	Medium to high	High	Poor	1.75	1.00
	Clays (inorganic) of high plasticity, fat clays	CH	CH	Practically impervious	Very slight	High	Poor to very poor	1.70	0.70
	Organic clays of high plasticity	CHO	OH	Practically impervious	Very slight	High	Very poor	1.50	0.50
Fibrous organic soils with very high compressibility	Peat and other highly organic swamp soils	Pt	Pt	Fair to poor	Slight	Very high	Extremely poor	1.40	0.40

*British Soil Classification System.
†Does not apply to in situ surface soils.

381

25

Water in Soils

In outlining the processes of formation of soils, the products were visualised in their complexity as composed largely of solid particulate matter. While this is usually so, the presence of water in the pores between the solid particles may have a great influence on the behaviour of the composite material. Much of the thinking of the geotechnical engineer is directed towards the understanding of this behaviour. The porous nature of soils permits the storage and flow of groundwater in the void spaces and these voids may be wholly or partly water-filled. In the coarser-grained soils this water may be a basic resource for industry, agriculture and human consumption to be drawn off from wells and recharged naturally or with man's intervention. Many situations arise in engineering design and practice where the flow, or even simply the presence, of water must be taken into account in analysis. The coarser soils generally do not experience much volume change as a result of movements of groundwater but in the finer-grained soils swelling or shrinkage will occur when some change is made in the environment of the soil. The presence of water has important effects on the states of stress in soils and on their compressibility and strength. In discussing water in soils, therefore, a distinction is usefully made in behaviour between the coarser-grained soils (silts and larger sizes) and the finer-grained soils (clays).

25.1 Transfer of stress through soil

The engineer is concerned with the influence of *stress* on his materials: stress changes may produce volume changes, changes in strength and may lead to yield or failure of materials. Although concretes, ceramics and timbers are porous materials it is usual in practice to deal with them as if they are *isotropic* single-phase materials such that definition of the *total normal stress* σ on a plane through a point in the material is all that is required to describe that normal stress. This is not sufficient in soils which, as we have seen, are compressible particulate assemblies comprising the three phases of solid particles, water and air or gas. The stresses in the pore spaces in the water and the air are measurable as u and u_a respectively where, owing to surface tension, u is always less than u_a. The total stress σ is also determinable. It is one of the problems in soil mechanics to determine what part of σ is supported by the soil particle assembly or skeleton.

For saturated soils it was shown by Terzaghi in 1923 that

$$\sigma' = \sigma - u$$

where σ' is termed the *effective stress*. It is changes in σ' which produce the changes in the volume and strength of such soils. This expression has been studied theoretically and experimentally and for practical engineering usage has stood the test of time. The effective stress σ' is closely related to the stress transmitted through the soil skeleton and is often called the intergranular stress. The two terms are usually used interchangeably. The unit of stress in soils is usually kN m^{-2} or kPa (1 kN m^{-2} = 1 kilopascal).

25.2 Coarser-grained soils – static groundwater

It will usually be possible in the coarser-grained soils, by inserting a borehole casing tube or standpipe down into the soil, to determine a water surface level. This water level at the standpipe location is variously described as the *groundwater surface* in the soil (in static conditions defining the level of the groundwater table), the *piezometric surface* or the *phreatic surface*. The position of the water level in the standpipe can be readily determined using, say, an electrical 'dipper' with its connecting cable marked in length units. At this groundwater level the absolute porewater pressure is equal to atmospheric pressure. The *porewater pressure u* at any place in a soil is defined as the excess pressure in the porewater, above atmospheric pressure, thus, $u = 0$ at groundwater level. Where there is no flow of groundwater taking place through the pores of the soil, u increases linearly with depth z below the groundwater level as illustrated in figure 25.1. That is

$$u = + \gamma_w z$$

Capillary water

Above the groundwater level the soil may be saturated for some distance owing to *capillary rise* of water pushed up from the water table or retention of water percolating down to the water table. This capillary rise is held by surface tension at the air–water interface from where the water is continuous to the groundwater level. Surface tension at an air–water surface generally tends to keep the surface flat, but when the water is in contact with soil particles, to which it acts as a wetting liquid, the water surface meets the particle surfaces tangentially or at a very small contact angle. This gives an air–water interface of a most complicated form in the soil. A simpler picture is obtained by considering a single vertical open-ended fine-bore glass capillary tube (*capilla*, Latin for hair) whose lower end is dipped in water. The tiny meniscus takes up a hemispherical concave form as shown in figure 25.2 and a pressure difference is set up across the meniscus, which is pushed up the tube until an equilibrium position is reached where the weight of

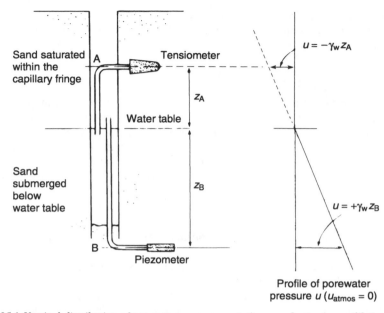

Figure 25.1 *Vertical distribution of porewater pressure – static groundwater in equilibrium*

the water column is just supported by the surface tension force. It is a simple matter to show that $z_s \propto 1/D$ where z_s is the height of capillary rise in a tube of diameter D. Within a real soil the meniscus or interface is of a complicated form with saddle-shaped masses of water forming at the points of contact of the interface and the soil particles. The greatest suction (pressure below atmospheric) which can be maintained at such an interface corresponds to the sharpest curvature the interface can form in the complex pore spaces. The sharpness of this curvature depends on the smallness of size of the pore spaces which in turn is related in some manner to the sizes of particles present in the soil and their closeness of packing. Thus, in a general way, the force raising the water in the soil capillaries above the groundwater level is inversely proportional to the particle size in the soil.

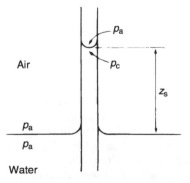

Figure 25.2 *Capillary rise of water in small bore tube.* p_a = *atmospheric pressure,* p_c = *pressure in capillary water,* $p_c < p_a$ *and as* $z_s \to 0$, $p_c \to p_a$

The weight of water hung up at the interface must tend to pull the supporting soil particles closer together, increasing the effective stress σ' and compressing the soil itself. In the coarser-grained soils the particles are already predominantly in physical contact with each other and little volume change occurs on capillary rise or retention, beyond that due to a slight rearrangement of the particles.

Capillary water that is continuous through the pores of the soil to the water table, will again show a linear variation of u with height z above the groundwater level, but with negative porewater pressures, that is, below atmospheric pressure. Thus

$$u = -\gamma_w z$$

These values of u are also illustrated in figure 25.1. The maximum height z_s of capillary rise or retention in the coarser-grained soils depends on a number of factors, but principally on the 'diameter' of the irregular pore spaces, the nature of the soil grain surfaces and the history of wetting of the soil (whether by capillary rise alone from the groundwater level or by percolation). Capillary rise and capillary retention are illustrated in the simple model in figure 25.3.

In a natural soil comprising a range of particle sizes a zone of full saturation may exist just above the groundwater level and this zone may merge into an upper zone of partial saturation. In such soils z_s is the combined height of these zones, sometimes called the *capillary fringe*. This height can only be measured approximately or estimated.

Since water to a height z_s is held in the soil pores, it will not drain out under gravity unless the groundwater level itself is lowered. An interesting phenomenon attributed to capillarity is described by Terzaghi and Peck (1948). Water pushed up by capillary action on one side of an impermeable barrier within soil may connect over the top of the barrier with pore spaces on the other side which are subject to gravity drainage, so forming a self-priming capillary siphon. Leakage over the impermeable core of the dyke on the Berlin–Stettin canal was apparently due to this phenomenon. When the core was 0.3 m above the free water surface in the canal an estimated loss of water of over 3500 m³ per day occurred along a canal length of some 19 km. When the height of the core was increased by 0.4 m the loss was reduced to less than 800 m³ per day.

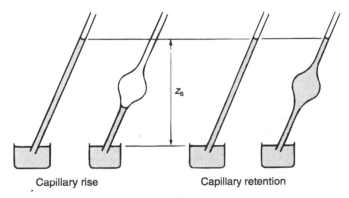

Capillary rise   Capillary retention

Figure 25.3 *Diagrammatic representation of capillary rise and capillary retention*

Contact moisture

Above the capillary zone the soil may contain moisture of a discontinuous nature in the form of traces of water at the points of contact of grains. This *contact moisture* produces surface tension forces which pull the soil grains together at points where they are bridged by a trace or droplet of water. This leads to the soil acquiring a strength which has a similar effect to that of cohesion in clays and this strength is sometimes described as *apparent cohesion*. Because of its origin, this apparent cohesion disappears completely on drying the soil or immersing it in water.

The phenomenon of 'bulking' of moist sands is well known in concrete technology and leads to gross inaccuracies in volume batching of moist sand aggregates. This is due to apparent cohesion from contact moisture enabling sands to remain stable at lower densities than would be practicable if they were dry or immersed.

Height of capillary rise

For most purposes it will be sufficient to estimate rather than attempt to measure the approximate height z_s of the capillary zones.

Taylor (1948) gives a relationship between z_s and the coefficient of permeability k (see section 25.5) of soils, and k in turn can be related approximately and empirically to D_{10}:

$$k = D_{10}^2 \times 10^4 \ \mu\text{m s}^{-1}$$

where D_{10} is in millimetres.

$$\frac{700}{\sqrt{k}} < z_s < \frac{2400}{\sqrt{k}}$$

where z_s is in millimetres. A coarse silty sand material with $D_{10} = 0.02$ mm would have a k value in the neighbourhood of 4 μm s^{-1} so that $0.35 < z_s < 1.2$ m.

It is only in the siltier types of coarse-grained soils that capillarity will have significant effects.

These soils are sometimes described as *non-swelling soils*, in that no significant volume change is observed as a result of water content changes.

25.3 Finer-grained soils – equilibrium of water contents

Clays come into the category of *swelling soils* and water content changes lead not only to changes in density of clays but also affect their strength, volume and compressibility. These changes occur mainly in the top metre or so of a swelling clay and can be important in the settlement behaviour of light shallow foundations – particularly roads and other paved areas resting on soil, dwelling houses and light commercial and industrial premises. *The water content of a clay soil* and its distribution with depth below the soil surface tend towards an equilibrium which *is essentially the result of competition for the water*: the force of gravity on the water induces drainage of the soil, the force of attraction or soil suction, if any, draws moisture into the soil, and the force of gravity on self-weight or surface loadings on the soil tends to expel water from the soil. On reaching this equilibrium the soil has experienced water content and volume changes. A further change in the environment produces another unsteady system tending with time towards a new equilibrium. For example, paving over a clay soil alters the environment by reducing or eliminating ground-surface evaporation, and if the soil is susceptible to volume changes it will swell or shrink, depending on the weight of the paving and the original seasonal condition of the soil, until a new equilibrium is reached. Other environmental changes producing similar water movements and volume changes can be visualised. The effects of these ground movements in expansive clay soils are often sufficient to cause damage to buildings and other engineering works of construction.

This competition for the porewater in a soil element may be expressed as an equation

$$s = \alpha\sigma - u$$

where s is the *soil suction*, u the porewater pressure with respect to the atmospheric pressure, σ the vertical total pressure due to overburden surface loading and α the fraction of this pressure transmitted to the porewater.

Soil suction

This is a measure of the capacity of a given clay at a given water content to retain its porewater when the soil is free from external stress. A range of equipment has been developed for measuring soil suction (Croney and Coleman, 1961). α can be measured by a simple loading test on a laboratory specimen but is mainly dependent on the plasticity characteristics of the soil.

$$I_P < 5 \qquad \alpha = 0$$
$$I_P > 40 \qquad \alpha = 1$$
$$5 \leqslant I_P \leqslant 40 \qquad \alpha = 0.027 I_P - 0.12$$

Soil suction is expressed as a pressure, namely the pressure s (below atmospheric pressure) which has to be applied to the porewater to overcome the capacity of the soil to retain the water. The pressure is often referred to a pF scale where the common logarithm of the suction expressed in multiples of 10 mm head of water is described as its pF value. Thus suction s of

$$1 \times 10 \text{ mm of water equals } \text{pF } 0$$
$$10 \times 10 \text{ mm} \qquad \text{equals } \text{pF } 1$$
$$100 \times 10 \text{ mm} \qquad \text{equals } \text{pF } 2$$

and so on. Saturated soils will show a pF value corresponding to zero porewater tension. There is no zero on a logarithmic scale but this condition would be described as pF 0. When oven-dry the soil will have a pF in the region of 7 (that is, some 10 000 atmospheres of suction). One atmosphere of suction corresponds to nearly pF 3. An example of the relation between pF and w is given in figure 25.4.

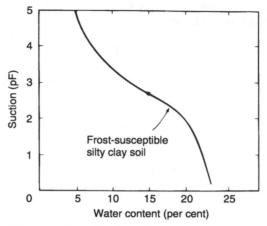

Figure 25.4 *Typical suction curve for a silty clay soil*

Strictly the relation is not a single curve and values measured differ on wetting and drying cycles. Except for salty materials, soils feel wet when below pF 2.5, moist in the range pF 2.5 to pF 4.5 and dry above pF 4.5, when they also appear lighter in colour.

Since a relationship of the form of figure 25.4 can be established between s (or pF) and w for a clay soil and in turn w can be related to soil strength and volume, it is possible by applying the above equation to explore the consequences of changes in the environment of a clay – changes in loading, in surface evaporation and transpiration, in position of groundwater table and so on. For example, in road pavement design the evaluation of the improvement in soil strength produced by lowering the groundwater table by drainage requires the application of soil suction characteristics. Prediction of shrinkage and swelling of clays which result from changes arising from construction works can be made with some success. Because of the impervious character of clays it may take some time to reach an equilibrium water content profile, indeed at shallow depths fluctuations in the climatic environment may cause continuous fluctuations in the water contents.

Plants, trees and general vegetation take part strongly in the competition for the porewater in soil, and roots exert a considerable suction. Unless the water drawn out by the vegetation is replaced, the clay will reduce in water content, with consequent shrinkage, until the wilting point of the vegetation is reached at about pF 4.18 (nearly 15 atmospheres of suction head). It follows that the removal of trees which draw off substantial quantities of porewater will produce swelling in such soils.

25.4 Frost susceptibility of soils

Another aspect of soil suction is shown when soils or other porous materials are subjected to freezing conditions. This is common in roads and paved areas resting on soil and in shallow foundations. When ice and water are present in a soil the soil suction experienced in the unfrozen water becomes dependent on the temperature alone and independent of water content. The relationship is shown in figure 25.5. Thus it is seen that high suctions may develop in the freezing zone in soils and so draw water into the freezing zone. This water accumulates in ice 'lenses' distributed through the freezing zone and the growth of these lenses may be

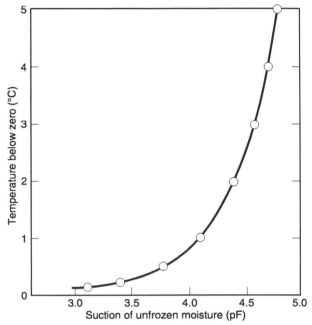

Figure 25.5 *Relationship between temperature below freezing point and suction*

(a)   (b)

Figure 25.6 *Frost heave in a silty soil: (a) before freezing; (b) after freezing*

accompanied by significant heaving pressures on pavements or foundations and by volume changes (heaving) of the supporting soil (figure 25.6).

Such suction is minimal in coarse sands and gravels, and in clays ($I_P > 15$ per cent) the pore sizes are so minute that the rate of flow is insufficient to build up ice lenses within the usual climatic period of freezing. Between these extreme particle size groups lies the silty range of soils and these are the materials broadly described as *frost-susceptible soils*. In the cases of shallow foundations mentioned, three factors must exist together if the frost is to occur:

(1) frost-susceptible porous material;
(2) supply of water (from neighbouring soil or water table); and
(3) penetration of freezing temperature into the soil.

If any one of these is absent, appreciable frost heave should not occur.

Improved drainage of foundation soils should reduce the growth of ice lenses. Very severe freezing conditions would be needed to give a frost penetration of 1 m and in Britain a value of 0.5 m is not often exceeded.

Heaving pressures

The direction of *frost heaving* of soil is usually vertical, corresponding to the direction of frost penetration and heat flow, and ice lenses form with their long axes practically horizontal. The heaving pressures are of two main types: those associated with the 'normal' expansion of water on freezing (ice having a volume about 9 per cent greater than the water from which it is formed), and those associated with the increase of water content (as ice) during freezing. Only the latter is regarded as the 'heaving' pressure in soils engineering. The process of heaving of freezing soil against a foundation or in lifting foundation piles continues to exert heaving pressures up to the maximum value for as long as the freezing conditions are maintained. These maximum heaving pressures can be considerable and substantial heaving displacements may therefore take place during a prolonged cold spell. Because particle size distributions and compositions are so variable precise values of pressure cannot be determined. Table 25.1 gives some values for illustration only. This pressure is also the pressure required just to prevent frost heaving.

TABLE 25.1
Heaving pressures exerted by freezing soils

Soil type	Heaving pressure P_i for $u = 0$ $(kN\ m^{-2})$
Coarse sands or coarser material only	0
Medium and fine sands or coarse silty sands	0 – 7.5
Medium silts or mixed soils with small amounts < 0.006 mm	7.5 – 15
Largely fine silts or silts with some clays	15 – 50
Silty clays	50 – 200
Clays	>200

From Williams (1967).

Thawing

The intake of water and growth of ice lenses produces problems when the soil thaws. A silty clay soil is broken up by the ice lenses to form a flaky structure and immediately after the thawing of the ice lenses the soil has a very high voids ratio and water content and its strength is substantially reduced. Thawing will usually occur from the top downwards so that initially the water is prevented from draining away. The overburden in time may reconsolidate the soil but in its initially weakened condition it may easily suffer damage, for example, by traffic movement over the soil.

25.5 Flow of groundwater

In practice the effects of capillary water are principally related to the vertical profile: water content and strength variations with depth below ground surface, vertical shrinkage and swelling of the top 2–3 m in clay soil.

Consider now the porewater below the groundwater surface in a continuously porous soil mass – the pore spaces are likely to be saturated or nearly so. If the porewater at a point in the soil mass possesses greater potential energy, or *total head H* (metres or other length units) above a datum level, than exists at an adjacent point in the mass, a flow will be induced towards the latter point. The flow of water in soils is accompanied by a loss of hydraulic head as energy is taken up in overcoming resistance to flow through the pores.

The total head H of water at a point in a soil is simply

$$H = \frac{u}{\gamma_w} + z$$

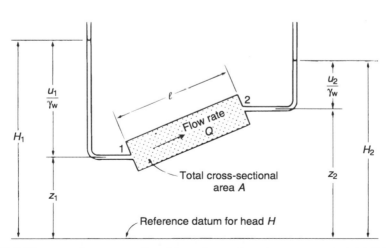

Figure 25.7 *Diagrammatic representation of flow of water through a porous material defining total head H, pressure head u/γ_w and elevation head z at two points 1 and 2. The flow rate is Q through a total area A normal to the direction of flow and the points 1 and 2 are a distance l apart in the direction of flow*

where u is the porewater pressure at the point, u/γ_w is the *pressure head* and z is now the height of the point above the reference datum level, and is the *elevation head* as in figure 25.7. This is the familiar Bernoulli equation from fluid mechanics in which the velocity head term $v^2/2g$ also appears, v being the velocity of porewater flow and g the acceleration due to gravity. In practical terms in groundwater flow, $v^2/2g$ is triflingly small numerically and is ignored.

In geotechnology the predictions of flow patterns in soils, and the resulting distribution of seepage forces and porewater pressures, form an important area of expertise. This lends itself to mathematical analysis and the study has long exercised mathematicians and physicists so that the flow patterns associated with many practical engineering problems now have a broad theoretical background. The successful matching of the theoretical to the practical continues to depend on the validity of the assumptions made in the analysis – and foremost among these is the description in quantitative terms of the drainage characteristics of the soils.

Permeability of soils

The first scientific investigation into the movement of water in sands is attributed to the famous railway engineer Robert Stephenson. During construction of the Kilsby tunnel, near Rugby, completed in 1838, extreme difficulties were encountered in waterlogged coarse sands. Stephenson had shafts sunk along the line of the tunnel and pumping from these took place. He took careful measurements of water levels and, among other observations, noted that the resistance which the water encountered in its passage through the sand was related to the angle of inclination of the groundwater surface down towards the pumped shafts. In 1856, Darcy reported on experimental studies of flow in saturated sands and noted that the flow rate (m^3 s^{-1}) per unit of total cross-sectional area of soil, at right angles to the direction of flow, was proportional to the rate of loss of total head H with distance l along the average flow path of the water. This is known as *Darcy's law* and may be expressed

$$\frac{Q}{A} = k\,\frac{(H_1 - H_2)}{l}\ (\text{m s}^{-1})$$

where Q is the rate of flow of water through a total area A, the conditions being illustrated in figure 25.7. Any elevation may be taken for the reference datum since the actual value of z, the elevation head, has little meaning: it is the difference in elevation head that is of interest. The quantity Q/A has units of velocity (m s^{-1}) and is sometimes referred to as the *superficial velocity* or specific discharge v. The ratio $(H_1 - H_2)/l$ is dimensionless and gives the rate of energy loss in flow through the soil and is called the *hydraulic gradient i*. Thus the above equation is commonly written

$$v = ki$$

Since flow only occurs through the voids spaces in the soil, v will not be a measure of the actual velocity of flow of water particles. The average water particle velocity can be estimated from $v_s = Q/nA = v/n$ where n is the porosity of the soil. It is usual, however, to work in terms of v. The coefficient k, termed the *coefficient of permeability*, is considered to be a soil property. Strictly it is influenced by variations in the porewater or pore fluid since $k = K\gamma_w/\mu$ where K is the permeability (unit, m^2 or equivalent) and is the soil property; μ is the coefficient of viscosity of water. For fresh ground water at ordinary temperature ranges, γ_w and μ are practically constant and the use of k as the soil property is acceptable and usual in practice.

Of much more significance and importance is the appreciation that the value of k appropriate to the analysis of groundwater flow problems is that representative of the soil mass as it exists in place. A review has already been given of the complex nature of soils, their stratified, fissured and otherwise anisotropic character, and the influence of their state of stress on their engineering behaviour. This has pointed to the importance of careful and perceptive sampling and of testing soils in place where practicable. This property of permeability is strongly influenced by the effects of anisotropy and the state of stress.

The validity of Darcy's law v=ki

When water flows through a pipe or in an open channel the relationship between the average flow velocity and the hydraulic gradient is often expressed as $v \propto i^n$. Students of fluid mechanics will know that such flows are well described by putting $n = \frac{1}{2}$ and are classed as *turbulent* flows. When water flows through the pore spaces in sands, however, the flow is usually classed as *laminar* for which $n = 1$. Further, in pipe and channel flows the velocities at the walls or boundaries are measurably less than at the centre of the flow, whereas in an ideal porous material of sandy texture which is subject to uniform hydraulic gradients, the average velocities of flow at the boundaries of the material are no different from those at the centre. It is seen then that there are two kinds of flow, turbulent flow and laminar flow. Turbulent flow is usual in pipe and open channel situations in civil engineering practice; laminar flow is usual in groundwater situations. In groundwater flow the flow conditions in the coarser soils may occupy a transition stage between laminar and turbulent flow when Darcy's law will not apply. A clear separation between laminar and turbulent conditions cannot be defined and a conservative limit for laminar flow is usually adopted. Taylor (1948) proposed that, under a hydraulic gradient of 100 per cent, uniform sands with an effective size D_{10} of 0.5 mm or less always have laminar flow. This means that in soils coarser than coarse sands or fine gravels the validity of Darcy's law should be considered questionable. It would be exceptional and in any case undesirable to place a water-retaining hydraulic structure on such pervious soil.

Many investigations following those of Darcy have given results substantially in accord with the equation $v = ki$ for sands. In soils containing fines there is some evidence that particle migration within the soil may cause departures from Darcy's law. Departures have also been observed when sands have been tested at very high hydraulic gradients. However, most problems of drainage of sands, flow of water beneath foundations or floors of hydraulic structures, where the hydraulic gradient is usually less than 100 per cent, fall well within the range of validity of Darcy's law.

Factors influencing the coefficient of permeability k

Assuming laminar flow, the relation $v = ki$ applies to groundwater flow in a given direction in a soil. We have already noted that k is a parameter which is a function of the pore fluid as well as of the soil. When the pore fluid is water and its temperature range not large, k is considered to describe fully the permeability of the soil in a given direction of flow.

Other factors influence k, principally

(1) particle size, shape and grading;
(2) arrangement of particles, and stratification and fissures;
(3) density of packing;
(4) composition;
(5) consolidation pressure;
(6) migration of particles; and
(7) presence of air or gas.

Particle size, shape and grading

These characteristics of the soil and, to a lesser extent, the density of packing of the particles, largely determine the size, shape and continuity of the pore spaces in the soil. Since k is a measure of the ease with which water can flow through a soil, these characteristics largely determine the value of k. It is reasonable that in a graded soil the coefficient of permeability should be mainly influenced by the finer sizes present. From experimental work on filter sands of medium permeability, Hazen showed that k could be roughly estimated from

$$k = D_{10}^2 \times 10^4 \ \mu m \ s^{-1}$$

where D_{10} is the effective size of the sand in millimetres. This expression contains no parameter describing either density of packing or particle size distribution, but for estimation purposes gives good results.

The coefficient of permeability is a parameter set apart from most other soil parameters in engineering usage in that it has a great range of numerical values.

Table 25.2 lists the terms describing degrees of permeability and tables 25.3 and 27.1 give values of k typical of certain soils. Hazen's relation of k and D_{10} appears to fit quite well.

TABLE 25.2
Classification of degrees of permeability

Degree	$k \ (\mu m \ s^{-1})$
High	Over 1000
Medium	10–1000
Low	0.1–10
Very low	0.001–0.1
Practically impermeable	below 0.001

TABLE 25.3
Typical values of coefficient of permeability

Soil type	Particle size range		Effective size	Measured coefficient
	D_{max}	D_{min}	D_{10} (mm)	of permeability
	(mm)			k ($\mu m\ s^{-1}$)
Uniform, coarse sand	2	0.5	0.6	0.4×10^4
Uniform, medium sand	0.5	0.25	0.3	0.1×10^4
Clean, well-graded sand and gravel	10	0.05	0.1	100
Uniform, fine sand	0.25	0.05	0.06	40
Well-graded, silty sand and gravel	5	0.01	0.02	4
Silty sand	2	0.005	0.01	1
Uniform silt	0.05	0.005	0.006	0.5
Sandy clay	1.0	0.001	0.002	0.05
Silty clay	0.05	0.001	0.0015	0.01
Clay (30 to 50 per cent of clay sizes)	0.05	0.0005	0.0008	0.001
Colloidal clay (over 50 per cent finer than 2 μm)	0.01	10 Å*	40 Å	10^{-5}

*1 Å = 10^{-10} metre.

Arrangement of soil particles

The processes of soil formation have been commented on several times in this section, the resulting soils often showing complex and erratic variations. Alluvial soils are likely to show less resistance to horizontal flow of water along the strata than to vertical flow, particularly where the particles are flake-shaped and tend to lie with their largest area nearly horizontal. Particle arrangements may even be such that coefficients measured in two horizontal directions at right angles to each other may differ, particularly in sloping strata.

A particle arrangement of importance to the engineer is that of *open-work gravel*. This term is used for the more uniform gravels, and layers of this type of highly pervious material may occur among deposits of less pervious sand–gravel mixtures. The presence of seams of such gravels, even of small thickness, enormously increases seepage flows and if undetected may lead to serious and unexpected water losses and water pressures in hydraulic construction works. Detection is difficult and trial pits, borings, pumped wells and observations during construction may all contribute information.

Problems of a similar kind may occur in *fissured clays* and *jointed rocks*, the fissures and joints often rendering the material considerably more pervious than the intact material between the fissures. It is obvious therefore that tests for permeability on the intact material would be misleading and that the importance of representative sampling and testing in place should again be recognised.

Density of packing of soil

As the pore volume in a given soil decreases as a result of densification or compaction the coefficient k also decreases. A number of empirical or semi-empirical expressions have been proposed for expressing the relation between permeability and porosity, or voids ratio. For example Terzaghi and Peck (1948) quote a formula of Casagrande

$$k = 1.4k_{0.85}\, e^2$$

where k is the coefficient of permeability at voids ratio e in a clean sand whose permeability is $k_{0.85}$ at voids ratio of 0.85. From this it would appear that $k \propto e^2$. There are considerable experimental data however to show that a plot of e against log k for a soil is a straight line for nearly all soils.

Composition of soil

Groundwater flow studies are mainly applied to the size range silts to small gravels and in this range the mineral composition of the soil has little significance as regards permeability. In clays, however, something has been seen of the importance of mineral composition on a range of properties – swelling, consolidation and strength – and permeability is also dependent on this composition. Montmorillonite clays exist at high voids ratios yet their permeability is very low. A sodium montmorillonite has a permeability less than 10^{-3} μm s^{-1} at a voids ratio of about 15. Kaolinite clays show k-values of about 10^{-1} to 10^{-2} μm s^{-1} at lower voids ratios of less than 2. Much depends too on micro-structure and it is observed that for a given voids ratio the higher permeabilities are associated with the more flocculated clays having the larger flow passages and the lower values with the more dispersed clays.

When clayey soils are used as fill materials the water content at the time of mechanical compaction appears to influence the micro-structure formed and hence the permeability – material compacted on the wet side of optimum (see Chapter 28) being significantly less pervious than material compacted on the dry side.

Consolidation pressure

It was stated earlier that it is desirable for the testing of soils to be carried out on specimens subjected to the same state of stress that exists in the soil in place. This requirement aimed to prevent disturbance of the soil. Stress changes may also occur *in situ* during or as a consequence of construction works. Increases in effective stress lead to volume decreases in soil and hence to voids ratio reductions and reductions in k. These effects are not usually of much significance in coarser-grained soils but in plastic clays, particularly if organic or if fissured, the influence of consolidation on the coefficient of permeability is quite marked owing to a reduction in voids ratio and a closing or tightening of fissures.

Migration of fine particles

The drag or seepage force exerted on the soil particles by the flowing water may be sufficient to transport some of the fines through the void spaces in the coarser fraction. In most soils the particle size distribution will be such that this does not happen but in coarse uniform sands containing fines such migration is possible.

Once the fines have been removed the remaining material is much more permeable. The phenomenon is known as internal erosion and may occur locally within a soil causing a 'pipe' of high permeability to develop. If this is allowed to grow in or beneath a hydraulic structure, the high flows may lead to undermining of the structure and eventual failure. This migration can be controlled to a large extent by preventing the loss of fines at the outlet boundary of the soil by means of filters (see section 25.6). Foreign matter in the flowing water, from solution, suspended solids and organic matter may also be present to clog the pore spaces and so reduce the permeability.

Air or gas in the soil

If the sand is not fully saturated, substantial reductions occur in the coefficient of permeability, a drop from 100 per cent to 85 per cent in the degree of saturation being accompanied by a drop of around 50 per cent in the value of k. The permeability will also change with changes in the pore pressure.

Determination of the coefficient of permeability k

In relatively uniform coarse-grained soil formations satisfactory estimate of k may be made based either on the particle size, shape and size distribution of the soil or on the results of laboratory permeability tests. Because of the variable and anisotropic nature of soils and because of sampling disturbance in obtaining core samples it is preferable to determine average values for k using tests carried out on the material *in situ*.

Indirect methods

The dependence of k on the particle size of a soil has been discussed and the simple and approximate expression

$$k = D_{10}^2 \times 10^4 \ \mu m \ s^{-1}$$

presented, in which D_{10} is the effective size of the soil in millimetres. The expression was originally limited in application to clean fairly uniform sands having uniformity coefficients C_u less than 5, but later work indicates that it is useful over a much wider particle size range and for non-uniform, well-graded sands.

The Kozeny–Carman expression (Loudon, 1952)

$$k = \frac{g}{k'S^2\mu} \ \frac{e^3}{(1 + e)}$$

is a development which takes account of the particle size distribution and the density of packing of the soil particles, although not the orientation and arrangement of the particles, and gives values of k to better accuracy than the previous equation. k' is a factor which accounts for particle shape and hence pore shape and varies somewhat with porosity. S is the specific surface (particle surface area per unit *volume*) of the sand, μ the viscosity of water, e the voids ratio of soil and g the acceleration due to gravity.

Indirect estimates of k in the finer-grained soils are also deduced from information obtained in the oedometer or consolidation test in which the porewater is induced to flow out of the parent soil by squeezing or loading (Chapter 26).

Direct methods in the laboratory

The more pervious soils can be directly tested for permeability using either the constant head or the falling head methods.

The *constant head permeability test* (figure 25.8(a)) is suitable for soils having k in the range 10^{-2} to 10^{-5} m s^{-1} and a test procedure is detailed in BS 1377: Part 5. In this test, a suitably prepared representative specimen of the soil to be tested is placed in the cell of the permeameter. After saturating the specimen with de-aired water (the permeant) a difference H in total head is applied between the ends of the specimen or between two points l apart in the direction of flow and a steady state of flow is given time to develop.

The volume Q of water then passing through the specimen of cross-sectional area A in time t is measured.

Hence from Darcy's law a value of k can be calculated since

$$v = ki$$

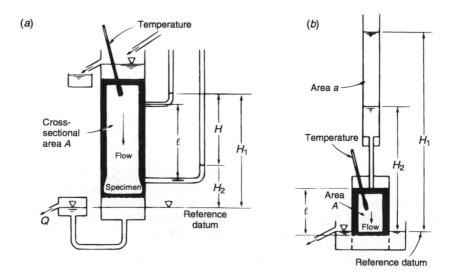

Figure 25.8 *Permeameters: (a) constant head and (b) falling head*

$$\frac{Q}{At} = k \times \frac{H}{l}$$

or

$$k = \frac{Ql}{HAt}$$

Tests run at several hydraulic gradients enable confirmation of laminar flow by demonstrating that $v \propto i$.

The *falling head permeability test* arrangement is shown in figure 25.8(b). Again the soil specimen is placed in the permeameter and saturated. A standpipe is connected to the specimen and filled with water. The water is allowed to drain from the standpipe through the soil specimen and the rate of fall of water level in the standpipe is measured. Usually several diameters of standpipe are available from which one is selected as giving a suitable rate of fall for measurement purposes with the particular soil under test. From continuity of flow in the standpipe and in the specimen

$$dQ = -a\ dH = k\frac{H}{l}A\ dt$$

with symbols as defined on figure 25.8(b), from which

$$k = \frac{al}{At}\log_e\frac{H_1}{H_2}$$

or

$$k = \frac{2.3\ al}{At}\log_{10}\frac{H_1}{H_2}$$

The preparation of specimens for these tests requires careful work to ensure homogeneity of the material. A thin segregated layer of finer grained material within or at the boundary of a specimen can have a large effect on head loss and hence on the computed k-value. Core samples of cohesionless soil are quite difficult to obtain and it is likely in any case that they are in a disturbed condition.

The values of k are usually adjusted to a standard temperature of 20°C, using

$$k = \frac{\mu_t}{\mu_{st}}k_t$$

where μ_t and μ_{st} are the viscosities of water at the test temperature and the standard temperature respectively, the adjusted values being used for comparison of k-values among soils. If specimens are tested for permeability at several densities or voids ratios, the value of k may also be adjusted to correspond to the density of the soil in the field.

For soils of low and intermediate permeability, BS 1377: Part 6 defines test procedures for the measurement of k in association with *consolidation tests* in a hydraulic cell and in a triaxial cell (Chapter 26).

Direct methods in the field

Because of the shortcomings of the sampling methods and methods of specimen preparation, none of the above tests can give results reliably representative of the *in situ* soil. The hydraulic gradient in the laboratory test is usually much greater than that typically experienced in actual hydraulic structures and hence untypical internal erosion may occur in the laboratory test specimen.

Where the significance of the property justifies the cost of *in situ* measurements, average values and local values of the coefficient of permeability can be derived from measurements of changes in water levels in boreholes and observation wells as water is drawn from or injected into the soil formation. Several procedures of measurement and analysis are in use (see, for example, BS 5930).

Coefficient of permeability of a layered soil

In an aquifer of total thickness H composed of n distinct horizontal layers of thickness H_1, H_2, ..., H_n having isotropic coefficients of permeability k_1, k_2, ..., k_n, the equivalent coefficient of horizontal permeability of the whole aquifer is

$$k_h = \frac{1}{H} \int_0^H k \, dH$$

$$= \frac{1}{H} (k_1 H_1 + k_2 H_2 + ... + k_n H_n)$$

since for horizontal flow the hydraulic gradient of the flow is the same along each layer.

The coefficient of vertical permeability of the aquifer is

$$k_v = \frac{H}{\int_0^H dH/k}$$

$$= H \text{ recip } (H_1/k_1 + H_2/k_2 + ... + H_n/k_n)$$

since, for a steady state with no volume change in the soil, vertical flow continuity gives the same flow rate across each layer. The permeability ratio k_h/k_v exceeds unity and may reach high values of a hundred and more.

25.6 Soils as drainage filters

When groundwater emerges from a soil through an exposed soil face or passes from a soil into a drainage system, the groundwater is likely to carry with it some of the fines from the soil. As noted, this is called *internal erosion* or *piping* and if it is allowed to continue over a period of time it may lead to instability in the soil, because the loss of fines produces zones or 'pipes' of increased permeability and flow rate extending in upstream directions. The hydraulic gradient is increased in the approaches to these permeable zones and so the tendency to erosion is aggravated. Flow from an open outlet face of soil, for example, at a side slope of an earth dam, may cause *surface erosion*, also leading to instability by sloughing of the surface of the slope. These situations are commonplace in engineering and for safe and durable earthworks and for their efficient drainage the erosion must be largely prevented.

By placing a layer or layers of selected natural soil at the outflow surface, protection from erosion may be obtained. These selected soils are called *filters* and their use is important in the control and proper functioning of drainage works, wells and indeed in any case where seepage water emerges from soil. Some examples of their use are given in figure 25.9. A suitable filter material will have pore spaces too small to admit the larger particles of the soil it is protecting and these particles will then partially block the pores of the filter, so reducing the size of the passages into the filter. This prevents the entry of successively finer particles from the soil and soon a zone is developed at the soil–filter boundary which largely prevents the erosion of the soil. The size fractions most likely to migrate through and from soils are the silts and fine sands. These are coarse enough to allow fairly high seepage velocities and yet fine enough to be readily transported. Coarser soils are usually too large to be transported by flow through void spaces and the fine silts and clays do not permit sufficiently high seepage velocities to cause appreciable internal erosion.

The *selection of a filter* is therefore based on ensuring that the pore spaces in the filter are small enough to prevent the migration of the coarser sizes of the soil being protected. For efficient conveyance of the water emerging from the soil the filter should nevertheless be as permeable as possible. These two requirements are in conflict and a compromise is necessary. A number of criteria exist to guide in the selection of filter soil gradings. The U.S. Waterways Experiment Station criteria are much used. These require that a filter should satisfy the following:
(1) Piping requirement

$$\frac{D_{15} \text{ filter}}{D_{15} \text{ soil}} \leq 20$$

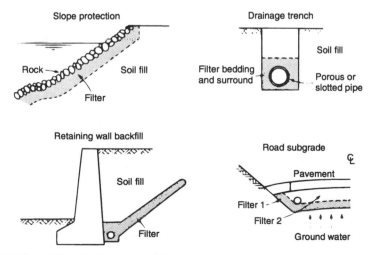

Figure 25.9 *Use of filters in the control of drainage*

$$\frac{D_{15} \text{ filter}}{D_{85} \text{ soil}} \leqslant 5$$

$$\frac{D_{50} \text{ filter}}{D_{50} \text{ soil}} \leqslant 25$$

except that for uniform soils ($C_u \leqslant 1.5$)

$$\frac{D_{15} \text{ filter}}{D_{85} \text{ soil}} \leqslant 6$$

and for well-graded soils ($C_u > 4$)

$$\frac{D_{15} \text{ filter}}{D_{15} \text{ soil}} \leqslant 40$$

(2) Permeability requirement

$$\frac{D_{15} \text{ filter}}{D_{15} \text{ soil}} \geqslant 5$$

The filter material should not be gap graded. It is sometimes recommended that the particle size distribution curves of filter and soil should be approximately 'parallel'. An example of the range of filter soils for the protection of a base soil is given in figure 25.10.

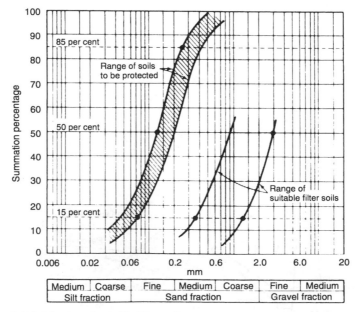

Figure 25.10 *Selection of suitable filter soils*

Often a single filter is sufficient but two or more layers may be used to form a *graded filter* with greatly increased effectiveness in drainage. Each filter layer should meet the above criteria with respect to the next layer upstream. It has also been seen that seepage exerts a force on soil. If that force is upwards, countering the weight of the soil, instability may occur. Filters are sometimes used to improve safety in such cases by acting as a surcharge on soil. Since a filter is by design relatively highly permeable the seepage forces in it will be low and its weight will be effective as a surcharge. Such filters are called *loaded filters*.

Where filter flow is collected in perforated or porous pipes for conveyance, the openings in the pipes must also be selected to prevent inflow of fines and clogging. The following criterion has been found satisfactory

$$\text{circular hole diameter} \;\leqslant\; D_{85} \text{ filter}$$

$$\text{slot width} \;\leqslant\; 0.83\, D_{85} \text{ filter}$$

There is a fast growing use of man-made fabrics, *geotextiles*, developed to meet a range of needs in earthworks construction. In particular, these durable plastic fabrics can be specified to satisfy the filter requirements (piping and permeability) defined in this section, and so reduce or eliminate the need for scarce natural soil aggregate filters. Because of the range of geotextile products available, selection requires reference to manufacturers' literature and performance testing for compliance.

26

Compressibility of Soil

Buildings and engineering works resting on soil will settle to some extent owing to their weight causing a compression or deformation or both of the supporting soil. There is a limiting value to the support a soil can offer to the weight of a building and the engineer ensures that at working load, when the building is fully loaded, an adequate factor of safety exists. That is

$$\text{working load} = \frac{1}{F} \times \text{ failure or limiting load}$$

where F will usually have a value of about 2.5 for ultimate failure.

Settlements will occur at the working load and lesser loads and the engineer makes predictions of the amounts and rates of these settlements based on measurements or estimates of the compressible quality of the soil. To do this he requires a stress–strain relationship for soil. In principle this is extremely complex and has not yet been fully developed for soil. For practical usage therefore some simplifications are made to allow predictions of vertical settlement. One simplification is to make use of the theory of elasticity for isotropic materials. In the application of the theory of elasticity it is not necessary that a material should show elastic rebound characteristics but simply that over a range of a stress increase (or decrease) the stress is proportional to strain.

The apparatus generally used in the laboratory for the study of the stress–strain behaviour of soils is the *triaxial compression apparatus*. The features of this are shown in figure 26.1 and these will receive further reference in Chapter 27. The test specimen of soil is cylindrical in form and of diameter usually 100 mm or more. The length is commonly twice the diameter, but if lubricated end platens are used, as is preferable, the length may equal the diameter. The specimen is enclosed in a rubber membrane and the whole placed in a cell which is filled with a liquid (water, usually) and the latter put under pressure σ_3. In this condition the specimen is subjected to a *hydrostatic state of stress*, σ_3 all round. An additional vertical stress ($\sigma_1 - \sigma_3$), termed the *deviator stress,* is applied axially on to the end faces of the cylindrical specimen. The stress system is then one of biaxial symmetry, as in figure 26.2.

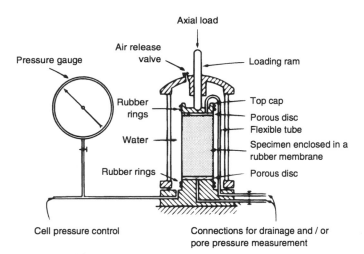

Figure 26.1 *Arrangement of the triaxial compression cell, showing the environment of the test specimen*

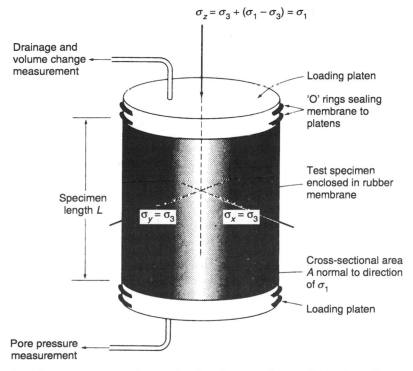

Figure 26.2 *Stress system on specimen in the triaxial compression test ('hidden' reaction vectors σ_x, σ_y and σ_z omitted)*

26.1 Compression of sands

Triaxial compression

In figure 26.3 the form of the stress–strain relationship for a dry or freely draining soil tested in triaxial compression is shown and, at low strains up to the maximum working load or stress, it is seen that there may be an acceptable linearity to the relationship of deviator stress ($\sigma_1 - \sigma_3$) and axial strain ε_1. The slope of this initial part defines *Young's modulus of elasticity* or *the elastic modulus E*, a constant for an ideal isotropic elastic material. In soils the linearity is likely to be imperfect and a tangent or a secant modulus is determined for E as shown in figure 26.3.

Predictions of settlements, assuming elastic theory and isotropic material, are based on equations of the form

$$S_i = \Delta q\, B \left(\frac{1 - \nu^2}{E} \right) I_S$$

where S_i is now the vertical settlement of a loaded area on the horizontal surface of the material, Δq the increment of vertical stress causing the settlement, ν Poisson's ratio, the other elastic constant of an elastic material, B the least dimension in plan of the loaded area and I_S an influence coefficient containing all the geometric proportions of the case under study; taken together with B, I_S defines the actual size, shape and stiffness of the loaded area.

In most practical cases, settlements in sands and gravels are relatively small with a high proportion occurring during construction (loading). It has already been emphasised that these soils are inherently variable both laterally and vertically as well as from site to site. Experience and a limited amount of field data on settlements, however, indicate that with careful and cautious selection of the parameters, good predictions are possible.

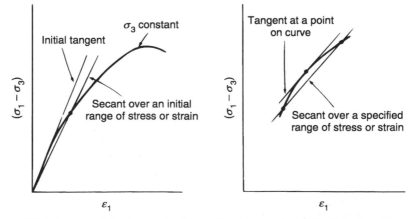

Figure 26.3 *Estimation of elastic modulus E from the stress–strain relationship of a soil*

Confined or constrained compression

If the cylindrical test specimen is constrained such that $\varepsilon_x = \varepsilon_y = 0$ the ratio of axial stress to axial strain is now called the *constrained modulus* E^*. These conditions are met in the oedometer test described in section 26.3.

The use of a Young's modulus E or a constrained modulus E^* in predicting settlements in an isotropic material would be quite straightforward. The value of the modulus, however, depends on sampling disturbance and on many other factors, not all clearly understood. The soil itself is unlikely to be isotropic and, even if homogeneous in appearance, will have a history of stressing which exerts an influence on its compressibility. In sands, provided crushing of the grains does not occur, the modulus E or E^* tends to increase with increasing relative density I_D; with increasing applied and confining pressures and hence with increasing depth; and with successive cycles of loading (the increase per cycle being large on the first cycle and diminishing on successive cycles until a fixed stress–strain curve is reached). Grain shape, grading and composition all have an influence on the modulus. The major difficulty is in obtaining undisturbed specimens for test.

The stress–strain data from confined compression tests on soils are also commonly presented graphically as the voids ratio e plotted against the corresponding effective vertical stress σ_z' ($= \sigma_z$ when $u = 0$) or against $\log_{10} \sigma_z'$. These plots for a sand and a clay are shown in figure 26.4, (a) and (b). In dry or freely draining sands the voids ratio change which accompanies the application of a stress change occurs almost without delay so that for these soils the data in figure 26.4 are not time dependent. The slope of the e–σ_z' curve in figure 26.4(a) at any stress σ_z' or stress range $\Delta\sigma_z'$ is described as the *coefficient of compressibility* a_v.

$$a_v = -\frac{de}{d\sigma_z'}$$

or

$$a_v = -\frac{\Delta e}{\Delta\sigma_z'}$$

and for that stress or stress range a_v is a parameter defining a soil property (compressibility). In figure 26.4 it should be noted that the unloading curve does not coincide with the preceding limb of the loading curve and the reloading curve does

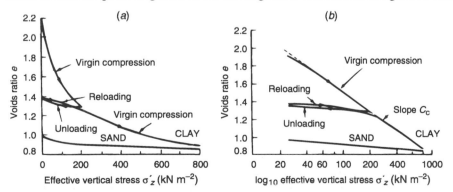

Figure 26.4 *Confined compression test data plotted as (a) voids ratio e against effective vertical stress σ_z' and (b) voids ratio e against $\log_{10} \sigma_z'$*

not coincide with either of these until at higher stresses it follows the extension of the preceding loading curve.

Another parameter in common use is the *coefficient of volume compressibility* m_v where

$$m_v = \frac{a_v}{1 + e_0} = \frac{1}{E^*}$$

where e_0 is the initial voids ratio and E^* the constrained modulus of elasticity. The plot in the form of figure 26.4(b) enables the soil compressibility to be displayed over a wide range of stresses. Usually this plot becomes linear at the higher stresses and the slope of the curve, usually its linear part, is described as the *compression index* C_c for the soil:

$$C_c = - \frac{\Delta e}{\Delta(\log \sigma_z')} = - \Delta e_c$$

where Δe_c is the change of voids ratio on the linear portion of the plot, in one logarithmic cycle of stress. The semi-logarithmic plot of figure 26.4(b) is particularly useful in presenting the compressibility of clay soils. The slope of the unloading (swelling) limb of the plot of figure 26.4(b) is a *swelling index* C_s.

26.2 Compression of clays – short term

In clays the problem of estimating settlement is especially difficult and the amount can be quite large. The amount of ultimate settlement is usually considered as comprising two principal parts – an *immediate* or *undrained settlement* occurring as the loading is applied to the saturated clay, and a long-term settlement called *consolidation settlement* occurring under sustained loading.

Immediate settlement

The first part, immediate settlement, is the result of a change of shape of the clay under loading, without any significant change in volume. This implies that Poisson's ratio ν_u is 0.5 for the undrained conditions. The modulus E_u of a clay for undrained conditions is very difficult to determine with any conviction as the value is affected by many factors associated with stress history, sampling disturbance and test procedure. Predictions of immediate settlement are based on the equation for S_i given in section 26.1.

If a natural stratum of saturated clay is relatively thin and is subjected to a stress increment over a large area in plan, that stratum is fully constrained and practically incompressible on initial loading. In this case the constrained modulus E^* is infinite and immediate settlements negligible. Where this constraint does not exist, that is, in a thicker stratum or one stressed over a small area, estimates of immediate settlement based on E_u and ν_u may be required. For a saturated clay ν_u is taken as 0.5 and E_u may be estimated from the results of laboratory triaxial compression tests on specimens from carefully extracted block samples or large diameter piston samples. Until a generally agreed test procedure is available it would seem reasonable first to restore the *in situ* stress system on each test specimen as closely as possible and to allow the specimen to stabilise (consolidate). The specimen is then subjected to loading up to the working stress and then unloading. This is repeated for as many as five cycles of loading and unloading and what should be a reasonable estimate of E_u is then estimated from σ_z/ε_z on the last loading. Use of a value of E_u from the first loading appears to give excessive values of immediate settlement and the effect of applying one or two cycles of loading may be to offset sampling disturbance.

In situ *tests for immediate and short-term compression*

The difficulties of obtaining reliable elastic parameters in the laboratory to define soil behaviour together with the non-linear and anisotropic behaviour of real soils, have led to the use of *in situ* tests to determine these elastic parameters or to classify soils sufficiently closely to allow predictions of immediate settlement to be made. These *in situ* tests include plate loading penetration tests and pressuremeter tests (BS 5930).

26.3 Compression of clays – long term

The compression producing *consolidation settlement* in a clay is due to a gradual expulsion of porewater giving a reduction in the volume of the voids accompanied by a compression of the assembly of solid particles forming the clay. The water and the solid matter from which the particles are formed are themselves relatively incompressible whereas the assembly of solid particles forming the clay is relatively compressible.

For example, Costet and Sanglerat (1969) list the changes in volume under a pressure of 100 kN m^{-2} as follows

water	1 : 22 000
solids comprising the particles	1 : 100 000
assembly of clay particles	1 : 100

Any air or gas in the voids spaces would be highly compressible.

When a sustained increment of loading, a stress increment, is applied to a saturated clay this increment is initially supported by the porewater – which is relatively incompressible, as just discussed. The porewater in the stressed volume of the clay is thus at a greater pressure than the porewater in the surrounding clay and a hydraulic gradient now exists. If drainage is free to take place, porewater flows from the stressed zone with a reduction in volume occurring in the voids of the clay in that zone, and the stress increment is gradually transferred to the particle assembly as an increment of effective stress. Since clays are relatively impervious this flow and stress transfer take some time. This *primary consolidation* process is

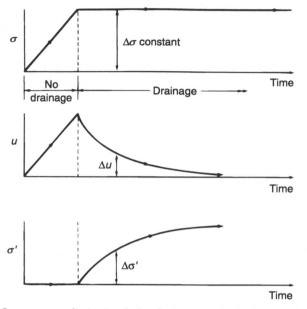

Figure 26.5 *Pore pressure dissipation during drainage, producing increase in effective stress (and also a decrease in volume)*

illustrated in its simplest form in figure 26.5 where at any time t, $\Delta\sigma = \Delta\sigma' + \Delta u$. At $t = 0$, the start of the drainage period, $\Delta\sigma = \Delta u$ and as $t \to \infty$, $\Delta\sigma \to \Delta\sigma'$. Predictions of the rate of settlement require a theoretical model of the process which, through appropriate parameters, applies to any given clay.

In addition to settlements arising from this drainage and pore pressure dissipation, the more organic clays may show a delayed or *secondary consolidation*, not considered further herein.

One-dimensional consolidation

The laboratory apparatus used for studying consolidation behaviour in constrained compression is the *oedometer* and two forms of this are shown diagrammatically in figure 26.6. The detailed methods of test are given in BS 1377: Parts 5 and 6. A specimen of the saturated clay to be tested is accurately trimmed to cylindrical form and confined in a metal ring or cell, as illustrated, to prevent strains taking place laterally when a vertical external stress $\Delta\sigma$ is applied. This stress is applied to and sustained on the specimen through rigid porous discs which permit vertical drainage of porewater from the clay. The vertical compression of the clay is measured at suitable intervals of time until the compression is completed. This process is described as *one-dimensional consolidation* and is often explained by reference to the *spring analogy* (figure 26.7).

The spring represents the compressible assembly of particles forming the soil, sometimes called the soil skeleton; the water in the cylinder represents the porewater in the soil; the stopcock on the piston represents the permeability of the soil.

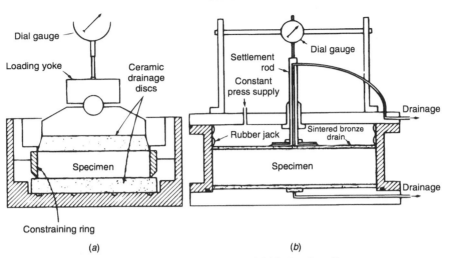

Figure 26.6 *Oedometers: (a) conventional fixed ring and (b) hydraulic cell types*

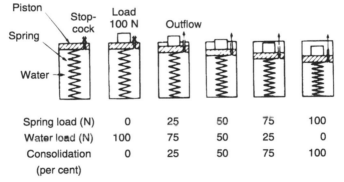

Spring load (N)	0	25	50	75	100
Water load (N)	100	75	50	25	0
Consolidation (per cent)	0	25	50	75	100

Figure 26.7 *Piston-and-spring analogy*

Thus if no drainage is permitted as a stress increment $\Delta\sigma$ is applied, this stress increment is carried by the porewater, as in the first stage of figure 26.5. Because the water is incompressible none of the stress $\Delta\sigma$ is carried by the spring at this stage and an equal stress increment Δu is experienced in the porewater, that is, $\Delta u/\Delta\sigma = 1$. Opening the stopcock permits drainage to occur and as the pore pressure dissipates the external stress is transferred to the spring as in the second stage of figure 26.5, the rate depending on the stopcock opening. The rate diminishes asymptotically with time since with dissipation of pore pressure the hydraulic gradient sustaining flow also diminishes.

In the one-dimensional oedometer test there is a similar time-dependent transfer of stress to the soil skeleton. It has been noted that in saturated soils

$$\Delta\sigma' = \Delta\sigma - \Delta u$$

and that consolidation depends on $\Delta\sigma'$. This serves as a reminder that consolidation (or indeed, swelling) may follow a change in σ or in u, or in both.

Amount of settlement

In one-dimensional consolidation there will be no volume change at the instant of applying $\Delta\sigma$. That is, E^* (or E_u) is practically infinite. It was also noted in section 26.1 that if consolidation of a particulate assembly is allowed time to complete itself under free drainage of porewater a voids ratio e is eventually reached for any applied stress σ (when $\sigma' = \sigma - u_0$, where u_0 is the porewater pressure before the applied stress was changed to σ'; $u_0 = 0$ for the present illustration). The relationship of e and σ' was expressed through the soil parameters a_v, m_v, C_c and C_s and is determinable from the results of the oedometer test. A knowledge of these parameters enables the engineer to answer questions about how much settlement is likely to occur.

Table 26.1 lists approximate values of m_v and C_c for a range of soils. The values should be directly measured and the purpose of the tabulated values is to alert the engineer to grossly untypical results.

It is seen from figure 26.4 that the unloading and reloading limbs of the curve differ from the first loading. This feature is significant to the engineer in that a given clay, in geological history, may have experienced unloading and exists today in a state represented by point a in figure 26.8. An increase of loading would need to exceed that represented by point b before settlements in the clay become appreciable. The straight-line portion of the $e-\log\sigma'$ plot for a clay is therefore called the *virgin compression line* and the other limbs are *unloading (swelling)* and *reloading curves*.

Terzaghi and Peck (1948) have assembled data on the slope C_c of the virgin compression line in relation to the liquid limit of clays to show that

$$C_c = 0.009 \, (w_L - 10) \text{ for undisturbed soil}$$

TABLE 26.1
Compressibility of clays

Soil description	Compressibility	m_v $(MN\ m^{-2})^{-1}$	C_c
Heavily over-consolidated clays	Very low	<0.05	<0.10
Very stiff to hard clays	Low	0.05 – 0.10	0.10 – 0.25
Medium clays	Medium	0.10 – 0.30	0.25 – 0.80
Normally consolidated clays	High	0.30 – 1.5	0.80 – 2.50
Very organic clays and peats	Very high	>1.5	>2.50

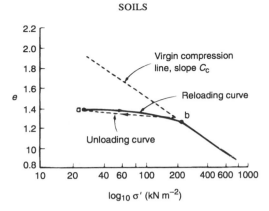

Figure 26.8 *Consolidation settlement on reloading a clay*

and

$$C_c = 0.007 \, (w_L - 10) \text{ for remoulded soil}$$

where w_L is the liquid limit in per cent. This usefully supplements values of C_c measured in the oedometer test. Because of its importance in the prediction of settlements, the virgin compression line has received considerable study. Two principal items arise: first, that the slope C_c is appreciably affected by sampling disturbance and, to be representative of field behaviour of the clay, some adjustment of the laboratory data is required; second, since the settlements on reloading are much less than those on virgin compression it is necessary to classify the present state of the actual clay which may be:

(1) *Normally consolidated* (NC): that is, the clay is consolidated under its present state of stress in place and this has never been exceeded. Its present voids ratio is therefore represented by a point on the virgin compression line.
(2) *Over-consolidated* (OC): that is, the clay has been consolidated in the past under loading greater than the loading today. Its present voids ratio is represented by a point on a reloading curve.
(3) *Incompletely consolidated*: this means that excess pore pressure exists in the clay and has not yet dissipated. Future settlements will therefore comprise the remaining settlement associated with the existing stress conditions together with settlement arising from new loading.

The *maximum past consolidation pressure* σ'_c of a clay is the maximum vertical effective stress experienced by the clay in its geological history. If σ'_0 denotes the current vertical effective *overburden stress*, then for a normally consolidated clay $\sigma'_c/\sigma'_0 = 1$ and for an over-consolidated clay $\sigma'_c/\sigma'_0 > 1$. For an incompletely consolidated clay $\sigma'_c/\sigma'_0 < 1$. The ratio σ'_c/σ'_0 is known as the *over-consolidation ratio* (OCR) of the clay. The maximum past pressure is commonly estimated using a method proposed by Casagrande. Referring to figure 26.9 the method consists of determining the point X of minimum radius of curvature on the laboratory e–log σ' curve. A line is then drawn through X parallel to the log σ' axis and through X a tangent is also drawn to the curve. The bisector from X of the angle between these lines is drawn and the point where the bisector cuts the extension

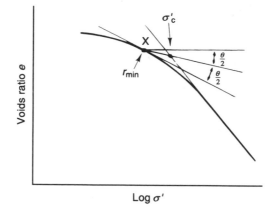

Figure 26.9 *Estimation of the maximum past consolidation pressure σ'_c experienced by a clay*

of the virgin compression line is considered to represent the state of the clay at the maximum past pressure σ'_c. There are other methods but the above is simple, if approximate.

For field conditions which satisfy the assumptions of one-dimensional consolidation, the *total vertical settlement S* of a clay layer of thickness H is given by

$$S = \sum_0^H m_v \, \Delta\sigma' \, \Delta z$$

where m_v, $\Delta\sigma'$ may vary from element to element Δz of H. If conditions of stress and soil property at the middle of the layer are representative of the whole layer, this may be written

$$S = m_v \Delta\sigma' \, H$$

where m_v and $\Delta\sigma'$ are now the values at the middle of the layer. Alternatively, $S = H(e_0 - e)/(1 + e_0)$ where e_0 is the voids ratio at the initial state of stress σ' and e is the voids ratio after complete consolidation under the increment $\Delta\sigma'$. This can be written

$$S = \frac{H}{1 + e_0} \, C_c \, \log_{10} \frac{\sigma' + \Delta\sigma'}{\sigma'}$$

Rate of settlement

Questions relating to the rate and the duration of the settlement process often have to be answered by the engineer but these aspects of settlement are beyond the scope of this book.

27

Shear Strength of Soil

A soil may be considered to have failed to support a built structure if the soil compresses or settles (or swells) to an extent which causes damage to the structure. This aspect was the subject of the previous chapter. When reference is made to *failure of a soil*, however, its failure in shear is usually meant, that is, the state of stress in the soil is such that the shearing resistance of the soil is overcome and a relative and significant displacement occurs between two parts of the soil mass. If this shearing resistance, or *shear strength*, of the soil is measured or predicted the geotechnical engineer is then able to analyse problems of stability of soil masses, estimate margins of safety against the occurrence of failure by shearing within such masses, and if necessary make design adjustments.

From the emphasis already placed on the complex nature of soil formations it will be understood that the measurement and representation of shear strength present difficult problems to the engineer. The strength is likely to differ from one major stratum to another in a natural soil formation and will differ also at any one location with the direction of measurement of the strength. The fabric of laminations and fissures in some clay soils makes it necessary to test quite large elements of the mass in order to obtain a representative value of strength. In a fissured material it is once more the defects which decide its behaviour, not the quality of the intact material between fissures.

A further complication lies in predicting the future behaviour of the soil. It has been noted that the clay soils are the principal group affected by time-dependent changes. If a clay is subjected to an increased sustained loading it will in time consolidate and become stronger. This behaviour is acknowledged in methods of improving (stabilising) clays by preloading or controlled loading – that is, by applying a loading for the purpose of consolidating the soil in advance of building construction, or by controlling the rate of increase of loading, from storage tanks say, to allow the increasing strength of the clay to keep pace with the increasing loading upon it. If the clay is subjected to a sustained relief of loading the soil will in time swell and become softer. In a relatively uniform clay this may take a number of years but in a clay formed with a fabric of fissures and drainage channels the strength may be affected quite quickly, even within the construction period. A relief of loading would occur in cases such as cuttings for side slopes in highways, trench and retaining-wall excavations.

Some aspects of the shear strength behaviour of soils are introduced in this chapter but the linking of shear strength and consolidation presented in the numerical models of 'critical state soil mechanics' is beyond the scope of this book. Doubtless the use of such models based on elastic–plastic soil behaviour, such as the Cam clay model, will become more conventional in application. Data from oedometer and triaxial tests, in principle, yield descriptive parameters for soils for adoption in this model, and research continues to bring model predictions closer to practical observations (Wood, 1990).

27.1 Graphical presentation of stress

A full description of the state of *stress at a point in a material*, soil in the present case, is given by the components of the stress tensor (see, for example, Schofield and Wroth, 1968) and from this the actual stress on any plane through the point may be deduced. The point may be any point within the mass of soil or it may simply be representative of conditions in the laboratory test specimen. The three mutually perpendicular planes at the point on which the shearing stresses τ are zero are defined as principal planes and the (normal) stresses σ on them are principal stresses. In soil mechanics literature these are conventionally described as *positive if compressive*. The largest of these stresses is called the major principal stress σ_1, the smallest is the minor principal stress σ_3 and the remaining principal stress is the intermediate principal stress σ_2.

The influence of the intermediate principal stress σ_2 is obscure and in this chapter the study is limited to the two-dimensional stress state, this being wholly defined by σ_1 and σ_3.

The *Mohr diagram* is much used to display one given state of stress at a point in a material. Figure 27.1 shows the components of two-dimensional stress at a point whose stress tensor is (σ_1, σ_3). The Mohr circle is the locus of the values of (σ, τ) on all possible planes at the point.

Sometimes it is desired to follow the state of stress through a series of controlled changes of stress and, to avoid obscuring the pattern of change in a large number of Mohr circles, a *p–q diagram* may be used. In this diagram only the points $p = \frac{1}{2}(\sigma_1 + \sigma_3)$, $q = \frac{1}{2}(\sigma_1 - \sigma_3)$ are plotted throughout the change of stress. The locus of points p, q for a stress change is called a *stress path* and at any point on this locus

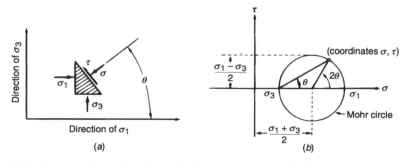

Figure 27.1. *Representation of two-dimensional stress at a point: (a) directions of stresses at a point; (b) Mohr diagram for stress at a point*

the coordinates p, q enable the Mohr diagram to be drawn for that particular state of stress at the point. For illustration, the stress path in a simple triaxial compression test (figures 26.1 and 26.2) is shown in figure 27.2, the initial application of cell pressure σ_3 being followed by an increasing deviator stress $(\sigma_1 - \sigma_3)$ until a maximum (failure) condition is reached. It is the failure state of soils that is the main concern in this chapter on shear strength.

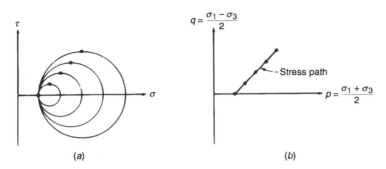

Figure 27.2 *Representation of a changing state of stress: (a) Mohr circles for successive states; (b) p − q diagram. The corresponding points • on the two diagrams represent the same stress conditions*

27.2 Shear-strength testing in the laboratory

The triaxial compression apparatus, including unconfined compression, is the most common laboratory apparatus for determining the stress–strain and shear-strength properties of soils. The simpler direct shear apparatus retains a place in the laboratory for testing the coarser sands, gravels, crushed rock and similar materials. Methods of test using these forms of apparatus are described in BS 1377: Parts 7 and 8. Because they are usually limited to the testing of rather small and sometimes rather disturbed specimens these tests may not be capable of giving results of value to the engineer. The obvious procedure to minimise disturbance is to test soils *in situ*. Such tests still only test rather a small volume of soil and the advancing of pits or bores to the point of test subjects the soil to some disturbance. The control of tests and boundary conditions are often uncertain so that problems remain in the interpretation of results. An outline of some *in situ* tests is given in section 27.3.

Test conditions

The principal features of the *triaxial test apparatus* were shown in figure 26.1. Good control of the state of stress is attained in this apparatus and several types of test condition are practicable. The three main types in common use are classified according to the conditions of drainage during a test and are:

(1) *The unconsolidated undrained* (UU) *test*, in which drainage is not allowed to take place, either on application of the cell pressure σ_3 or with increases in the

deviator stress ($\sigma_1 - \sigma_3$).

(2) *The consolidated–undrained* (CU) *test*, where the specimen is first allowed to consolidate (drain) on application of σ_3 and when excess pore pressures are fully dissipated, that is primary consolidation is complete, the deviator stress is applied under conditions of no drainage. It is usual to measure pore pressure during the second, undrained stage, enabling the stresses to be expressed also in terms of effective stresses.

(3) *The drained or consolidated–drained* (CD) *test*, in which drainage is allowed throughout the test and no excess pore pressure is allowed to build up at any time.

In the triaxial compression test the specimen is cylindrical and usually 38 mm or 100 mm in diameter, as described in Chapter 26. Sizes up to 250 mm in diameter are sometimes required to obtain sufficiently representative specimens.

The limited control of drainage available in the *direct shear test* means that this method is usually used for drained tests on sands, gravels and similar coarse-grained soils. The direct shear apparatus is simple in concept. The soil to be tested is placed in a square box-shaped container which is split horizontally across at mid-height, as shown diagrammatically in figure 27.3. A vertical stress σ is applied and shearing is induced on a horizontal plane in the soil. The maximum shearing stress τ_f is measured.

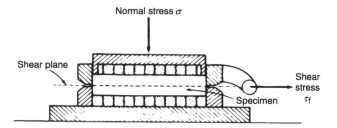

Figure 27.3 *Direct shear box, vertical section showing stressing arrangement. The top half of the box and the normal stress platen are free to move up or down as volume changes occur during shearing of the soil specimen*

The soil particle size determines the appropriate size of the shear box. For tests on fine to medium sands a box giving a specimen size 60 mm or 100 mm square by about 20–25 mm thick is often used, but for testing gravels, broken rock and similar materials a much larger box is required. For example, for testing particle sizes up to 20 mm a shear box specimen 305 mm square and 150 mm thick may be adopted.

Coarser-grained soils

When their strengths are expressed in terms of *effective stresses* there is a great similarity between the behaviour of sands and gravels on the one hand and clay on the other. There is also, however, a great difference in the permeabilities of these two extremes of particle size, as shown in table 27.1, and this leads to widely differing rates of response to the application of stress. It has been noted that a change

$\Delta\sigma$ in total stress applied to a freely draining sand produces an immediate response in a corresponding effective stress change $\Delta\sigma'$ and also a volume change. In a saturated relatively impervious clay, however, the response to a total stress change is not complete until sufficient time has elapsed to allow the clay to drain or absorb water and reach a new equilibrium under the changed stress.

It is convenient therefore to look first at the stress–strain and strength characteristics of the coarser-grained soils straining under drained conditions in which $\sigma = \sigma'$ and $u = 0$.

TABLE 27.1

Permeability and shear strength parameters of typical soils

Material	I_P (per cent)	k ($\mu m\ s^{-1}$)	c' ($kN\ m^{-2}$)	ϕ' (degrees)
Rockfill: tunnel spoil	—	5×10^4	0	45
Alluvial gravel: Thames Valley	—	5×10^2	0	43
Medium sand: Brasted	—	—	0	33
Fine sand	—	1	0	20–35
Silt: Braehead	—	3×10^{-1}	0	32
NC clay of low plasticity (undisturbed samples)	20	1.5×10^{-4}	0	32
NC clay of high plasticity (undisturbed samples)	87	1×10^{-4}	0	23
OC clay of low plasticity (undisturbed boulder clay samples)	13	1×10^{-4}	8	$32\frac{1}{2}$
OC clay of high plasticity (undisturbed London clay samples)	50	5×10^{-5}	12	20
Quick clay (undisturbed samples)	5	1×10^{-4}	0	10–20

From Bishop and Bjerrum (1960).

The abbreviation NC indicates normally consolidated, and OC over-consolidated.

Internal friction and cohesion

The shear resistance of an assembly of particles is essentially frictional in character. The over-simple analogy to a sliding block is commonly made (figure 27.4). When a normal compressive stress σ is applied through the block to the supporting surface, a relative displacement in shear between the block and the surface will not occur until the shear stress τ reaches a value τ_f at which the available shearing resistance between the material surfaces is fully mobilised and the obliquity of the resultant stress is then ϕ, as shown. A change in the value of σ would lead to a change in τ_f but in every case the obliquity of stress at the failure condition would remain at ϕ. The angle ϕ is thus the parameter which determines the shearing resistance and $\tau_f/\sigma = \text{constant} = \tan\phi$ for all values of σ. Tan ϕ is the coefficient of friction familiar to students of engineering mechanics. When a shear failure occurs within a soil, a zone or surface of shearing exists between the parts experi-

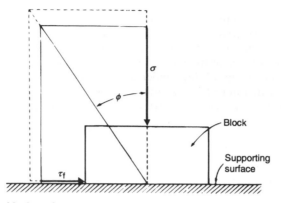

Figure 27.4 *Sliding block analogy*

encing relative displacement and in this zone the shearing stress τ_f is proportional to the normal stress σ as in the sliding block analogy. If there is still a shearing resistance c in a soil when this normal stress is reduced to zero that resistance is described as *cohesion*. The shearing resistance is then generally expressed by the Mohr–Coulomb failure criterion as

$$\tau_f = c + \sigma \tan \phi$$

At this stage a coarse-grained soil is being considered in which $u = 0$, where the value of c is denoted by c' and is insignificant so that

$$\tau_f = \sigma' \tan \phi'$$

in which ϕ' expresses the frictional characteristic of the soil and is known as the *angle of shearing resistance* of the soil in terms of effective stresses σ'. It may also be noted that in the failure state the principal stress ratio

$$\sigma'_1/\sigma'_3 = \frac{1 + \sin \phi'}{1 - \sin \phi'} = \tan^2 [45° + (\phi'/2)]$$

where $[45° + (\phi'/2)]$ is the angle between the major principal plane and the theoretical plane of shear at the point.

It should be understood throughout this introduction to shear strength that the paired parameters c and ϕ for a soil are not unique to the soil alone but are functions of both the soil and the conditions of testing.

The angle of shearing resistance ϕ' of a coarse-grained soil is more than simply a measure of the friction between two sliding surfaces; it embraces effects of interlocking of particles and their rolling and sliding actions when subjected to shear. It is apparent then that while the value of ϕ' is dependent on the soil minerals, it is also a function of the particle shape, size and grading, and the initial density of packing and confining pressure of the soil.

Maximum stress and volume change on shearing

Stress–strain curves for loose and dense sand specimens tested in triaxial compression have the general form shown in figure 27.5. The dense specimen shows a peak resistance in terms of deviator stress, thereafter continuing to shear at a lower stress. The loose specimen shows an increasing resistance until a maximum deviator stress is reached and thereafter the deviator stress remains more or less constant.

When the volume changes are examined it is found that the dense specimen expands on increasing the axial compressive strain while the loose specimen decreases in volume at first, followed by a partial recovery in volume. These volume changes, particularly in the loose sands, have implications in practice in their effect on the behaviour of saturated sands under shock loading. There is apparently a density, or voids ratio, at which a given sand will neither expand nor decrease in volume on shearing. This voids ratio has been termed the *critical voids ratio* and is a function of the sand itself and also of the effective confining pressure on the sand.

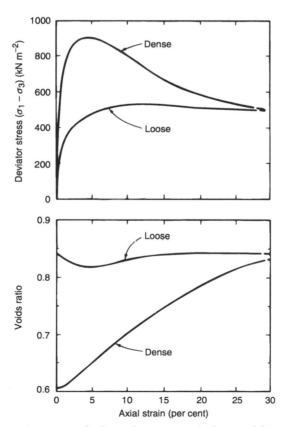

Figure 27.5 *Stress–strain curves and volume change curves for loose and dense specimens of a medium fine sand tested in triaxial compression at* $\sigma_3 = 200 \ kN \ m^{-2}$

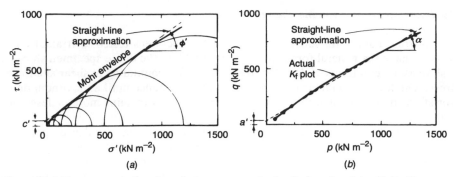

Figure 27.6 *Shear strength data for a dense coarse-grained soil plotted on (a) a Mohr diagram and (b) a p–q diagram*

The peak values of deviator stress for tests run at several values of cell pressure on dense specimens, at a given initial density, can be represented by a series of circles on a Mohr diagram, as in figure 27.6(a), each circle representing a state of stress at a point in the material, at failure. Consideration of the Mohr–Coulomb failure criterion shows that this will be represented by a straight line *tangential* to these effective stress circles. The line drawn tangential to the circles, in figure 27.6(a), is called the *Mohr failure envelope*; this may be nearly a straight line passing through the origin of the diagram and having the equation

$$\tau_f = \sigma' \tan \phi'$$

where ϕ' is the slope of the tangent and is the measured angle of shearing resistance of the dense sand. If the envelope is curved, a straight-line approximation is usually required for practical usage, covering a specified range of direct stress associated with a particular design problem. For this approximation

$$\tau_f = c' + \sigma' \tan \phi'$$

and usually c' is small. A similar plot using the ultimate value of deviator stress (that is, at high axial strains) gives the value of ϕ' for the loosened material, equalling the value from tests on initially loose material.

The data for the failure conditions may also be plotted on a p–q diagram as in figure 27.6(b), again giving a line or curve through the points. This line is called the K_f line and is sometimes found to be easier to plot than the Mohr envelope. If the best straight-line approximation on the p–q diagram has slope α and intercept a, then

$$\phi' = \sin^{-1} (\tan \alpha)$$

and

$$c' = a/\cos \phi'$$

The effect of the initial voids ratio e_0 on ϕ' is illustrated for a given sand in figure 27.7. The angle of friction ϕ_μ between the soil minerals is about 26° in the case of rough quartz grains and is little affected by wetting. Shearing within a granular mass will not however follow a planar surface with an angle of friction ϕ_μ and the many directions of particle contacts lead to a higher angle of shearing resistance ϕ_{fr} at the critical voids ratio. The remaining part of ϕ' derives mainly from particle interlocking, which depends on grain shape and grading and on density of packing of grains.

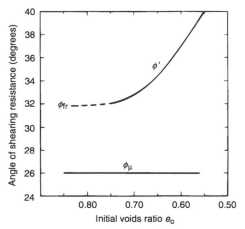

Figure 27.7 *Effect of initial voids ratio of a sand on the measured angle of shearing resistance*

Data from the direct shear test are also easily plotted to give estimated ϕ' values and volume change behaviour. The same pattern emerges as has just been seen for the triaxial compression test.

Finer-grained soils

When clays are tested in triaxial compression under consolidated drained (CD) conditions, that is, with $\sigma = \sigma'$ and $u = 0$, some of the general characteristics just seen in the tests on sands again emerge. Heavily over-consolidated clays are stronger than normally consolidated clays of the same material, just as dense sands are stronger than loose. Heavily consolidated clays have peak drained shear strengths just as for dense sands, and at large axial strains their drained strengths approach those of the normally consolidated materials. After an initial decrease in volume, the over-consolidated clays decrease in volume with increasing axial strain whereas the normally consolidated clays decrease in volume. These volume changes during shearing will be recalled when the conditions during the undrained tests are examined.

The Mohr failure envelope for the results of a set of drained tests on a normally consolidated clay passes near the origin, and in the case of an over-consolidated clay may be slightly curved. It is not usually difficult to fit a straight line to represent the envelope giving $\tau_f = c' + \sigma' \tan \phi'$ as before. The value of ϕ' for the clay is then established and also c' is estimated. The intercept c' is usually rather small. Table 27.1 shows permeability k and shear strength parameters c' and ϕ' of typical soils. The range of k-values is seen to be large whereas the range of ϕ' is not.

Drained or CD tests in clays must be run at very slow rates in order to keep pore pressures negligibly small as the test proceeds. For this reason they are sometimes called slow tests.

Residual or ultimate strength

When the shearing deformation in a clay soil is continued well beyond the relative displacement at which the peak resistance is mobilised, a lower ultimate resistance remains and this is termed the *residual strength*. It has been noted that the peak resistance is usually pronounced in the more heavily over-consolidated clays, less so in the normally consolidated and remoulded soils. The loss in strength is partly attributable to a change in water content but also to a reorientation of the minute clay platelets in the zone of shearing to a parallel, dispersed arrangement aligned in the direction of the deformation and of notable smoothness. A clay containing a proportion of silt sizes and coarser, which inhibits this reorientation, shows an ultimate strength in terms of ϕ' which is closer to that for non-platy particles, say about 30°. If, however, a clay is almost wholly composed of material less than 2 μm in size the reorientation dominates the strength and the reduced value of ϕ' for the residual state (now termed ϕ'_r) may be less than 15° and the cohesion intercept c' virtually disappears. The large deformations required to attain this loss of strength *in situ* may be part of a long-term process currently taking place unnoticed, or they may have occurred in the geological past in the course of ground movements which have since stabilised. Ground which has experienced past movements may be in a delicate state of stability today and any slickensided shear surfaces observed during site investigation or subsequent works should be noted as indicators that the residual strength may already be the maximum available resistance.

Undrained tests

While CD tests are sometimes made on fine-grained soils it is much more common practice to carry out triaxial compression tests in which drainage is prevented. These undrained tests may be made with or without the measurement of porewater pressures and are usually associated with the finer grained soils of relatively low permeability.

In the UU test, specimens as sampled are brought to failure under undrained conditions (Bishop and Henkel, 1962; BS 1377: Part 7). If a set of specimens from a point in a stratum of saturated clay is taken, each specimen has been consolidated *in situ* to the same pressure. Removal of the clay from the stratum reduces to zero the external total compressive stresses on each specimen and this stress change results in an equal reduction in the porewater pressure since undrained (constant volume) conditions prevail. Thus there is no change in the effective stress in the specimen. Application of a cell pressure to a specimen in the triaxial apparatus now results in an equal increase in the pore pressure, but again no change occurs in the effective stress if drainage is prevented.

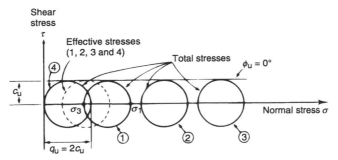

Figure 27.8 *Undrained test data, for specimens of a structured clay, plotted on a Mohr diagram*

When the deviator stress is applied and increased until failure takes place changes occur in the porewater pressure which depend both on the initial effective stress in the clay and on the consolidation history of the clay. The maximum deviator stress $(\sigma_1 - \sigma_3)_f$ and hence the *undrained shear strength* c_u is a function of these two factors so that even if each specimen of the set is tested at a different cell pressure, the maximum deviator stress should be constant. This is found to be so in the test conditions described and the Mohr circle plot would have the form of figure 27.8. This is the $\phi = 0°$ condition where $\sigma = \phi_u$ and $\tau_f = c_u = \frac{1}{2}(\sigma_1 - \sigma_3)_f$ for the undrained tests. It is interesting to note that specimens tested at zero cell pressure (atmospheric pressure) should also yield the same result, $\tau_f = c_u$, as in circle 4. This simple version of the triaxial test is called the *unconfined com-*

TABLE 27.2
Undrained shear strength of clays. From BS 8004

In accordance with BS 5930	Consistency		Undrained shear strength c_u ($kN\ m^{-2}$)
	Widely used	Field indications	
Very stiff	Very stiff or hard	Brittle or very tough	greater than 150
Stiff	Stiff	Cannot be moulded in fingers	(75) 100 to 150
	Firm to stiff		75 to 100
Firm	Firm	Can be moulded in the fingers by strong pressure	(40) 50 to 75
	Soft to firm		40 to 50
Soft	Soft	Easily moulded in the fingers	20 to 40
Very soft	Very soft	Exudes between the fingers when squeezed in the fist	less than 20

pression test and for saturated clays provides a rapid way of measuring c_u. The unconfined test may be carried out in the laboratory in the triaxial compression apparatus but for use in either laboratory or field a simple and inexpensive form of portable apparatus is available, which is suitable for testing saturated non-fissured clays. One such apparatus is described in BS 1377: Part 7. Specimens may be obtained from a borehole, trial pit or open excavation and tested on the spot. The *unconfined compressive strength* q_u of a soil is sometimes given as its measure of strength, where $q_u = 2c_u$ as shown.

If pore pressures are measured during the tests at the various cell pressures the total stress data at failure may be expressed in terms of effective stress and it is found, as expected from the above argument, that only one effective stress Mohr circle is obtained from the entire set of specimens. The total stress data cannot therefore be interpreted to give values of c' and ϕ'.

If a soil is only partially saturated the strength measured in undrained testing is not independent of the total stress conditions and, although the test results may be expressed by a curved Mohr failure envelope, and values of c_u and ϕ_u may be selected, the interpretation is rather difficult.

Table 27.2 lists a classification of consistency and undrained shear strength of clays from BS 8004.

Sensitivity

Disturbance of saturated clay soils usually leads to a rapid loss of undrained shear strength and at large amounts of disturbance a remoulded strength is reached. In clays with a flocculated structure this is often much smaller than the peak undisturbed strength. Any disturbance leading to remoulding, for example, by kneading or rolling, will have the effect of changing the whole structure of the clay to a more oriented, dispersed particle arrangement. An initially dispersed clay would show little change in strength on remoulding.

The ratio of the undisturbed (peak) undrained strength of a soil to its fully disturbed and remoulded strength at the same water content is known as its *sensitivity* S_t. A classification of degrees of sensitivity is given in table 27.3. The low sensitivities would be associated with the more highly consolidated clays.

TABLE 27.3
Degrees of sensitivity in soils

S_t	Classification	S_t	Classification
1	Insensitive	8–16	Slightly quick
1–2	Slightly sensitive	16–32	Medium quick
2–4	Medium sensitive	32–64	Very quick
4–8	Very sensitive	over 64	Extra quick

Recent marine clays deposited with flocculated structures would be expected to be quite highly sensitive (see Chapter 21, section 21.2). Clays of extreme sensitivity are termed quick clays. Some of the leached marine clays of Norway and the Leda clays of Canada are striking examples of clays which are altered by disturbance from soft cohesive soils to viscous liquids. These clays commonly have sensitivities exceeding 100 and pose very difficult geotechnical problems in dealing with landslides and also during site investigation and the construction of works.

Some partial recovery of strength with time after completion of remoulding has been observed in clays. This characteristic is termed *thixotropy* and reasons for the behaviour are still speculative.

Consolidated undrained tests

The consolidated undrained (CU) test extends the information from the undrained test for wider application to stability problems, particularly when pore pressures are also measured at the undrained stage (BS 1377: Part 8). Several sets of specimens may be sampled from a point in a saturated clay stratum and tested in triaxial compression. Each set is first allowed to consolidate, that is drain, at a selected cell pressure σ'_c; different sets are consolidated at different σ'_c values. Each specimen is subsequently brought to failure under undrained conditions by increasing $(\sigma_1 - \sigma_3)$. As with the undrained tests, previously outlined, each *set* of specimens will yield a single value of c_u related to the consolidation pressure (σ'_c) selected for that set. When the test results are plotted in the form of figure 27.9(a) it would appear that the relationship has the form of a strength c_u–depth profile (with σ'_c representing depth). Laboratory measurements of the slope c_u/σ'_c, however, give higher values than those estimated in the field from *in situ* measurements. This is attributable to a number of factors including sampling disturbance, and differing consolidation conditions in laboratory and field. Figure 27.9, (a) and (b), shows alternative ways used to convey the results of the CU test in terms of c_u and σ'_c. The consolidated undrained behaviour may be presented by giving the value of $\tan^{-1}(c_u/\sigma'_c)$ or alternatively the value of ϕ_{cu}. The two values are not equal, although they convey the same information. Both enable c_u to be predicted for selected values of σ'_c.

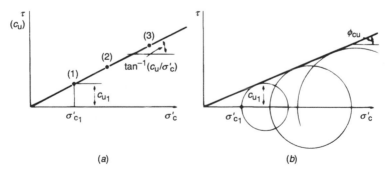

Figure 27.9 *Alternative ways of presenting results of CU tests in terms of c_u and σ'_c*

If *pore pressures are measured* in the CU tests, the effective stress Mohr circles may be plotted as shown in figure 27.10. The failure envelope enclosing the circles is usually closely linear like that for the coarse-grained soil in figure 27.6(a). The intercept and slope give values for c' and ϕ', the strength parameters in terms of effective stress. This envelope expresses the same information as the values of

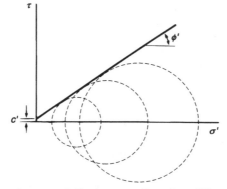

Figure 27.10 *Mohr diagram in terms of effective stress. Data from CU tests, on specimens of a saturated clay, with pore pressure measurement*

c_d and ϕ_d that are obtained in slow, drained (CD) tests on the same soil. For practical purposes the values of c' and ϕ' from the two types of test are the same.

It is worth reiterating that the values of c' and ϕ' are not unique for a given soil but depend on whether they are referred to total stresses ($c_u \phi_u$) or effective stresses ($c' \phi'$ and $c'_r \phi'_r$). The quality and direction of sampling, the test specimen sizes and the method of testing should all be stated when presenting the values of measured shear strength.

27.3 Shear-strength testing in the field

The difficulties of sampling soils without seriously disturbing their fabric and structure have led to the development of a number of field or *in situ* methods of assessing their quality. Such tests may be carried out at the ground surface, in pits or shafts, in boreholes or by probing from the surface. By eliminating the need to extract samples, eliminating handling, transport and laboratory preparation, some of these tests enable properties of relatively undisturbed soil to be assessed. In some ground conditions it may be difficult and expensive to obtain quality samples. Usually both sampling for laboratory testing and *in situ* testing will be undertaken.

The 'standard penetration test' is a long established, if rather crude, dynamic penetration test carried out in boreholes. It involves driving a sample tube, or a solid cone, into a soil below the bottom of a borehole and observing the blow count. Using the tube, a disturbed sample of soil may be retrieved for identification. The test procedure has been standardised and must be precisely followed.

The 'cone penetration test' is a 'static' method in which the resistance to penetration of a standard cone pushed downwards into the ground is measured. There is no borehole required and no sample of soils is obtained. The measured resistance has been correlated with other soil parameters for application in design. The test has advantages in being relatively cheap and in providing a continuous record of soil variations. Determination of the resistance of soils to dynamic probing with a driven cone has also been standardised.

In uniform saturated clays the 'vane test' has found much application. A cruciform section vane is lowered in a borehole, pushed some distance below the bottom and rotated in the clay. The required torque is measured and the undrained shear strength of the clay deduced.

'Plate loading tests' and 'California bearing ratio tests' are quite commonly adopted, the former over a range of plate diameters. The plate etc. is loaded on to the soil under prescribed conditions, in the manner of small foundation, until a certain penetration is reached. From the data obtained a measure of shear strength is determined.

In the 'pressuremeter test', an inflatable probe is inserted through a borehole, or by self-boring, into the soil. The probe is expanded laterally against the soil by gas pressure and the radial deformation measured. The observations enable undrained shear strength and other soil parameters to be estimated. Further information on *in situ* tests, and further references, are to be found in BS 1377: Part 9 and BS 5930.

28

Soil Stabilisation by Compaction

In the wider sense, soil stabilisation is a term used for the improvements of soils either as they exist *in situ* or when laid and densified as fill. The purpose of stabilisation is to make a soil less pervious, less compressible or stronger, or all of these. This is often achieved by injecting a fluid into the pore spaces in soil, the fluid gelling or hardening in the pores. In suitable soils stabilisation may be achieved by introducing admixtures, for example in improving clay soils by mixing in a few per cent by weight of quicklime, hydrated lime or Portland cement before mechanical compaction. For further information see Ingles and Metcalf (1973)

28.1 Mechanical compaction

In this chapter the study of stabilisation is limited to processes of densification or *compaction by mechanical equipment* such as vibrators or rollers. The treatment of the subject is necessarily brief but the subject itself is important. Extensive developments have occurred and are continuing in excavating, transporting, spreading and compacting machinery enabling large quantities of material to be economically handled in earthworks. Use of such plant is seen in the construction of highway embankments, earth dams and general infilling. Suitable fill, when placed in loose layers and compacted under controlled conditions, can be made up into a relatively uniform strong, incompressible and impervious material with predictable properties. As such it will be preferred as an engineering material to randomly dumped fill and even to some natural soils if these are erratic or poor in quality and distribution. Some natural soils and existing fills may be improved by compaction in place – although usually to very shallow depths of under 1 m – the process often being used to improve road foundations. Greater depths have been successfully compacted using very high tamping energies, the final densities being measured using *in situ* tests.

Consolidation and compaction have similar meanings but in geotechnical engineering the term *consolidation* is usually reserved for the time-dependent reduction in voids volume accompanying an expulsion of pore water due to the application of a loading. The term *compaction* is used for the rapid reduction in

430

voids volume in a soil arising from the application of mechanical work – the reduction being due to an expulsion of air from the voids. The state of compaction of a soil is usually described quantitatively in terms of the *dry density* of the soil. That is, although the soil itself is moist, a greater or lesser compaction means that a greater or lesser mass of solids is packed into a unit total volume, with a consequent influence on the engineering quality of the soil. The compacting energy is applied to the soil in the field by one of three principal methods: static (smooth-wheeled roller, pneumatic-tyred roller, grid roller and sheeps foot roller), dynamic (falling weight, rammers) and vibrating (vibrating smooth-wheeled or pneumatic-tyred roller, vibrating plate). Choice of field compacting plant depends on the soil to be compacted – kneading methods work best in the more clayey soils, vibration in the sandy soils.

The selection of earthworks materials and their suitability and acceptability for placing as fill is based on a number of practical and economic criteria, such as availability, ease of excavation and transport, and strength when placed and compacted at an acceptable water content. Unsuitable materials include highly organic soils such as peats, hazardous material, material susceptible to spontaneous combustion, highly plastic clays and frozen soils. For an introduction to earthworks practice, see Horner (1981).

28.2 Compaction properties of soil

Laboratory testing for the compaction behaviour and condition of soil is detailed in BS 1377: Part 4. This includes determinations of the dry density/moisture content relationship for a soil, maximum and minimum dry densities for granular soils and the Moisture Condition Value (MCV) of a soil. The first of these involves the application of standard compaction energies to the soil, which is contained in a standard mould. Batches of soil are prepared to cover a range of water contents. Three tests are described, differing mainly in their compaction method, two in which dynamic compaction is applied through blows from standard rammers falling on to the soil and one, for cohesionless soils, in which a vibrating hammer is used. For a given compacting effort, the dry densities ρ_d corresponding to several water contents w in the soil are determined. The relationship of these is shown typically for a cohesive fill in figure 28.1. The water content at which the maximum value of dry density is obtained is called the *optimum water content* for the compacting energy used. With a given soil, an increase in the compacting energy leads to a shift of the entire curve, upwards and to the left – that is, giving higher maximum density and lower optimum water content. With different soils, the more cohesive tend to have lower maximum dry densities and higher optimum water contents, when compacted at a given energy level, than the more granular. Air voids lines on the ρ_d–w graph can be calculated from

$$\rho_d = \rho_w \frac{\left[1 - \dfrac{V_a}{100}\right]}{\left[\dfrac{1}{\rho_s} + \dfrac{w}{100}\right]}$$

CIVIL ENGINEERING MATERIALS

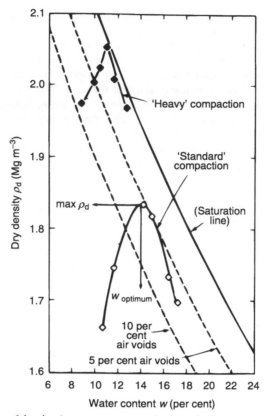

Figure 28.1 *Relation of dry density to water content for a sandy clay compacted at two different compaction energies*

Lines corresponding to air voids contents V_a of 0, 5 and 10 per cent are shown in the figure. Lines of equal degree of saturation S_r are sometimes given. It is seen that the maximum dry density occurs at an air voids content of 5 per cent or less. Investigations reported by Sherwood (1970) suggest that with skilled operators following the BS 1377 procedures, the tests for maximum dry density and optimum water content may be expected to give reproducible results just within acceptable requirements for design purposes. For control of compaction, however, the variation in results among test operators was found to be high. This means that for effective control on site a considerable number of tests must be undertaken even for the rare case of a uniform soil.

Although the optimum water content is not in itself a unique parameter there are some broad differences in the characteristics of clayey soils between those which are compacted on the dry side and those on the wet side of optimum. In UK climatic conditions soil fills tend to exist on the wet side of optimum. If just on the wet side they should compact well and form stable soils.

In the UK it is usual to specify either the required performance to be achieved in compaction, or the required method of compaction. In the first, the requirement is to achieve a certain product, for example a compacted soil having 90 per cent of its density when measured in the 'Heavy' compaction test. In the second, the method to be used in compacting the soil is specified, for example type of compaction plant, thickness of layer of soil after compaction and number of passes of plant.

The laboratory test fulfils a useful function in *comparing* the compaction behaviour of fills. It does not necessarily reproduce the compaction characteristics of any given mechanical plant in the field. There is no unique optimum water content and density – as we have seen they depend on the compacting energy applied. This means that tests should be run in the field, if possible at several water contents, using the selected mechanical plant to give a portion of the density–water-content curve of the type shown in figure 28.1. Methods for the determination of soil density *in situ* are detailed in BS 1377: Part 9. Because of the quantities of material involved it is not usually practicable to manipulate the water content in the field. The question usually is to select the most stable fill material available and the most suitable plant, sometimes operating with more than one stage of compaction to give finally a high percentage of the maximum soil density in an acceptable number of passes of the plant. As the plant makes the initial passes over the loosely placed fill, the greatest changes in density occur.

The increase of density per pass becomes smaller with each successive pass, as shown in figure 28.2, and a stage is reached at which further rolling produces little further compaction.

28.3 Moisture condition value

Material properties to be determined for judging the acceptability of soils for use in earthworking in the UK – earthmoving and compaction as fill – now include the Moisture Condition Value (MCV). An apparatus, the Moisture Condition Apparatus, and test procedures have been developed for assessing soil suitability, for cohesive and most granular soils, in an environment which broadly simulates field compaction.

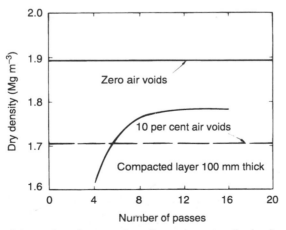

Figure 28.2 *Effect of the number of passes of a roller equipment on the dry density achieved in a sandy clay (w = 16 per cent) under field compaction*

The apparatus consists of a frame containing a drop rammer and a mould to hold the soil test sample. In use the the level of compaction is determined by the penetration of the rammer into the mould as the sample is compacted by repeated drops of the rammer. The weight and height of fall of the rammer are kept constant. Since both water content and sample weight also remain constant during the test, the sample volume reflects the bulk density ρ and degree of compaction achieved. Slots are incorporated in the base of the mould and any build-up of porewater pressure is observed by the appearance of water at these slots. Testing is normally stopped when water appears. Details of apparatus and test procedure are given in BS 1377: Part 4. Again, a good standard of quality control and operator training is necessary.

The MCV is defined in terms of the effort required to compact a 1.5 kg sample of the soil being evaluated to a level where the *change* in penetration of the rammer, calculated between B blows and $4B$ blows, is 5 mm. A graph as shown in figure 28.3 is plotted to give this value B, from which the MCV is expressed as $10 \log_{10} B$ – and this can be read directly from the same graph. Thus an MCV of 12 would indicate a very dry condition and 3 a very wet and weak condition. If $B = 5.5$, then the MCV is $10 \log_{10} 5.5 = 7.4$.

Each MCV relates to a specific soil water content w and a calibration line can be set out at the ground investigation stage for the soil type being assessed. Figure 28.4 shows relevant parts of such lines for several soil types.

Once a particular soil is calibrated, MCV testing allows its suitability in its sampled condition to be assessed on site in 6–10 minutes without the need for water content determination. The MCV also allows the selection and use of earthworking plant in highway and similar construction to be optimised for soil conditions pertaining at the site. An MCV of between 7 and 9 may be judged by the engineer as a lower limit of acceptability for a range of general cohesive fills and earthworking plant.

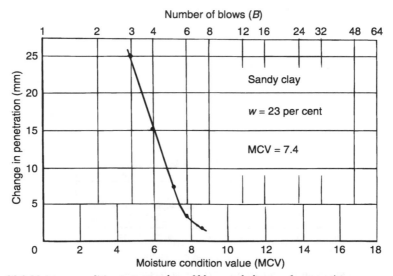

Figure 28.3 *Moisture condition test – number of blows and change of penetration*

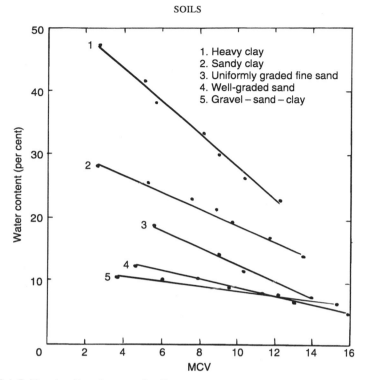

Figure 28.4 *Calibration lines for several soil types*

29

Engineering Applications

The study of soil as a material has led to a codified description of soils and soil behaviour, including description by number values – bulk density, voids ratio, index limits, coefficients of permeability and volume compressibility, shear strength and others. Once these values are determined it becomes practicable to introduce them into conjectural models of the mechanics of behaviour of soils, sometimes to prove the worth of the models and more often to make predictions of margins of safety or acceptability of built structures which include soils. For details, reference must be made to texts in *soil mechanics*, but the following are a few simple illustrations of the application of some material properties discussed in the preceding chapters.

29.1 Safe bearing capacity – foundations

For, say, column foundations on isotropic saturated clays, theoretical analysis indicates that a vertically loaded spread foundation, square in plan, will overcome the capacity of the clay to offer support when the imposed bearing pressure

$$q_{ult} = 6c_u$$

where c_u is the undrained shear strength of the clay (Chapter 27).
   A safe value of bearing pressure could thus be

$$q_{safe} = \frac{6c_u}{F}$$

where F is a factor of safety against failure in bearing.
For $F = 3$

$$q_{safe} = 2c_u = q_u$$

where q_u is the *unconfined* compressive strength of the clay – a useful 'rule of thumb'. Thus a firm clay with a measured c_u value of 75 kN m^{-2} would offer a safe bearing capacity of 150 kN m^{-2} for a proposed building foundation.

Even at q_{safe}, the foundation will experience some settlement with the passage of time. In Chapter 26 (figure 26.5) it was noted that this is accompanied by some increase in shear strength in the supporting clay. The above value of F, on completion of foundation construction and loading, is thus the critical, minimum value, and F will increase with time.

29.2 Settlement

Suppose the square foundation is of side length B and the supporting ground is not influenced by any other construction or other loading nearby. The settlement of the foundation due to the load intensity q_{safe} is partly immediate elastic settlement and partly long-term consolidation settlement.

From Chapter 26, the immediate settlement S_i is estimated from an expression of the form

$$S_i = \Delta q B \left(\frac{1 - \nu_u^2}{E_u} \right) I_s$$

where Δq is the applied load intensity, q_{safe} in this example.

Suppose the foundation is supported on a thick stratum of the firm saturated clay (say, of $c_u = 75$ kN m^{-2}), for which $\nu_u = 0.5$ and $E_u = 22.5$ MN m^{-2}, and that $B = 4$ m and $I_s = 0.5$. Then

$$S_i = \frac{150 \times 4 \times 0.75 \times 0.5}{22.5} = 10 \text{ mm}$$

In addition, the long-term consolidation settlement S is estimated (Chapter 26) from

$$S = \sum_0^H m_v \, \Delta \sigma' \, \Delta z$$

This may be modified by a multiplying factor μ accounting for some geometrical and geological aspects (allow $\mu = 0.8$ in this example). $\Delta \sigma'$, the representative vertical stress increment *through* the clay arising from the foundation loading of 150 kN m^{-2} is, say, $0.6 \Delta q$. For a clay having $m_v = 0.25$ (MN m^{-2})$^{-1}$ and a thickness 4 m, say, affected by the foundation loading, the estimated primary consolidation settlement will be

$$S = \mu m_v \Delta \sigma' H = 0.8 \times 0.25 \times 0.6 \times 150 \times 4$$

$$= 70 \text{ mm, nearly}$$

The estimated total settlement is thus

$$S + S_i = 70 + 10 = 80 \text{ mm}$$

It would be for the designer now to judge if this amount is tolerable as a displacement or tilt in the proposed building, and to adjust the design as required.

29.3 Stability of soil slopes

Slopes in dry or submerged cohesionless soils will just be stable at slope surface angles ϕ' to the horizontal, 'where ϕ' is the effective stress angle of internal friction of the soil in a loose density (Chapter 27).

For a flatter slope surface of angle β, that is $\beta < \phi'$, the factor of safety F of the slope against failure by shearing of the soil is

$$F = \frac{\tan \phi'}{\tan \beta}$$

Thus a bank of dry sand with a surface slope of 1 vertical to $2\frac{1}{2}$ horizontal, the sand having $\phi' = 30°$ in a loose state, has an estimated factor of safety of

$$F = \frac{\tan 30°}{\left(\dfrac{1}{2.5}\right)} = 1.44$$

This applies irrespective of the height of the bank.

However if the bank is subject to groundwater flow parallel to and up to the surface of the slope

$$F = \frac{\gamma_w}{\gamma_{sat}} \frac{\tan \phi'}{\tan \beta} \doteq 0.5 \frac{\tan \phi'}{\tan \beta} = 0.72$$

which is unsafe, since F must exceed unity. This demonstrates the importance of including groundwater in any review of the safety and stability of soil slopes.

Analyses of cut slopes in isotropic saturated clay soils, stressed under undrained conditions, show the factor of safety against shear failure to be dependent on height H of the slope, as well as on β and on the clay strength c_u and density ρ_{sat}. For example, the short-term stability of a steep cutting in a saturated isotropic clay is described theoretically by

$$\frac{c_{um}}{\rho_{sat}gH} = 0.2 \text{ for } \beta = 65°$$

where c_{um} is the actual, mobilised maximum shear stress.

Thus, for a 65° slope of height $H = 7.5$ m cut in a clay of $c_u = 75$ kN m^{-2} and unit weight $\gamma_{sat} = \rho_{sat} g = 20$ kN m^{-3}, the actual shear stress is

$$c_{um} = 0.2 \times 20 \times 7.5 = 30 \text{ kN m}^{-2}$$

and $$F = \frac{c_u}{c_{um}} = 75/30 = 2.5 \text{ which is satisfactory.}$$

29.4 Stresses due to self-weight of soil

Consider the horizontal plane ZZ in figure 29.1 at a depth of 5 m below the level
surface of a cohesionless soil which is saturated with groundwater from 1 m below
the ground surface. There is no flow of the groundwater.

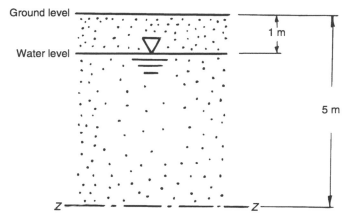

Figure 29.1 *Vertical section in a soil*

The soil properties are

$$G = 2.7 \quad e = 0.55$$

In the top metre, $w = 11$ per cent

Hence in the top metre

$$\rho = \frac{(1 + w)G}{1 + e} \rho_w = \frac{(1 + 0.11)\,2.7}{(1 + 0.55)}$$

$$= 1.93 \text{ Mg m}^{-3}$$

$$\gamma = \rho g = 18.97 \text{ kN m}^{-3}$$

and in the saturated zone

$$\rho_{sat} = \frac{G + e}{1 + e} \rho_w = \frac{(2.7 + 0.55)}{(1 + 0.55)}$$

$$= 2.1 \text{ Mg m}^{-3}$$

$$\gamma_{sat} = \rho_{sat}\, g = 20.57 \text{ kN m}^{-3}$$

Hence the stresses on the horizontal plane ZZ at 5 m depth are

Total vertical stress $\quad \sigma \quad = \quad 1 \times 18.97 + 4 \times 20.57 \text{ kN m}^{-2}$
$$= \quad 101.25 \text{ kN m}^{-2}$$

Porewater pressure u $=$ 4×9.81 kN m^{-2}

                                  $=$ 39.24 kN m^{-2}

Effective vertical stress σ' $=$ $\sigma - u = 62$ kN m^{-2}

Alternatively, σ' could be obtained direct from unit weight using the submerged or buoyant unit weight:

$$\rho' = \frac{G - 1}{1 + e} \rho_w = \frac{(2.7 - 1)}{(1 + 0.55)}$$

$$= 1.097 \text{ Mg m}^{-3}$$

$$\gamma' = \rho' g = 10.76 \text{ kN m}^{-3}$$

and $$\sigma' = 1 \times 18.97 + 4 \times 10.76 \text{ kN m}^{-2}$$

$$= 62 \text{ kN m}^{-2}$$

It may be noted that σ and σ' – in this example vertical stresses – vary with direction at any point (the Mohr circle gives the locus of all values at the point). On the other hand u is a water pressure and has the same value in all directions at the point.

29.5 Lateral earth pressure

Analysing the safety of walls that support and retain earth masses is one of the oldest and most common applications of soil mechanics. Because of the interaction between the soil and the wall in contact with and supporting it, there is a range of earth pressures to be considered in any situation. The lateral pressure may arise only from the self-weight of the supported soil, though commonly other loadings are also in action.

In the simple, idealised case of a vertical frictionless wall retaining soil with a level ground surface, the vertical effective stress in the soil σ'_v at any depth will be a principal stress, as will be the horizontal lateral stress σ'_h on the wall. In practice

$$\sigma'_h = \sigma'_3 \lesseqgtr \sigma'_1 = \sigma'_v$$

Where conditions of soil deposition and vertical consolidation are close to one-dimensional, the ratio

$$\frac{\sigma'_3}{\sigma'_1} = K_0 \quad \text{where } K_0 \text{ is an 'earth pressure coefficient'.}$$

This is called the 'at rest' state – it is not a failure state – and for normally consolidated soils

$$K_0 = 1 - \sin \phi' \quad \text{closely}$$

The magnitude of the lateral earth pressure on a retaining wall depends on a number of factors in addition to soil weight and other loadings on the soil. Principal among these is the translation or rotation of the wall away from or towards the retained soil.

For the present purpose of introducing the use of soil parameters consider simply the horizontal pressure, at a depth, on a vertical plane within a dry cohesionless soil with a level ground surface. The vertical and horizontal effective stresses, σ'_1 and σ'_3, are principal stresses. In the 'at rest' state we noted relative values for these, namely

$$\frac{\sigma'_3}{\sigma'_1} = \frac{\sigma'_{30}}{\sigma'_1} = K_0 = 0.5 \text{ for } \phi' = 30°$$

This state is shown in the Mohr diagram presentation in figure 29.2.

For a given depth in the soil, that is a given σ'_1, a large range of values of $K = \sigma'_3 / \sigma'_1$ is possible, the lower and upper limits of this range being when the soil is brought to a state of shear failure by lateral stretching and compressing respectively (through movement of the vertical plane, and by analogy the retaining wall). For cohesionless soils we have seen this state to be described sufficiently by ϕ', the effective stress value of the angle of internal friction.

If the soil is allowed to stretch laterally from the 'at rest' state, it is seen from figures 29.2 and 29.3 that an effective stress ratio

$$\sigma'_3 / \sigma'_1 < K_0$$

is reached where the Mohr circle of stress becomes tangential to the line defining stress ratios of failure states in shear within the soil – the line being at angle ϕ' as shown.

Equally, if the soil is compressed laterally, a different effective stress ratio is reached where the Mohr circle again displays failure states.

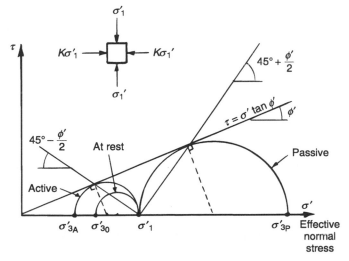

Figure 29.2 *Mohr diagram of stresses*

These two stress ratios – failure states – define the *active* and *passive* Rankine states respectively. From figure 29.2 in the active state

$$\sin \phi' = \frac{\frac{1}{2}\left(\sigma'_1 - \sigma'_{3A}\right)}{\frac{1}{2}\left(\sigma'_1 + \sigma'_{3A}\right)}$$

Hence

$$\frac{\sigma'_{3A}}{\sigma'_1} = \frac{1 - \sin\phi'}{1 + \sin\phi'} = \tan^2\left(45° - \frac{\phi'}{2}\right) = K_A$$

For $\phi' = 30°$, $K_A = 0.33$

By the same reasoning

$$\frac{\sigma'_{3P}}{\sigma'_1} = \frac{1 + \sin \phi'}{1 - \sin \phi'} = \tan^2\left(45° + \frac{\phi'}{2}\right) = K_P$$

For $\phi' = 30°$, $K_P = 3$

The ratios K_A and K_P are also 'earth pressure coefficients'.

The result is that the effective stress ratios K_0, K_A and K_P are readily determined from a knowledge of ϕ', but intermediate values of the ratio are not. Experiments demonstrate that K_A is attained with quite small stretching deformations in cohesionless soils and that larger movements in compression are needed to attain K_P, as indicated in figure 29.3.

In the example, section 29.4, the limiting total horizontal pressures on a vertical plane at 5 m below the ground surface can be developed. The soil is not dry and the pressures are derived by separating effective or grain-to-grain, stresses from porewater pressures

$$\sigma_h = \sigma'_{3A} + u = \sigma'_1 K_A + u \qquad \text{in the 'active' state}$$

$$= 62 \times 0.33 + 39.24 \text{ kN m}^{-2}$$

$$= 59.91 \text{ kN m}^{-2} \text{ in the 'active' state}$$

The 'passive' resistance to compression, and the water pressure combine to give a total

$$\sigma_h = \sigma'_1 K_P + u$$

$$= 62 \times 3 + 39.24 \text{ kN m}^{-2}$$

$$= 225.24 \text{ kN m}^{-2} \text{ laterally at the same depth}$$

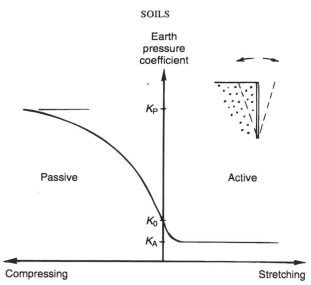

Figure 29.3 *Variation of earth pressure coefficient*

29.6 Groundwater flow

In Chapters 25 and 26 some aspects of submergence of soil and flow of ground-water through the pore or void spaces between the particles in coarser and finer grained soils were presented.

Flow of water through the voids in free draining soils often requires the engineer's attention. For example, the question may be one of predicting the temporary lowering of a groundwater surface by pumping from wells in order to facilitate construction work, or it may be to predict sustained patterns of flow through or below dams as a result of differences in water levels between upstream and downstream. The theoretical study of the patterns of groundwater flow, convergences and divergences, transient or steady state, has produced much published literature and forms an essential element in the study of soil mechanics.

Consider two simple cases of steady-state groundwater flow.

(1) *Flow in a confined aquifer*
For the vertical cross-section of ground shown in figure 29.4, as in figure 25.7, the uniform hydraulic gradient is

$$i = \frac{H_1 - H_2}{l}$$

The superficial velocity of laminar flow is

$$v = ki$$

The flow rate

$$Q = v \times D \text{ per unit length of aquifer normal to the cross-section}$$

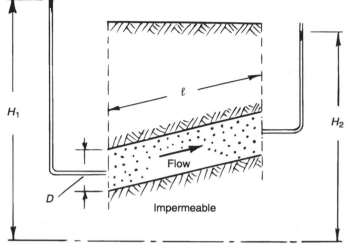

Datum for head H

Figure 29.4 *Groundwater flow in a confined aquifer*

Thus for such an aquifer of thickness $D = 3.5$ m, comprising an isotropic sand having $k = 10^{-4}$ m s^{-1}, and subject to water flow under a hydraulic gradient of 0.2

$$v = 10^{-4} \times 0.2 = 2 \times 10^{-5} \text{ m s}^{-1} = 72 \text{ mm per hour}$$

and $Q = 2 \times 10^{-5} \times 3.5 \qquad = 7 \times 10^{-5}$ m^3 s^{-1} per metre width of aquifer

$$= 0.25 \text{ m}^3 \text{ per hour per metre width}$$

(2) *Vertical flow*
For the conditions shown in figure 29.5, the uniform hydraulic gradient of upward flow is

$$i = \frac{\Delta H}{z}$$

At the horizontal plane ZZ at depth z the total vertical stress

$$\sigma = z \, \gamma_{sat}$$

The porewater pressure

$$u = (z + \Delta H) \, \gamma_w$$

Hence the vertical effective stress

$$\sigma' = \sigma - u$$

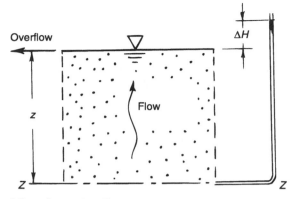

Figure 29.5 *Upward flow of water in soil*

$$= z\, \gamma_{sat} - z\, \gamma_w - \Delta H\, \gamma_w = z\gamma' - \Delta H\, \gamma_w$$

Thus if ΔH is sufficient, $\sigma' \to 0$ and no frictional strength can be mobilised at plane ZZ, namely when

$$i \to \frac{\Delta H}{z} = \frac{\gamma'}{\gamma_w}$$

The cohesionless soil is then said to be in a state of liquefaction or 'quick', as in 'quicksand', and has no strength and hence no bearing capacity. The phenomenon of 'quicksand' is the result of the upward flow of water within the sand.

As noted earlier

$$\frac{\gamma'}{\gamma_w} = \frac{\rho'}{\rho_w} = \frac{G-1}{1+e}$$

This value of $i = i_c$ is called the *critical hydraulic gradient* and its value for usual values of G and e is approximately unity.

It is part of the engineer's skill to anticipate locations in and around construction works where such undesirable hydraulic gradients are likely to be approached as flow emerges from the ground, and to take timely preventive action.

It follows from the above that generally with vertically upward flow

$$\sigma' = z\gamma' - \Delta H\, \gamma_w$$

$$= z\gamma' - \frac{\Delta H}{z}\, z\, \gamma_w$$

$$= z\gamma' - i\, z\, \gamma_w$$

The force effect of seepage on the soil can be expressed as a seepage force j per unit total volume of the soil from

$$j = \frac{iz\gamma_w A}{zA} = i\gamma_w$$

where A is the total area normal to the direction of flow, so that j is proportional to i and acts as a force on the soil grain structure.

In *isotropic* soil these seepage forces act in the direction of flow. Thus if there is a downward component to the flow, the value of σ' vertically on a horizontal plane is increased over the stress owing to submerged self-weight, thereby increasing the mobilisable frictional shear strength of the soil.

In the particular case of vertically downward flow, the above reasoning gives

$$\sigma' = z\gamma' + iz\gamma_w$$

and when $i = 1$ in free vertical drainage

$$\sigma' = z\,(\gamma' + \gamma_w)$$

$$= z\,\gamma_{sat}$$

These concepts have to be developed for flow in any direction, steady state or otherwise, and in more complex soil formations, and enable the engineer to answer many geotechnical questions relating to performance and safety.

References

Barnes, G.E. (1995) *Soil Mechanics – Principles and Practice*, Macmillan, Basingstoke.

Berry, P.L. and Reid, D. (1987) *An Introduction to Soil Mechanics*, McGraw-Hill Book Company (U.K.) Ltd.

Bishop, A.W. and Bjerrum, L. (1960) The relevance of the triaxial test to the solution of stability problems, *Proceedings of the Research Conference on Shear Strength of Cohesive Soils*, American Society of Civil Engineers, Boulder, Colorado.

Bishop, A.W. and Henkel. D.J. (1962) *The Measurement of Soil Properties in the Triaxial Test*, Edward Arnold, London.

BS 812: Part 1: 1975 Methods for determination of particle size and shape.

BS 1377: 1990 Methods of test for soils for civil engineering purposes.

BS 5930: 1981 Code of practice for site investigations.

BS 6031: 1981 Code of practice for earthworks.

BS 8004: 1986 Code of practice for foundations.

Costet, J. and Sanglerat, G. (1969) *Cours Pratique de Mécanique des Sols*, Dunod, Paris.

Croney, D. and Coleman, J.D. (1961) Pore pressure and suction in soils, *Proceedings of the Conference on Pore Pressure and Suction in Soils*, London. pp. 31–37.

Fookes, P.G. and Vaughan, P.R. (1986) *A Handbook of Engineering Geomorphology*, Blackie, Glasgow.

Horner, P.C. (1981) Earthworks, *I.C.E. Works Construction Guides*, Thomas Telford, London.

Ingles, O.G. and Metcalf, J.B. (1973) *Soil Stabilisation, Principles and Practices*, Butterworth, London.

Loudon, A.G. (1952) The computation of permeability from simple soil tests, *Géotechnique*, Vol. 3, pp. 165–183.

McGown, A., Marsland, A., Radwan, A.M. and Gabr, A.W.A. (1980) Recording and interpreting soil macrofabric data, *Géotechnique*, Vol. 30, No. 4, pp. 417–447.

Schofield, A. and Wroth, C.P. (1968) *Critical State Soil Mechanics*, McGraw-Hill, New York.

Sherwood, P.T. (1970) *The reproducibility of the results of soil classification and compaction tests*, Road Research Laboratory Report LR 339, Ministry of Transport, London.

Site Investigation in Construction (1993) *SISG. Part 1. Without Site Investigation Ground is a Hazard*, Thomas Telford, London.

Skempton, A.W. (1953) The colloidal activity of clays, *Proc. Int. Conf. Soil Mech.*, Vol. 1, p. 57.

Taylor, D.W. (1948) *Fundamentals of Soil Mechanics*, Wiley, New York.

Terzaghi, K. and Peck, R.B. (1948) *Soil Mechanics in Engineering Practice*, Wiley, New York.

Williams, P.J. (1967) *The Nature of Freezing Soil and its Field Behaviour*, NGI Publication No. 72, pp. 90–119.

Wilun, Z. and Starzewski, K. (1975) *Soil Mechanics in Foundation Engineering, Vols 1 and 2*, Intertext, London.

Wood, D.M. (1990) *Soil Behaviour and Critical State Soil Mechanics*, Cambridge University Press.

VI POLYMERS

Introduction

Polymers are now well established alongside metals and ceramics as one of the major classes of manufactured materials. Over the last thirty years, a great variety of such materials has been developed, based on about fifty individual synthetic polymers. They are used widely throughout industry and engineering. These materials have a spread of engineering properties very different from those of metals and ceramics, and a distinct manufacturing and processing technology centred largely on moulding, extrusion and fibre-forming operations. The relatively low elastic modulus and high permissible strains of many polymers frequently allow a 'softer' approach to design and product manufacture. However, at the other extreme, there are high-performance polymer fibres which are stiffer and stronger than steel wire.

Building construction is one of the main volume markets for polymers, taking about one-fifth of total production in the USA and Western Europe. However the use of polymers in civil engineering is less conspicuous, because these materials do not generally compete directly with the traditional load-bearing materials, the structural metals, concrete and masonry. Nonetheless a number of polymers have important and established civil engineering uses. One of the most straightforward and prominent applications is in pipework, where longer established materials like ductile iron, heavy clay and concrete face severe competition from several polymer materials. Polymers also play important supporting roles as surface coatings, membranes, adhesives and jointing compounds, roofing materials, claddings and thermal insulants. Fibre-reinforced plastics have a limited role in light structures. In addition, new uses are emerging in various types of polymer concrete and for textiles both in fabric structures and in ground engineering. Some polymer fibres are now used as a reinforcement for cement-based materials.

30

Materials Science of Polymers

A polymer is a large molecule containing hundreds or thousands of atoms formed by combining one, two or occasionally more kinds of small molecule (monomers) into chain or network structures. The *polymer materials* are a group of carbon-containing (organic) materials which have *macromolecular* structures of this sort. Polyethylene, polystyrene and the epoxy resins are well known examples. Table 30.1 lists the main polymer materials which are used in civil engineering. The table includes the internationally accepted standard abbreviations, the use of which greatly reduces the difficulties of polymer nomenclature. For a fuller listing of polymer materials and discussion of structures and of nomenclature the reader should refer elsewhere (for example, Hall, 1989).

Macromolecules occur in huge abundance and variety in biology: these *natural polymers* have complex and often very precise molecular structures. Almost all the polymers used in engineering are wholly synthetic and have much simpler molecular structures. The only natural polymers of any major importance in engineering are timber and natural rubber. Timber is discussed separately in Part II. Natural rubber is included in this Part, together with the synthetic rubbers; it finds little application in civil engineering.

30.1 Molecular structure

Chain polymers

Polyethylene (PE) is an important hydrocarbon polymer which has the simple molecular structure shown in figure 30.1. The molecular chain is built of carbon atoms each covalently bonded also to two hydrogen atoms, and thus the PE polymer has the formula $(CH_2)_n$, where n measures the chain length. In commercial

Figure 30.1 *Molecular chain structure of polyethylene, showing covalent bonding of carbon and hydrogen atoms*

451

TABLE 30.1
The main polymer materials used in civil engineering

Hydrocarbon polymers

PE	polyethylene
PP	polypropylene
PB	polybutylene
NR	natural rubber
PS	polystyrene
IR	polyisoprene rubber

Other carbon chain polymers

PVC	poly(vinyl chloride)
CPVC	chlorinated poly(vinyl chloride)
PTFE	polytetrafluoroethylene
PVDF	poly(vinylidene fluoride)
PMMA	poly(methyl methacrylate), *acrylic*
PAN	polyacrylonitrile
PVAC	poly(vinyl acetate)
CR	polychloroprene rubber, *neoprene*
CM, CSM	chlorinated, chlorosulphonated polyethylene rubbers

Heterochain polymers

PA	polyamide, *nylon* (aromatic polyamide, *aramid*)
PETP	poly(ethylene terephthalate), *polyester*
PBTP	poly(butylene terephthalate)
PC	polycarbonate
POM	polyoxymethylene, *acetal*
PPO	polyphenylene oxide

Network polymers

PF	phenol–formaldehyde resins
UF	urea–formaldehyde resins
MF	melamine–formaldehyde resins
EP	epoxy (epoxide) resins
UP	unsaturated polyester resins
PUR	polyurethanes

Copolymers, alloys and hybrids

EPDM	ethylene–propylene–diene rubber
ABS	acrylonitrile–butadiene–styrene copolymer
NBR	acrylonitrile–butadiene rubber, *nitrile*
SBR	styrene–butadiene rubber
IIR	isobutene–isoprene rubber, *butyl*
TPE	thermoplastic elastomer
HIPS	high impact polystyrene
FEP	fluorinated ethylene–propylene copolymer
PPO/PS	poly(phenylene oxide)/polystyrene alloy

PE materials n is generally in the range 1000 to 10000, so that the linear chains have a length-to-thickness ratio similar to a piece of household string a few metres long. In PEs produced by the high-pressure polymerisation of ethylene (ethene) gas, some degree of chain branching invariably occurs during synthesis: this is largely suppressed in the low-pressure catalytic route to PE. The absence of chain branching permits closer packing in the solid so that the catalytic PE has a higher density (about 960 kg m^{-3} rather than about 920 kg m^{-3}). The two forms are commonly known as high- and low-density polyethylene (HDPE and LDPE).

Other important linear polymers with carbon chain molecular backbones are polypropylene (PP), poly(vinyl chloride) (PVC) and polytetrafluorethylene (PTFE). PP differs from PE in that one of the pair of hydrogen atoms attached to alternate carbon atoms along the chain is replaced by a methyl CH$_3$ group. PP is therefore also composed of only carbon and hydrogen and like PE is a hydrocarbon polymer, $\{CH_2CHCH_3\}_n$. A useful engineering polymer is produced only if each of the CH$_3$ side groups in PP is attached to the chain in the same spatial configuration: that is, all the CH$_3$ groups must lie on the same side of the chain backbone. This stereoregular form of PP is described as *isotactic*. A requirement for stereoregularity is common to several synthetic polymers and can be achieved by catalytic polymerisation.

In poly(vinyl chloride) chlorine atoms are attached to the main carbon chain, replacing one in four of the hydrogen atoms in the PE structure. In PTFE, all the hydrogen atoms are replaced by fluorine (F), giving a linear polymer $\{CF_2\}_n$. Other carbon chain linear polymers include polyacrylonitrile (PAN) (from which carbon fibre is made), poly(methyl methacrylate) (PMMA, the leading acrylic polymer) and poly(vinyl acetate) (PVAC).

Other important materials are based on linear chain polymers in which the chain backbone itself contains other atoms besides carbon (heterochain polymers). For example, the polyamides (or nylons) incorporate the amide link –NH–CO– at intervals in the chain and in polyesters such as poly(ethylene terephthalate) (PETP, used mostly as polyester fibre) the ester link –O–CO– appears periodically along the chain.

Network polymers

Several major families of materials consist of polymer networks rather than these essentially linear chains, figure 30.2. Polymer networks may arise either by cross-linking preformed linear chains, in a two-step polymerisation; or directly in a single step to form a more or less randomly interconnected three-dimensional macromolecule of indefinite extent. Examples of the first type are the epoxies and unsaturated polyesters, extensively used with glass fibre-reinforcement, in which relatively short linear *prepolymer* chains are cross-linked at the curing stage. Polyethylene itself may be converted into a network polymer by chemical or radiation cross-linking.

The phenol–formaldehyde resins (PF) (and other formaldehyde resins such as MF, UF and resorcinol–formaldehyde) are network polymers of the second type. Such unorganised networks are not common in natural materials although very similar structures are found in the lignin component of wood, and indeed the formaldehyde resins are widely used as timber adhesives.

Hybrids: copolymers and polymer alloys

The properties of a polymer may often usefully be modified by incorporating a second chain-building unit to form a *copolymer*. Thus ethylene may be copolymerised with

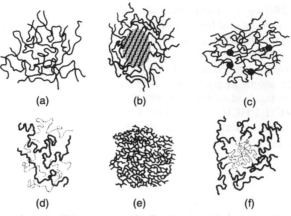

Figure 30.2 *Polymer chains in different types of polymer material: (a) amorphous linear thermoplastic; (b) semi-crystalline linear thermoplastic; (c) lightly cross-linked network polymer (elastomer); (d) copolymer; (e) highly cross-linked network polymer; (f) two-phase polymer hybrid*

propylene or with vinyl acetate to produce a variety of ethylene–propylene or ethylene–vinyl acetate copolymers. These have significantly different properties from the PE homopolymer, particularly in respect of elasticity, toughness and transparency. Catalytic polymerisation of ethylene with small amounts of hydrocarbons such as octene produces an unbranched LDPE known as linear low-density polyethylene (LLDPE): an important technical advance of the late 1970s.

Copolymers are examples of hybrid polymer materials; other hybrid types include blends and alloys. Hybridisation of polymers represents an important area of polymer development by means of which controlled changes in polymer properties may be achieved and new materials introduced. An important example of a hybrid is ABS (acrylonitrile–butadiene–styrene), a two-phase graft copolymer. This is an established engineering polymer, extensively used in pipework. In ABS the styrene–acrylonitrile (SAN) random copolymer is modified by incorporating finely dispersed particles of styrene–butadiene rubber, the molecular chains from the two phases being chemically grafted at the surface of the dispersed particles. The main effect is to toughen the otherwise rather brittle SAN. Polymer alloys are produced by blending two compatible or miscible polymers, often with a third polymer as a separate dispersed phase. Examples include ABS/PVC and PPO/PS.

30.2 Thermoplastics, thermosets, elastomers and gels

Those polymer materials (such as PE, PP, PVC and the PAs) which are composed of independent linear polymer chains melt to form liquids which, although highly viscous, may normally be extruded and injection moulded to fabricate polymeric products. Melting and solidifying occur reversibly according to temperature; hence these chain polymer materials are often described as *thermoplastics*. Most thermoplastics materials melt at relatively low temperatures in the range 100 to 250°C.

In contrast the cross-linked polymers once formed cannot melt because the segments of the macromolecular network cannot move freely relative to one another. Fabrication must be carried out before the cross-links are formed. In the early days of polymer technology the cross-linking or curing stage was often brought about by heating and these materials were therefore called *thermosets*. The term is now somewhat misleading: many such materials can be cross-linked without heating, or *cold-cured*.

In the thermosets such as PF, UF, UP and EP the degree of cross-linking is very considerable, to the extent that the polymer is really a three-dimensional random network. In other materials a much lighter cross-linking produces materials of very different properties. Light cross-linking of certain linear polymers serves simply to restrain the long-range movements of the chains and produce rubbery rather than liquid-like properties. Natural rubber (NR) cross-linked or vulcanised with sulphur was the first material of this kind but a number of synthetic rubbers now exist. The earliest of these (introduced between 1930 and 1960) are carbon chain polymers like NR itself, but later several rather different polymer materials with rubbery properties came into use, notably the heterochain polyurethanes. Generically all these materials are called *elastomers*.

The distinction between rubbers and plastics is now thoroughly blurred, both in terms of composition and of properties. For example rubbers may be made from polyethylene by chemical modification (CM and CSM types) and by copolymerisation (EPM and EPDM types); the polyurethane family contains both rubbers and thermosets (for example rigid and flexible foam formulations). Indeed there are now several commercially important *thermoplastic elastomers* (TPEs) which melt and may be processed like thermoplastics but which are rubbery solids at normal temperatures. A leading example is the styrene–butadiene block copolymer.

Cross-linking reactions which join separate polymer chains into extended networks can also be used to convert a dilute polymer solution into a soft solid or *gel*. Practical gelling polymers are often water soluble to avoid the use of organic solvents. Polyacrylamide is one widely used synthetic water soluble polymer. A 3 per cent solution of polyacrylamide in water can be chemically cross-linked after a delay controlled by addition of a chemical inhibitor. Such fluids have low initial viscosity for easy pumping but gel *in situ* to form a soft solid, for example for soil consolidation.

30.3 The polymer solid state

Polymers which have simple and regular molecular structures, such as PE, isotactic PP, PTFE and PETP, can form crystallites in the solid state, within which the molecular chains are aligned in a regular manner. The efficient packing of the molecules increases the density and maximises the forces of attraction acting between adjacent chains. Nevertheless the long molecular chains do not easily organise themselves into a crystalline lattice during cooling from the melt, and the solid polymer materials invariably comprise mixtures of crystalline and amorphous regions.

The proportion of crystalline material is about 50 per cent in LDPE but may be as high as 90 per cent in HDPE. In solid polymers the crystalline regions are roughly spherical, and grow out from central nuclei during solidification from the melt. The spherulitic morphology (figure 30.3) and hence the physical properties of the solid are considerably influenced by the rate of cooling.

Several of these semi-crystalline polymers (notably PE, PP, the polyamides PA and the polyester PETP) have civil engineering uses as fibres. The polymers are drawn to high strains to form the fibres, producing a fibrillar rather than a spherulitic morphology, in which the molecular chains are orientated roughly parallel to the fibre axis. The primary chemical bonds of the chain lie along the fibre axis, which enhances the strength and modulus of the materials in this direction.

Chain orientation is also used to enhance stiffness in orientated PVC. As applied to extruded pipe, the process involves expanding the pipe diameter after initial extrusion to achieve some preferential chain orientation in the direction of hoop stresses.

A number of linear chain thermoplastics however are completely amorphous (non-crystalline); examples are polystyrene (PS) and the acrylics such as PMMA. The failure of these polymers to form any crystalline regions may be attributed to

Figure 30.3 *Polyethylene, a semi-crystalline polymer, observed in transmitted polarised light, showing the spherulitic microstructure (Courtesy of V. F. Holland, Monsanto)*

the presence of bulky side groups along the chain. The solid polymer is composed of a myriad of randomly entangled polymer chains, which at normal temperatures (say 20°C) are locked together wherever the chains cross. The long chains are not cross-linked by chemical bonds but by physical intertwining. At these temperatures, the material is stiff (with an elastic modulus very similar to that of network thermosets) and generally brittle. As the temperature rises the thermal molecular motion gradually increases to allow chains to free themselves and move under an applied stress; thus the modulus falls and the tendency to creep increases. It is possible to identify a characteristic temperature, *the glass transition temperature T_g*, at which large-scale molecular movement becomes possible and in the region of which the mechanical and other properties show major changes. The value of T_g for PMMA is about 105°C and for PS about 95°C. As the temperature rises further beyond T_g the material becomes progressively more mobile, turning without obvious discontinuity into a viscous melt.

In the partially crystalline polymers the individual polymer chains pass from amorphous to crystalline regions and back again. The crystallites serve as cross-links in an amorphous matrix. Since the T_g of the amorphous material may be below or around the normal temperature of use, the crystalline regions stiffen the material, producing generally tough, leathery solids. The forces acting between the molecules within the crystallites themselves are not very strong and the melting temperatures are invariably rather low.

The thermoset polymers and gels have essentially random network structures and this rules out the possibility of regularly organised domains within the solid. Such materials are therefore inherently amorphous and share some of the characteristics of inorganic glasses which are similarly amorphous and macromolecular.

Elastomers are also amorphous in the unstrained state, but in some elastomers sufficient alignment of the molecular chains may occur at high strain to permit some temporary strain-induced crystallisation. This has an effect on the stress–strain diagram, producing a marked increase in modulus at very high strains.

30.4 Summary of polymer types

The linear chain thermoplastics fall into two main groups. First, there are those like PE, PP, PA, PC and PETP which are crystalline and are used either above or not far below their glass transition temperatures. They are generally tough, versatile materials, and several have important uses as solid polymer, film and fibre. Second, there are those like PMMA and PS which are wholly amorphous and which are used below their glass transition temperatures. They are generally stiff, transparent and brittle. Rubber modified thermoplastics and most polymer alloys have two-phase microstructures (figure 30.4), each phase having different physical properties. Alloys and blends are made both from crystalline and from amorphous polymers.

The elastomers are network polymers based on linear chains above their glass transition temperatures, lightly cross-linked to develop rubber elasticity. The thermosets are heavily cross-linked or random network polymers. Their mechanical properties are much less sensitive to changes in temperature than either thermo-

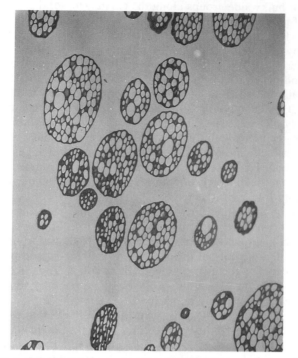

Figure 30.4 *Rubber-modified high-impact polystyrene (HIPS): electron micrograph showing dispersed SBR elastomer particles, each about 2 μm in diameter, in a PS matrix (Courtesy H. Keskulla, Dow Chemical Company)*

plastics or elastomers. They are generally stiff and strong, but usually brittle, and are widely used in fibre-reinforced composites, adhesives and surface coatings.

31

Polymer Technology

This chapter surveys the very diverse technology of polymer materials, with particular emphasis on aspects which are of interest to the design engineer.

31.1 Polymerisation reactions

The polymerisation reactions by which the solid thermoplastics are produced are mostly specialist matters for the chemist and the chemical engineer. These materials are manufactured from hydrocarbon feedstocks by the polymer sector of the petrochemical industry, remote from the processor and the consumer.

On the other hand, the reactions by which the thermoset resins (and the adhesives, surface coatings and composites based on them) are cured are of some importance to the civil engineer. The polymerisation of these materials (or at least its completion) must be deferred to the final stages of manufacture or construction, and therefore may occur in the fabrication shop or on site.

The chemistry of these reactions is complicated; in most cases, the precursor of the ultimate polymer material is provided in the form of a liquid resin (occasionally a low melting solid or paste). The polymerisation is usually completed by the addition of one or more reactive chemical substances, known as catalyst, hardener or curing agent, as in the following examples.

Unsaturated polyesters

UP resins are widely used in producing glass fibre-reinforced composites, including for example filament-wound pipe up to a diameter of 3 m. Besides the glass fibre there are three components: the linear chain UP itself, the cross-linking agent (usually styrene, an organic liquid), and the catalyst or hardener (generally an organic peroxide). The uncured UP is soluble in the styrene, which are generally supplied premixed. The addition of catalyst (in carefully controlled quantity) initiates a chain reaction between the UP and the styrene, which links the UP chains through the styrene molecules into a solid macromolecular network. The reaction may be promoted by heat or by a cold-curing accelerator (for example a cobalt salt).

Alkyd paint drying

One of the classes of paint used in the protection of structural steelwork is made by copolymerising natural vegetable oils (such as tung oil and linseed oil) and organic esters to yield branched polyester resins known as *alkyds*. Surface coating alkyds of this kind form solid films on exposure to the air by the action of atmospheric oxygen, which acts as a cross-linking agent to produce a network polymer. The paints usually contain a small quantity of cobalt accelerator to assist the reaction.

Silicone elastomer curing

Silicone elastomers are widely used as sealants in the construction industry. 'One-pack' versions are formulated and packaged with silicone elastomer already mixed with cross-linker and catalyst. Curing is initiated by atmospheric moisture. Certain cyanoacrylate adhesives also make use of atmopheric moisture to initiate the bonding reaction.

Polymer impregnated concrete

The permeability of concrete may be greatly reduced and its strength enhanced by impregnation with a liquid monomer followed by *in situ* polymerisation. The monomer most successfully used is methyl methacrylate, which has a low viscosity and is readily absorbed, and which may be polymerised by a chemical initiator or more satisfactorily by gamma-radiation. The ionising radiation acts directly on the monomer to produce a small concentration of reactive molecules (free radicals), initiating the polymerisation reaction which propagates throughout the pores of the concrete, forming solid PMMA.

31.2 Compounding of polymer materials: polymer additives

In both thermoplastics and thermosets, the main polymer constituent is rarely the sole component. Almost all commercially available polymer materials are carefully formulated combinations of base polymer and property-modifying additives included to improve performance. The main types of additives are given below.

Fillers are finely divided minerals such as china clay, talc or short staple glass fibre included to increase elastic modulus, hardness, durability and resistance to wear and to reduce cost. Impact resistance and tensile strength may be simultaneously reduced. Inorganic and organic pigments, included primarily for colour, have similar effects on polymer properties.

Plasticisers are non-polymeric organic substances added to polymers (especially to PVC materials) to soften them and make them flexible.

Stabilisers are organic substances, tailored to the requirements of particular polymers, which interfere with thermal oxidation and photo-oxidation and slow down the degradation of polymers during processing and in use.

Flame retardants are organic and inorganic substances which impede the combustion of the polymer by poisoning flame reactions.

By varying the base polymer itself (particularly the chain length and chain length distribution), by copolymerisation and alloying, and by alteration of the compounding formulations, polymer manufacturers can produce many commercial materials of the same nominal type but of widely different properties. As an example, table 31.1 lists the members of the polyethylene family.

TABLE 31.1
Polymer materials of the polyethylene family

Homopolymers	
Low-density polyethylene LDPE	Pipe (BS type 32), sheeting
Linear low-density polyethylene LLDPE	
High-density polyethylene HDPE	Pipe (BS type 50), non-woven fabrics
Ultra-high molecular weight polyethylene UHMW PE	Abrasion resistant linings
Medium-density polyethylene MDPE	Water and gas pipe
Copolymers	
Ethylene–vinyl acetate copolymers EVA	Bitumen blends, sewer repair, cable insulation
Ethylene–propylene copolymers (including EPDM, EPM elastomers)	Industrial silo liners
Ethylene–carboxylate copolymers	
Modified polyethylenes	
Chlorinated polyethylene (including CM elastomer)	Reservoir covers
Chlorosulphonated polyethylene (CSM elastomer)	Industrial roofing, pond liners
Cross-linked polyethylene	Foams, pipe
Cross-linked ethylene–vinyl acetate copolymer	Foams
Fluorinated ethylene–propylene copolymer	Hot-melt adhesive

31.3 Processing methods for thermoplastics

The principal methods of forming and shaping thermoplastics are extrusion and moulding, especially injection moulding and rotational moulding. In all these methods, the polymer is processed in the molten state (figure 31.1). Extrusion of melt through a shaped die is a relatively simple operation well suited to the production of pipe, film, conduit and fibre. Co-extrusion of two or more polymers is used to form a composite extrudate, for example providing a more durable surface layer (or *capping*) on a less durable substrate. Injection moulding is similarly suited to the repetitive production of components of complicated shape: the polymer melt is pressure-injected into a mould from which the component is ejected after

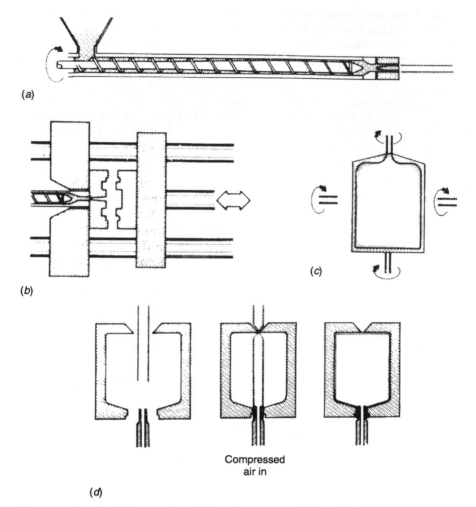

Figure 31.1 *Shaping thermoplastics: (a) extrusion; (b) injection moulding; (c) rotational moulding; (d) blow moulding*

rapidly solidifying. Rotational moulding is a low-pressure method of producing large thermoplastics products such as tanks; the thermoplastics material is melted within the mould and spread over the walls by biaxial rotation. Vacuum or pressure-assisted thermoforming of thermoplastics sheet heated above the softening temperature is another method of shaping that is widely employed. The methods of polymer processing are now highly developed and it is possible for example to produce pipe with complicated wall profiles (figure 31.2) to counteract buckling and bending in a way that would be prohibitively difficult with other materials. Likewise, complex one-piece components may be produced in long production runs by injection moulding, although the cost of tooling may be considerable and uneconomic for small production volumes.

Figure 31.2 *Large-diameter HDPE pipe of extruded spiral wall construction used for drainage (Courtesy of Bauku GmbH)*

The solubility of certain polymers in organic solvents is exploited in solvent welding and in fibre-spinning from solution. Direct heat welding is also used, particularly in joining sheet and pipe on site. For example, polyethylene pipe is commonly butt-welded in this way. Many polymers may also be satisfactorily shaped by conventional machining operations, suitably modified to take account of the deformability and low melting temperature of the polymer materials.

PTFE has an unusually high melting temperature (about 320°C) and special fabrication methods have been developed in which PTFE powders are sintered by hot pressing. Similar techniques are used for UHMW PE, which is too viscous for normal methods of melt processing.

31.4 Thermoset processes

Processes for the forming of products from unreinforced thermosets usually involve compression, transfer or injection moulding. In civil engineering the use of solid unreinforced thermosets is insignificant compared with various types of reinforced materials, for which special processing techniques have been developed.

Fibre-reinforced polymer materials

The bulk of fibre-reinforced composites are based on glass fibre with an unsaturated polyester UP or epoxy EP matrix. The term *glass fibre-reinforced plastics* (GRP) denotes these materials. The combination of a high-strength, high-modulus inorganic fibre with a polymeric matrix has proved highly successful; the polymer component protects the fibre surface from mechanical damage which would otherwise rapidly weaken it, and the adhesion between matrix and fibre ensures stress transfer within the composite. The non-metallic materials often provide resistance to environmental attack under conditions where electrochemical corrosion occurs in metals.

Furthermore, relatively simple production techniques are available. The glass fibre reinforcement incorporated into the UP or EP resin matrix may be in various forms – woven glass fibre fabric or woven rovings, chopped strand mat or continuous rovings. The simplest method of combining fibre and resin is by hand moulding (hand lay-up), but chopped fibre may be sprayed and the filament winding method for pipe and tube is to a high degree automated (figure 31.3). Forms of extrusion (pultrusion) and injection moulding (reinforced reaction injection moulding, RRIM) are also used. Fibre-reinforced plastics are strongest in the direction of the fibre and it is possible to orientate the reinforcement to support

Figure 31.3 *Production of GRP pipe by filament winding (Courtesy of Fibaflo Reinforced Plastics Limited, Poole, Dorset)*

particular service stresses. GRP sheet is often stiffened by lamination to PUR, PF or PVC foams to form sandwich structures.

Besides glass, several other high-performance fibres are available (table 31.2), some of which are themselves polymeric. Carbon fibre, produced by high-temperature treatment of polyacrylonitrile (PAN) fibre under stress, has a structure of linked carbon rings, highly oriented and strong in the direction of the fibre axis. Carbon fibre is especially compatible with EP resins and they are used together to produce high-performance composites. Another fibre with outstanding mechanical properties is the aromatic polyamide known as *aramid*. Aramid fibre has been used experimentally as a reinforcement in concretes.

Glass fibre fabric may also be used in combination with thermoplastics, notably with PTFE to produce a high-performance flexible sheet for air-supported and tension structures.

TABLE 31.2

Mechanical properties of reinforcing fibres

	Young's modulus (kN mm^{-2})	Tensile strength (kN mm^{-2})	Density (kg m^{-3})
PA 66 *(nylon 66)*	5	0.8	1100
E glass	70	3.5	2550
Carbon (high modulus)	420	1.8	1900
Aramid	125	2.9	1450

Other composites

While fibre reinforcement provides a highly controlled means of providing strength and stiffness where it is required in engineered products, other types of polymeric composites are also of some importance. For example, a variety of wood based composite materials depend on polymer binders. In plywood, peeled wood veneers are bonded in a cross-grained sandwich with PF or amino resin adhesives; wood particle boards utilise wood waste in a thermoset blend with UF resin. There are also all-polymer composites. An ingenious example is the polypropylene fibre with a surface layer of polyethylene from which a number of non-woven geotextiles are produced. PE melts at a lower temperature than PP and the textile is stabilised by heat-bonding.

31.5 Cellular polymers

Many polymers are readily converted into cellular or foamed materials. Foams may be made either by mechanical frothing or by the action of blowing agents, which produce gas bubbles within the polymer. Subsequently, the cell structure is stabilised by cross-linking, either by chemical means or by ionising radiation. PE foams are made by both of these methods.

Cellular polymers have outstanding thermal insulation properties, and densities as low as 10 kg m^{-3} are possible. Rigid foams have structural value, particularly in sandwich panel construction where a strong sheet material is bonded to one or both faces to support tension, and in integral skin structural foams. Polymer foams with closed cell structures (including syntactic foams formed with hollow glass microspheres) have oustanding long-term buoyancy in water.

Cellular forms of most major polymers are commercially available (including PE, PP, PVC, PS, UF and PF), but the rigid and flexible polyurethane PUR foams have an unusually wide range of properties and are extensively used in construction as insulating and semi-structural materials.

31.6 Impermeable membranes

Plain and fibre-reinforced membranes are made by extrusion, blowing and calendering from a variety of polymers, notably PE, PVC, EPDM and CSM elastomer. Plain sheet membranes may themselves be composite, for example, multilayer bonded sheets of EPDM and butyl rubber. The fibre-reinforced membranes may be all-polymer, as in bonded combinations of PE sheet and PA fabric, or the fibre may be non-polymeric as in the glass fibre-reinforced PTFE sheet developed for fabric structures.

31.7 Textiles: permeable membranes

Those synthetic polymers which are readily formed into continuous filaments or fibres may be converted into a variety of fibre assemblies or textiles. Woven textiles are formed from fibres or yarns interlaced in various patterns at right angles to produce a porous fabric with its greatest strength along fibre directions. Non-woven types include random mats of non-bonded fibres, melded or spun-bonded fabrics in which the fibres are partially melted and welded together at points of contact, resin-bonded and needle-punched types. In addition there are numerous open mesh thermoplastic webs, with aperture sizes as large as 75 mm. The physical properties of non-wovens are generally more isotropic than those of woven fabrics.

Figure 31.4 shows a process for making oriented polymer grids from polyethylene or polypropylene. The biaxial stretching achieves some polymer chain orientation and enhances stiffness.

31.8 Polymer emulsions

With very few exceptions, synthetic polymers are insoluble in water. In most respects this is a definite virtue, but it means that polymer solutions (for adhesives and paints for example) must be made up with organic solvents. Such solvents (commonly hydrocarbons and ketones) are flammable, need safe handling and cost more than water.

To avoid these difficulties, many polymers may be prepared in liquid form as aqueous emulsions. These are not true solutions, but dispersions of very fine par-

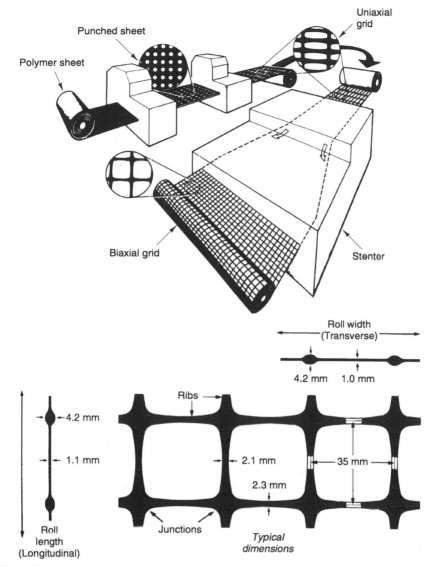

Figure 31.4 *Starting with a continuous polymer sheet, the grid is formed by punching followed by two stretching operations (Courtesy of Netlon Limited)*

ticles of polymer (usually about 1 μm diameter) in water. The dispersions normally contain 50 to 60 per cent of polymer and are stabilised by the addition of surfactants. Polymer emulsions, known also as latexes, are now widely used in surface coatings, numerous adhesives formulations, and to some extent in polymermodified Portland cement materials. Poly(vinyl acetate) (PVAC) and its copolymers, natural and synthetic rubbers and members of the PMMA family are the polymers most often encountered as emulsions.

32

Properties of Polymer Materials

The civil engineer is concerned chiefly with mechanical properties (in the widest sense, including not only deformation and strength but also impact, friction and wear characteristics) and with the physicochemical properties which determine durability. Electrical and optical properties are generally much less significant and are omitted from this chapter (see however Hall, 1989; Waterman and Ashby, 1991). Thermal properties and permeability are discussed briefly.

32.1 Density

Polymers are materials composed mainly of light elements, relatively open-packed at the molecular level; consequently solid polymers have low densities. Polymethylpentene (830 kg m^{-3}) and polypropylene (PP) (905 kg m^{-3}) have the lowest densities of commercial solid polymers; the density of PTFE is abnormally high (2150 kg m^{-3}) as a result of its high fluorine content. By comparison the steels have densities of about 7900 kg m^{-3}. When appropriate, allowance should be made for the large density factor in comparing the mechanical properties of polymers with other classes of materials.

32.2 Mechanical behaviour of polymers

As a class polymer materials show a great range of mechanical properties and in a single material the mechanical response may change appreciably over quite a small temperature range.

Some polymers (amorphous thermoplastics below T_g and the unreinforced network thermosets) are stiff materials and they fail by brittle fracture at relatively small strains (2 to 5 per cent) (figure 32.1). They show a reasonably elastic response to stresses below the failure stress. However the elastic modulus falls sharply in the region of the glass transition temperature and at higher temperatures creep effects become pronounced. The amorphous thermoplastics do not exhibit a sharp melting point, but as the temperature rises viscous flow becomes the dominant feature of the response to stress.

468

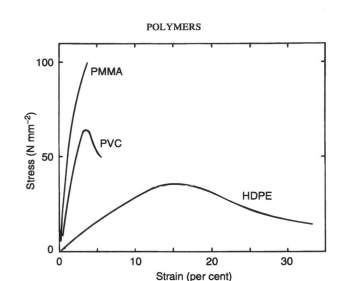

Figure 32.1 *Conventional short-term stress–strain curves of three polymer materials. (Note: the stress–strain curves are strongly influenced by the strain rate and the temperature – see text)*

The crystalline thermoplastics are tougher; they have lower elastic moduli and higher failure strains, but creep and stress relaxation are marked at normal service temperatures, as is a strong frequency-dependence in the dynamic stress response. Failure is preceded by yield; there may or may not be a well-defined yield point on the stress–strain curve. With increasing temperature the modulus falls and there is a fairly well-defined melting temperature, above which the polymer is a viscous liquid. At low temperatures the materials are stiffer and ultimately become brittle.

The elastomers exhibit uniquely low tensile and shear moduli and are capable of supporting very large strains (up to 500 per cent) completely recoverably. Cross-linking largely precludes viscous flow but at low temperatures the elasticity may be sluggish and *retarded*. A strong frequency-dependence may be observed in the mechanical response.

As a general rule the measured mechanical properties of all types of polymers vary with the rate of straining, and with temperature. Polymers as a class should be regarded as viscoelastic rather than elastic materials, in which the strain response to applied stress may develop only slowly with time, and may be only partially recoverable on removing the stress. The recoverable part of the creep strain may be regarded as a retarded elastic (or *anelastic*) component; the irrecoverable part as a viscous flow component. The time-dependence of the mechanical response introduces phase lag and hysteresis when the stress is cyclic, and cyclic stressing of polymers may cause mechanical energy to be converted to heat in these materials, producing undesirable temperature rise.

Tensile modulus and compliance of thermoplastics

Creep is pronounced in thermoplastics at normal service temperatures; thus for example the strain ε under constant tensile load or stress σ increases with time as shown in figure 32.2. A family of creep curves is one very useful way of presenting mechanical data for design. Creep strain is often approximately linear on a double logarithmic plot.

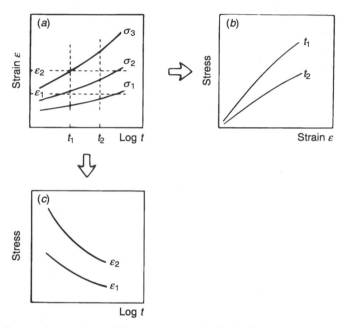

Figure 32.2 *Creep in thermoplastics: (a) creep strain curves; (b) isochronous stress–strain curves; (c) isometric stress–time curves, derived from the creep curves*

The data contained within a family of creep curves may be depicted in two other ways: as isometric stress–time curves ($\sigma(t)$ at constant ε), and as isochronous stress–strain curves ($\sigma(\varepsilon)$ at constant t). Isometric stress–time curves are useful in connection with maximum strain approaches to design.

For many thermoplastics, especially well below failure, the creep strain is proportional to the stress; that is, the time-dependent tensile creep compliance $D(t) = \varepsilon(t)/\sigma$ or the creep modulus $E(t) \approx 1/D$ is independent of the stress σ. For such *linear* viscoelastic materials, the family of creep curves may be consolidated into a single curve representing the apparent or viscoelastic modulus $E(t)$. Furthermore for these materials the isochronous stress–strain curve is a straight line.

Linear viscoelastic materials also approximately obey Hooke's law in short-term testing, so that the stress–strain curve obtained by direct tensile testing at constant strain rate is approximately linear. For those thermoplastics for which the stress–strain relation is not linear (non-linear viscoelastic materials) the short-time modulus is then taken as the tangent at the origin or the 0.2 or 1.0 per cent secant modulus.

In thermoplastics, most but not always all of the creep strain is progressively recovered on removal of the load. It is conventional to present recovery data in the form of a graph of the fractional recovered strain versus reduced time (see figure 32.3).

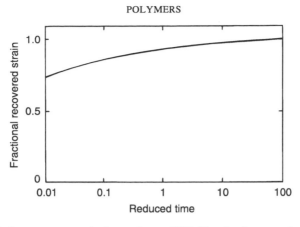

Figure 32.3 *Typical creep recovery of polypropylene at 20°C. (Fractional recovered strain = recovered strain/initial creep strain; reduced time = recovery time/duration of preceding creep)*

Stress relaxation

In polymer materials held at constant strain the tensile stress diminishes with time; such stress relaxation results from the molecular chain mobility in the solid, of which creep is itself another manifestation. Stress relaxation data may be obtained by direct test at constant strain; alternatively isometric stress–time data derived from creep tests are often used, although the stress history of the material is different in the two cases.

High stress response and tensile strength

At high stress the viscoelastic response of polymeric materials tends to become markedly non-linear, so that the creep modulus is no longer independent of strain (figure 32.4). At sufficiently high strains, failure occurs, although this may only be after prolonged creep. Polymers exhibit many modes of mechanical failure ranging from brittle fracture to ductile yielding (with or without the formation of a stable neck) leading ultimately to rupture. The mode of failure is greatly influenced by the rate of deformation (strain rate) and temperature. Thus the strength itself is a function of time (see figure 32.5) and also of temperature. For both thermoplastics and thermosets a linear double logarithmic plot may be used to estimate long-term strengths, but the extrapolation should be made with caution. For some materials the line takes a downturn at long times, and it is not wise to extrapolate beyond a factor of about 10 in time. The downturn in the creep rupture strength marks a change in the failure mode, often associated with long-term environmental stress cracking.

Impact strength

High stress creep testing carried to failure (creep rupture) represents an extreme condition of very low strain rate. At the other extreme is the impact test in which

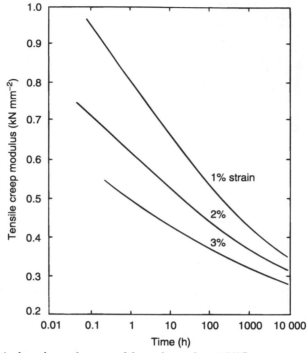

Figure 32.4 *Strain-dependence of creep modulus: polypropylene at 20°C*

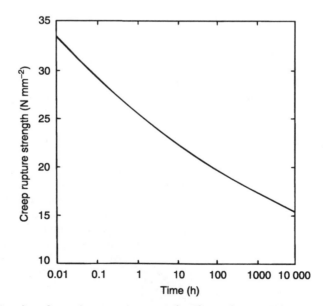

Figure 32.5 *Time-dependence of creep rupture strength: polypropylene at 20°C*

a very high strain rate is imposed. The impact strength of polymers may be measured by the energy absorbed during fracture using pendulum impact machines similar to those for metals (see Chapter 5). Charpy and Izod test arrangements are employed, normally with carefully prepared specimens with a radiused notch. Vincent (1971) introduced a simple threefold classification of impact performance in polymers (see table 32.1). Impact strength, like other mechanical properties in polymers, is strongly influenced by changes in temperature, impact strength generally falling as the temperature falls.

TABLE 32.1

Threefold classification of impact performance of polymers (20°C) after Vincent (1971)

Brittle*	Notch brittle*	Tough*
PMMA	PP	Wet nylon PA
Glass-filled nylon PA	PVC	LDPE
PS and HIPS	Dry nylon PA	ABS (some)
	POM	PC (some)
	HDPE	PTFE
	PPO	Some high impact PPs
	ABS (some)	
	PC (some)	
	PETP	

*Brittle: specimens break even when unnotched. Notch brittle: specimens do not break unnotched, but break when sharp-notched (notch tip radius $r = 0.25$ mm). Tough: specimens do not break completely even when sharp-notched ($r = 0.25$ mm).

Friction and wear

PTFE has the lowest coefficient of friction of any solid engineering material (0.03 to 0.10), whereas the rubbers have among the highest (1 to 3). The coefficient of friction is found to vary with sliding speed, so that the classical laws of friction do not apply to polymers in their simple form. Resistance to wear is complex, but is broadly related to friction and strength. Certain tough polymers are very resistant to wear; for example HDPE polymer pipes carrying mineral slurry have been found to wear at one-fifth the rate of cast iron pipes. Ultra-high molecular weight polyethylene has outstanding abrasion resistance combined with a very low coefficient of friction.

Thermoset materials

Creep is measurable in the structural thermosets UP and EP but at normal temperatures the effect is small in magnitude compared with thermoplastics. Without reinforcement both UP and EP thermosets are generally brittle, with breaking strains of 1 to 5 per cent. Some increase in flexibility can be obtained by plasticisation of UP, but durability is impaired. Epoxies may be toughened by copolymerising with butadiene–acrylonitrile elastomer to give a two-phase hybrid.

Temperature-dependence of mechanical properties

Polymers exhibit creep because molecular reorganisation occurs in the solid under the action of an applied stress, as a result of at least partial chain mobility. This mobility increases sharply at the glass transition temperature and above T_g creep increases rapidly with increasing temperature up to the melting or softening temperature. (Indeed in elastomers, for which normal service temperatures are well above T_g, retardation times may be so short that the materials are effectively elastic.) Creep may be regarded as a premelting phenomenon; it is a feature of the mechanics of polymers because they are used at temperatures much closer to their melting ranges than is normal for metals and ceramics. When metals are used at elevated temperatures creep likewise becomes a problem.

The measurement of mechanical response at elevated temperature provides a form of accelerated testing to obtain long-term performance data for use at somewhat lower service temperatures. Provided that wild extrapolations are avoided the approach is most useful and indeed may be reasonably quantitative. For example at constant stress the time to failure t_f in a creep rupture test is related to the temperature T by the equation $t_f = a \exp(K/T)$ where a and K are constants. $K = T \ln(t_f/a)$ is known as the Larson–Miller parameter.

Simplified time-dependent properties

For many design purposes, short-term and long-term values of the creep modulus and of the tensile strength provide an adequate mechanical specification of a polymer material. The moduli may be obtained from graphs such as those of figure 32.2. Alternatively creep strain data may be fitted to semi-empirical creep functions, such as that of Findley which has been widely used (see ASCE, 1984):

$$\varepsilon = \varepsilon_0 \sinh(a\sigma) + bt^n \sinh(c\sigma)$$

where ε_0, a, b, c and n are material constants which do however vary somewhat with temperature. The apparent creep modulus E_t can be obtained by calculation at any time and stress (t, σ).

Corresponding long-term strengths are obtained from creep rupture test data, extrapolated if necessary.

Cellular materials

Structural foams are of engineering interest because it is often possible to obtain higher stiffness from a given mass of material in cellular than in solid form. The elastic modulus of closed-cell thermoplastics decreases roughly as the square of the foam density (Iremonger and Lawler, 1980) but the increase in moment of inertia I may be such that the product EI increases and the deflection in simple bending may decrease. If, in addition, there is a dense integral skin or bonded sheet providing stiffness and strength on the tension face, then a useful structural foam composite or structural sandwich foam may be formed.

Fibre-reinforced materials

The mechanics of the fibre reinforcement lies outside the scope of this chapter (but see Hull, 1981; ASCE, 1984; Matthews and Rawlings, 1994). Generally the presence of a relatively stiff and elastic reinforcing-fibre reduces creep in the matrix, though measurable creep may still occur. Nevertheless for design purposes fibre-reinforced thermosets are frequently regarded as elastic materials, especially for stress parallel to the fibre direction. Creep at right angles to the fibre axis may be more pronounced. Chopped strand mat reinforcement produces approximately isotropic composites, but continuous rovings and woven reinforcements are markedly anisotropic. Fibre-reinforcement normally improves impact resistance because the fibre-matrix interfaces act as crack stoppers and crack deflectors.

The properties of some GRP materials, thermoplastics, unreinforced thermosets and non-polymer materials are compared in table 32.2.

Fibre-reinforced thermoset composites do not fail in a brittle manner but normally by progressive loss of adhesion at the fibre–matrix interface. First damage is microcracking, which may be detected by a change of slope in the stress–strain curve, by a whitening of the material (due to light scattering at microvoids) and by the development of porosity. In hydrostatic pipe testing microcracking is apparent as leaking or weeping of water from the pipe wall. This may occur at strains as low as 1 per cent in static loading and less in materials subject to cyclic fatigue.

32.3 Thermal properties

Thermal conductivity and specific heat capacity do not vary greatly from one polymer to another. Polymers do not conduct heat readily; most solid polymers have thermal conductivities in the range 0.15 to 0.45 W m^{-1} K^{-1}. Thermal conductivities as low as 0.024 W m^{-1} K^{-1} can be obtained in foamed polymers and such materials are widely used as thermal insulants.

The thermal expansivity of polymers is larger by a factor of about 10 than that of most metals and ceramics, and usually lies in the range 5 to 20×10^{-5} K^{-1}.

32.4 Permeability

Solid polymers do not usually contain interconnected pores and may generally be regarded as impermeable, except when used as very thin sheets or surface coatings, when they may transmit measurable amounts of gases, vapours and liquids by direct permeation through the solid. Considerable differences are found in the permeability of individual polymers to permeants of different kinds, as table 32.3 shows. The permeability data suggests that an LDPE sheet 0.2 mm thick would transmit about 0.3 g m^{-2} of water vapour per day.

Sheet membranes used for water retention and pipes for water and gas distribution are usually so thick for mechanical reasons that the permeance (permeability/thickness) to water and water vapour is effectively zero. However, plastic materials such as GRP and MDPE pipe may be vulnerable in contact with organic fluids. For example, chlorinated hydrocarbons will permeate through MDPE and can contaminate potable water carried in MDPE pipe.

TABLE 32.2
Physical properties of some thermoplastics, thermosets, composites and foams

	Density (kg m^{-3})	Short-term mechanical properties		Approximate coefficient of thermal expansion (× 10^6 K^{-1})	Specific heat capacity (kJ kg^{-1} K^{-1})	Thermal conductivity (W m^{-1} K^{-1})
		Tensile modulus (kN mm^{-2})	Tensile strength (N mm^{-2})			
PVC	1400	2.4–3.0	40–60	70	1.05	0.16
HDPE	960	0.4–1.2	20–30	120	2.3	0.44
LDPE	920	0.10–0.25	10–15	160	1.9	0.35
PP	900	1.1–1.6	30–40	60	1.9	0.24
PA-6	1130	1.0–2.8	50–80	60	1.6	0.31
UP	1200–1400	2.0–4.5	40–90	100	2.3	0.20
EP	1100–1400	3.0–6.0	35–100	60	1.05	0.17
GRP: polyester						
50 to 80% glass unidirectional	1600–2000	20–50	400–1000	10	0.95	0.3 – 0.4
45 to 60% glass woven rovings	1500–1800	12–24	200–350	15	1.0 – 1.2	0.2 – 0.3
25 to 45% chopped strand mat	1400–1600	6–11	60–180	30	1.2 – 1.4	0.2
GRP: epoxy, 60 to 80% glass unidirectional	1600–2000	30–55	600–1000	6	—	0.02
PU rigid foam	50	0.01	0.07	—	—	0.02
	25	0.003	0.02	—	—	0.02

TABLE 32.3

Permeability of common polymers to air, methane and water vapour

| | Permeability, 10^{-11} cm^{-3} STP/(cm s torr) | | |
	Air	Methane	Water vapour
HDPE	0.19	0.39	12
LDPE	1.36	2.9	90
Unplasticised PVC	0.02	0.03	275
PP	0.81	—	51
Natural rubber (NR)	12.2	30.1	2300

So-called permeable membranes for drainage and soil stabilisation are made from woven and non-woven fabrics of porous construction. The hydraulic permeability of these materials is determined by the weave, the fibre or yarn thickness, the porosity and the pore size, and fabrics are manufactured to meet specific engineering requirements for flow rate.

32.5 Durability

The changes which bring about environmental degradation in polymers and ultimately determine durability are complex and varied. The complexity arises from the conjoint action of a number of agents of degradation, notably ultra-violet radiation (from sunlight), heat, oxygen, ozone and water. Table 32.4 lists the main agents and modes of degradation in polymer materials. These agents produce physical and chemical changes at the molecular level; the changes may differ considerably from one polymer to another, producing polymer-specific behaviour. For example, polyethylene is subject to photo-oxidation in sunlight, unless deliberately stabilised by ultra-violet absorbers such as carbon black. In contrast, PMMA and PTFE are largely resistant to this form of degradation. As a result they are much more resistant to weathering.

Photo-oxidation and thermal oxidation by reaction of polymer with atmospheric oxygen may bring about cross-linking of polymer chains with accompanying embrittlement, or may break chains into small oxidised fragments. Often these fragments are water soluble and are washed out, producing surface erosion. In

TABLE 32.4

Main agents and modes of degradation in polymers

Oxygen at moderate temperatures	*thermal oxidation*
Oxygen at high temperatures	*combustion*
Oxygen + ultra-violet radiation	*photo-oxidation*
Water	*hydrolysis*
Heat alone	*pyrolysis*
Ionising radiation	*radiolysis*
Micro-organisms	*biological attack*
Atmospheric oxygen + water + solar radiation	*weathering*
	atmospheric degradation
Solvents, organic fluids	*softening/dissolution*

heavily filled polymer materials (paint coatings and unplasticised PVC for example) this causes chalking and loss of gloss.

Ionising radiation also causes molecular damage to polymers, resulting in rupture, cross-linking and degradation of polymer chains. As in other aspects of durability individual polymers differ considerably in their resistance to radiolysis.

Several polymers are known to fail by slow brittle fracture at stresses well below the normal failure stress when they are exposed to certain specific organic substances. *Environmental stress cracking* probably results from the action of these substances (ESC promoters) at points of local stress concentration. Polyethylene is susceptible to this form of failure and detergent surfactants may act as promoters. By increasing the chain length however PEs may be made considerably more resistant to stress cracking.

The processes which lead to degradation in polymers are generally not the same as those which promote corrosion in metals or deterioration in ceramics and cementitious materials. Many polymers are resistant to substances which are normally considered aggressive; thus while subject to photo-oxidation and pyrolysis, many (but not all) polymers are highly resistant to attack by acids, alkalis and aqueous salt solutions. This ensures that the durability characteristics of the different classes of engineering materials are frequently broadly complementary.

Many common polymers are also resistant to many organic fluids but some are also vulnerable to softening with degradation of performance properties. High temperatures usually exacerbate solvent attack. Fluids such as chlorinated hydrocarbons, petroleum hydrocarbons, esters and ketones are particularly aggressive to certain polymers (table 32.5). Test data should be checked carefully on individual materials where contact with contaminated ground, spillages or contact with adhesives and coatings is foreseen.

TABLE 32.5

Guide to risk of deterioration by contact with organic chemicals for selected polymers

	PE	PMMA	PVC	EP	PTFE	UP/GRP	EPDM elastomer
Alcohols	2	3	2	1	1	2	1
Esters	2	3	3	2	1	3	1
Ketones	1	3	3	2	1	3	1
Chlorinated hydrocarbons	3	3	3	2	1	3	3
Gasoline	2	2	2	1	1	2	3
Fuel oil	2	3	2	1	1	1	3

Note: This table is intended only to indicate in the very broadest terms the probable response of some major groups of polymers to prolonged contact with organic fluids. Practical materials selection should be based on careful assessment of test data.
1: Little or no loss of performance.
2: Some softening, expansion and loss of strength.
3: Severe softening or loss of performance.

Glass fibre-reinforced plastics

In fibre-reinforced materials, durability depends on maintaining good adhesion between fibre and matrix. Water penetration at the matrix–fibre interface may produce long-term deterioration in mechanical properties. In glass fibre-reinforced polyester mouldings a surface layer of resin (or *gel coat*) about 0.3 mm thick is normally provided to give protection to the component. The gel coat is sometimes itself reinforced with a fine glass or polyester fibre surfacing tissue. The weather resistance of isophthalic acid resins is somewhat greater than that of general purpose orthophthalic acid resins, but flame-retardant grades have poorer weather resistance.

Fire behaviour

Heat alone produces irreversible damage to the primary molecular chains in polymers at comparatively low temperatures (200 to 300°C) by pyrolysis. Some polymers are more resistant to thermal degradation than others, but in all polymers at sufficiently high temperatures pyrolysis produces small molecular fragments which appear as an organic vapour. A solid carbonaceous residue (incorporating any mineral filler present) may also be formed. Under fire conditions the vapours from the decomposing polymer support combustion and the heat from the combustion reactions itself promotes further degradation. Thus combustion becomes a self-supporting process and ignition occurs. The combustion of synthetic polymers (as of natural combustible materials) produces toxic substances which are a serious fire hazard. All polymers are combustible but vary somewhat in the ease with which they burn. Hydrocarbon polymers (such as PE, PP and PS) burn rather like paraffin wax and release about the same amount of heat per unit mass. Polymers containing oxygen release rather less heat. Halogen-containing polymers such as PVC and PTFE burn only with difficulty and will not burn freely in air. However PVC does decompose at relatively low temperatures releasing copious amounts of acidic hydrogen chloride gas. The fire behaviour of polymers may be modified by compounding with flame retardants, which interfere with the flame chemistry, but these additives cannot change the fundamental combustibility of organic polymer materials.

For use in building construction, polymer materials must meet the requirements of building regulations and codes, and must perform satisfactorily in standard fire tests. In the UK, the test procedures are described in the various parts of BS 476. Wide variation in fire properties exists among polymer materials, and fire performance may be considerably improved by formulation, including choice of base polymer, the incorporation of mineral fillers and of flame retardants. Surface spread of flame and fire propagation index test data are particularly important in assessing fire hazard. At present there is no British Standard test for the smoke yield of combustible building materials. The ASTM tunnel test (E84) assesses flame spread, smoke yield and heat contribution.

Besides methods of test for building materials in general, there exist a number of special small-scale procedures for polymers. These include the critical oxygen index test (BS 2782 Part 1; ASTM D2863), which provides a measure of the difficulty with which materials burn. The index (COI) is the minimum fraction of oxygen in an oxygen–nitrogen mixture in which the polymer, once ignited, burns freely under specified test conditions.

The assessment of fire hazard is complex and difficult. Polymer materials have obvious shortcomings when exposed to fire, and their use should follow a balanced assessment of risk.

Biological attack

Most synthetic polymers appear to be very resistant to bacterial and fungal attack and withstand soil burial without decay (Küster, 1979). This is a benefit for durability in service but also means that the major synthetic polymers are not biodegradable. The main exceptions are plastics derived from natural polymers, such as cellulose materials, natural rubber and surface coating alkyds incorporating natural drying oils. Heterochain fibre-forming polymers (polyesters and polyamides) show some long-term loss of strength in burial tests which may be caused by microbiological action. There is also some evidence that polymers degraded by photo-oxidation and other weathering processes are more susceptible to microbial attack.

A number of polymer additives, especially plasticisers, stabilisers and dyes, are undoubtedly vulnerable to biological degradation, and indeed this is the commonest cause of fungal and bacterial attack on plastics, notably on plasticised PVC.

32.6 Toxicity

Some of the organic monomers from which polymers are synthesised are recognised as toxic, and severe controls are placed on the handling of these substances. Residual free monomer levels in thermoplastics are generally extremely low and these materials are not normally considered hazardous. However if exposed to abnormally high temperatures, partial pyrolytic decomposition may occur, releasing monomer or other volatile and toxic substances.

There is no doubt that polymer materials contribute to the rapid production of toxic gases in the early stages of building fires.

Unpolymerised substances should be handled with strict attention. Materials in this category include epoxy and polyester components for thermosets, adhesives and coatings, acrylamide grouts for chemical soil stabilisation, and formaldehyde for PF and UF foams.

Further toxicity problems arise with certain polymer additives, and additives permitted in formulations for contact with potable water should be subject to tight control.

33

Polymers in Civil Engineering

The diversity of polymer materials has been emphasised in previous chapters. Here the main uses of polymers in civil engineering are considered from the view point of the materials specialist.

33.1 Structural plastics and composites

Except for pipe, large components of unreinforced solid polymers are unusual because of the relatively low stiffness of these materials. PMMA, PVC and PC are used for glazing, roofing and cladding panels, where loads are generally low, and component stiffness is achieved by vacuum forming to a domed profile, or by incorporating webs, ribs or folds. PP and PE tanks are produced up to dimensions of several metres. Innovative uses of thermoplastics exploit extrusions or mouldings of complicated shape, including numerous building components from window frames to reinforcement spacers.

Glass-reinforced plastics, usually polyesters, are more extensively used, for large tanks, conduit and building panels (ASCE, 1984, 1985; Hollaway, 1994), the latter often of sandwich construction with a polyurethane or other cellular polymer core. In all applications of structural plastics, fire performance is a paramount consideration.

Ropes, grids and rebars for structural concrete

A potentially important structural application of high-performance polymers is the use of continuous fibres (in various forms) as structural reinforcement and as prestressing tendons. Aramid and carbon fibre have stiffness and strength comparable with that of conventional steel materials and are beginning to find valuable application in situations where steel is liable to corrosion. Figure 33.1 compares the mechanical properties of such systems and shows that the polymer element can act in a fully load-bearing fashion. A number of approaches are currently in development (Clarke, 1993). These include various combinations of polymer and nonpolymer fibre and polymer matrix. Among these are:

481

(1) the use of unbonded aramid fibre bundles (*parafil* ropes) as prestressing
    tendons;
(2) the use of aramid, carbon or glass fibre bonded into solid rods with epoxy
    resin matrix and pultrusion-processed to form a polymer version of a
    conventional rebar;
(3) the use of carbon, glass and aramid fibres bonded with epoxy to form two-
    and three-dimensional grids.

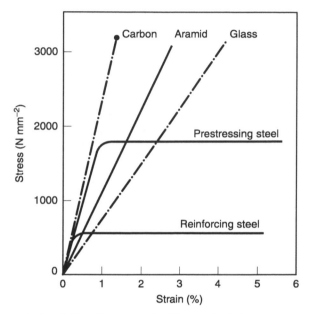

Figure 33.1 *Stress–strain relations for engineering fibres and reinforcements*

33.2 Pipe

Polymer pipes are now used very widely for water and gas distribution, drainage,
for carrying industrial effluents and sewage, and in chemical plant construction.
MDPE and unplasticised PVC are the main thermoplastics materials but there is
some use of HDPE, LDPE, PP, ABS and polybutylene. In addition, thermoset
pipes are manufactured from glass fibre-reinforced polyester and also from reinforced thermoset mortars containing silica sand fillers. There are also polymer
pipes of more complicated construction: for example PVC pipe wrapped with
epoxy-bonded glass fibre rovings to support hoop stress; and PVC pipe with
polyurethane foam insulation cased in a PVC outer skin.

All these materials are generally resistant to degradation from within the pipe
and from external attack in aggressive ground conditions unless there is contamination by organic fluids such as oils and solvents (De Rosa *et al.*, 1988). HDPE
pipes may be thermofusion welded or coupled with compression joints; PVC pipes
may be solvent welded although this is not recommended for underground pipes,
for which elastomeric compression fittings are commonly used. Thermoset pipes

may be jointed either with elastomeric compression rings or with resin-bonded overlays.

Polymer materials are also employed in combination with longer-established pipe materials. Composite polymer/steel pipes with polystyrene foam sleevings are produced for insulated pipework. Cast iron pipes are manufactured with thermoplastics and thermoset polymer linings to eliminate metal corrosion. A variety of plasticised PVC, UP, PUR and butyl rubber jointing compounds, gaskets and collars are used in jointing vitrified clay pipes.

Thermoplastics pipes may be subject to somewhat greater deflections than steel and concrete pipes. The design calculations must allow for creep or stress relaxation by using appropriate long-term (10, 20, or 50 year) values of tensile modulus or pipe stiffness. Maximum design stress is determined from creep rupture strength–time data (figure 32.5). In the UK, Type 32 and Type 50 polyethylenes as defined in British Standard CP 312: Part 2 have long-term stress ratings of 3.2 and 5.0 N mm^{-2} respectively. Creep effects are considerably smaller for thermoset pipes, as indicated by the smaller values of the ratio R ($= E_0/E_{10y}$) given in table 33.1.

TABLE 33.1

Long-term mechanical properties of pipe materials (from Chambers and McGrath, 1981)

	E_0 (kN mm^{-2})	R^*	Long-term strength (N mm^{-2})
PVC	3	2	30
PE	0.7	2	10
Reinforced thermosetting resin RTR	20	1.25	95
Reinforced plastic mortar RPM	14	1.25	45

$^*R = E_0/E_{10y}$ where E_0 is the short-term modulus and E_{10y} the extrapolated 10 year modulus.

Reported service defects in polymer pipes which are directly attributable to materials include environmental stress cracking in PVC and PE gas distribution pipes, cracking in large-diameter HDPE water pipes traceable to thermal oxidation during manufacture, and fractures caused by inclusions and voids in the wall of PVC water pipes, since PVC is generally somewhat notch-sensitive.

TABLE 33.2

Current usage of various materials for trunk water mains: UK practice (De Rosa et al., 1988)

Nominal bore (mm)	Trunk $\geqslant 300$	Distribution $300 - 50$	Service < 50
Asbestos cement	•	•	
Copper			•
Ductile iron	•	•	
Glass fibre-reinforced plastic	•		
Medium density polyethylene	•	•	•
Prestressed concrete	•		
Steel	•		
Unplasticised polyvinyl chloride	•	•	

However, failures more often arise from defective jointing or embedment. For use in drinking water supply, the formulation of polymer materials is stringently controlled to exclude toxic additives, such as certain antioxidants and heat stabilisers.

Water mains: pipe selection

Polymer pipe materials now compete with ductile iron and concrete for trunk and distribution piping of water and with copper for small-diameter service piping. Table 33.2 shows the application for the UK, patterns in other countries differ somewhat. For drainage, the main competing materials are unplasticised PVC and heavy clay. Polymer pipe materials have outstanding durability in a wide range of soil conditions but are subject to degradation in contact with certain organic fluids.

Table 33.3 summarises performance factors for materials selection.

33.3 Polymer membranes

Impermeable polymer membranes have long been used in building construction as barriers to capillary water movement in walls and floor slabs. The use of loose-fitting tubular polyethylene film as protective sleeving for buried ductile iron pipe is highly effective in reducing external corrosion and is well established (BS 6076). So also are the larger-scale civil engineering uses as linings for reservoirs, canals and industrial ponds and tanks and in roof construction, which have followed the development of proven methods for manufacturing, handling and joining large flexible sheets.

The main requirements placed on membrane materials are impermeability, flexibility, strength and tear resistance; and above all durability. Of the thermoplastics, polyethylene and poly(vinyl chloride) membranes are widely used. Below ground PE is damaged only by the boring action of animals and insects such as termites, a problem for all polymer membrane materials. Otherwise resistance to biological attack is excellent. Above ground, ultra-violet radiation promotes photo-oxidation in PE, which must be stabilised, usually with carbon black.

Other civil engineering membranes are mostly elastomers, generally thicker and stronger than PE and PVC. They include butyl rubber, chlorinated and chloro-sulphonated polyethylene elastomers (CM and CSM), and ethylene–propylene copolymer elastomer (EPDM). Some of the membranes used are composite and in certain cases they are laid on permeable fabrics, such as non-woven polyester fibre mats, to provide bedding. Membrane materials may be solvent welded or heat welded to form strong, water-tight bonds.

Permeable membranes: geotextiles

Fabrics are applied primarily for ground stabilisation and containment, for separation and for drainage and filtration purposes. For all these applications numerous fabrics are available, with controlled permeability and strength. Both woven and

TABLE 33.3
Polymer materials for trunk water mains: check list for materials selection
(De Rosa et al., 1988)

Pipe material	Advantages	Disadvantages	Applications Suitable	Applications Not advised
Glass fibre-reinforced plastic	Corrosion resistant Relatively light weight Ease of jointing Flexible joints tolerate some deflection	Susceptible to impact damage May suffer strain corrosion attack in soils/water of low pH Reliant on stable support from soil (2) Pipe location difficult Leakage detection complicated Retrospective installation of fittings/repair complicated Susceptible to permeation/ structural degradation by certain organic contaminants	Most low and medium stress applications including: • gravity mains • minor carriageways	• pumped mains/surge conditions (1), (3) • locations subject to third party interference • major carriageways • ground subject to significant ground movement and subsidence (5) • contaminated ground • soil/water pH <5.5
Medium density polyethylene	Corrosion resistant Relatively lightweight Flexible Alternative installation techniques possible (narrow trenching etc.) Can be welded to form leak-free system that will resist end load Out of trench jointing possible	Fusion jointing requires skilled installers and special equipment Reliant on stable support from soil (2) Need to pressure derate where risk of long-line fracture exists Susceptible to permeation/ degradation by certain organic contaminants Ultra-violet degradation of *blue* MPDE on prolonged exposure to direct sunlight Pipe location difficult Leakage detection complicated Retrospective installation of fittings/repair complicated	Most low and medium stress applications including: • gravity mains • minor carriageways • ground subject to movement and subsidence	• pumped mains/surge conditions (1) • major carriageways • contaminated ground • applications involving long-term exposure to direct sunlight (i.e. above ground outdoors)
Plasticised PVC (vinyl chloride)	Corrosion resistant Relatively lightweight Ease of jointing	Susceptible to impact damage Susceptible to poor installation practice Reliant on stable support from soil (2) Ultra-violet degradation on prolonged exposure to direct sunlight Susceptible to permeation/ degradation by certain organic contaminants Pipe location difficult Leakage detection complicated	Low stress applications (4)	• pumped mains/surge conditions • major carriageways • ground subject to movement and subsidence • contaminated ground (4)

Notes: (1) Unless a suitable surge analysis has been carried out and the range of pressures shown to be within acceptable limits.
(2) Dependent upon pipe stiffness.
(3) Resistance to negative surge pressures dependent upon pipe stiffness.
(4) Only recommended on a trial basis in certain low stress applications where the level of supervision is sufficient to ensure correct installation.
(5) Unless careful consideration given to joint performance and pipe design limits.

non-woven textiles are used. PP, PP/PVC, PP/PA and polyester spun-laid fabrics are used in temporary and permanent road construction, as sub-grade/sub-base separator, and for drainage. In a somewhat similar way fabrics can be incorporated in railway track construction to separate ballast from sub-grade. Geotextiles have proved valuable in controlling coastal and river erosion, and in providing soil stabilisation in the construction of causeways and in land reclamation projects. For drainage and irrigation, geotextiles may be matched to the particle size distributions and permeabilities of soils to optimise flow. Non-woven matting can provide very efficient drainage at the face of retaining walls and abutments, taking off water directly to slit polymer pipes.

Fabric structures

Fabric structures provide a technical solution to the problem of enclosing or roofing large areas (figure 33.2). Strong fabrics with long-term (say, 20 year) durability capable of meeting building regulations for permanent rather than temporary structures are based on woven glass fibre cloth coated with PTFE. PTFE has outstanding resistance to weathering. PTFE/glass fibre fabrics are used both in mast and cable-supported tension structures and in air-supported structures. Less durable but cheaper PVC coated polyester (PETP) fabrics are available for temporary or short-life structures.

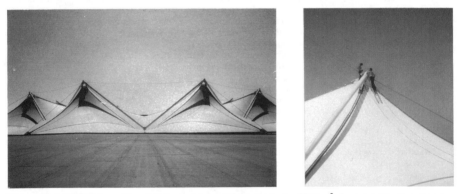

Figure 33.2 *PTFE/glass fibre fabric structure: fabric roof of a 6500 m² department store; tension structure supported on steel A frames (Courtesy of Birdair Structures, division of Chemical Fabrics Corporation)*

The PTFE/glass fibre materials developed for these uses are based on a plain weave of glass fibre yarn (the yarn having about 200 individual filaments, each 3.8 μm in diameter) coated with PTFE to produce an impermeable fabric 0.8 to 1.0 mm thick, weighing 1.25 to 1.5 kg m^{-2}. Since they are based on woven reinforcements, the strength and elongation characteristics vary somewhat with direction. Fabric design stresses are normally about 15 per cent of the tensile strength, which is about 140 N mm^{-2} along the warp. Tear and flex strength are satisfactory. The fabric is produced in continuous lengths 5 m wide, and panels are joined by heat welding. Overlaps of 75 mm are bonded by a fluorinated ethylene–propylene copolymer FEP hot melt adhesive, with which the fabric is precoated.

The combination of non-combustible glass fibre and PTFE (which has a critical oxygen index of 0.95) produces a composite which has performed well in standard fire tests to assess surface spread of flame and smoke yield. PTFE/glass fibre fabrics have a high solar radiation reflectance (68 to 75 per cent of incident radiation); a further 5 to 11 per cent of incident light is diffusely transmitted and the rest is absorbed. The translucency may be reduced if necessary by opaque pigments. Considerable thermal control may be achieved by the use of inner fabric liners with a separating air space.

33.4 Coatings

Surface coatings are used to control corrosion on metal structures of all kinds and to a lesser extent on wood and concrete also. While many different types of surface coating are manufactured, those widely used in the protection of steel structures are the alkyds and oleoresinous paints, and epoxy EP and chlorinated rubber paints. The selection and specification of coating materials for this purpose are fully discussed in BS 5493.

All paint coatings are mixtures of a polymeric binder (which forms the coating film) and solid fine-particle pigments. Pigments act as fillers from the point of view of mechanical properties. They serve also to protect the polymer film from photo-oxidation and thus improve weathering. Pigments may also be active in corrosion control; for example, zinc metal pigments in *zinc-rich paints* act electrochemically at the steel surface, and phosphate and chromate pigments are corrosion inhibitors. All these pigment functions are in addition to the obvious ones of colorant and opacifier. Non-pigmented coatings, particularly epoxy resins, are used to protect steel reinforcement in concrete against corrosion under aggressive conditions (Clifton *et al.*, 1975).

33.5 Adhesives

Several thousand adhesive formulations based on numerous polymers are in industrial use, but within civil engineering the most significant are the high-performance adhesives based on epoxy resins and the adhesives available for constructional timber.

In laminated timber structures and to an increasing extent in concrete structures, adhesive bonds may play a fully structural role. For example, external plate reinforcement (in which the plate may be steel or a carbon fibre-reinforced polymer sheet) may be bonded to concrete structures with epoxy adhesives, particularly in structural repairs.

Epoxy resins are a family of related polymeric adhesives which may be formulated in many ways according to substrate, curing time, viscosity and gap-filling properties. They are invariably two-pack thermoset adhesives in which the prepolymer, catalyst and activator are mixed immediately before use. Certain epoxies are compatible with fresh Portland cement and may be emulsified with water to form epoxy grouts and mortars. Such materials are widely used in concrete jointing, bonding and repair.

Most high-strength durable adhesives for timber are based on phenolic or amino formaldehyde thermoset resins, although PVAC emulsion adhesives have become established as excellent wood adhesives when water resistance is not required.

33.6 Polymer concretes

Polymers may be combined with concrete in several ways, with the aim of modifying the properties of the concrete, particularly increasing its strength and reducing its permeability, thereby improving durability. One form, polymer impregnated concrete (PIC), has been described briefly in Chapter 31; the polymer is combined with the concrete after the concrete has hardened. On the other hand, *polymer cement concrete* (PCC) is produced by mixing organic and inorganic constituents at the time of hydration. The organic constituent may be introduced as a polymer emulsion, as a monomer dispersion or as a resin/hardener dispersion to polymerise during the setting of the cement.

Broadly, PIC has superior properties because the pore system within the concrete is more or less fully occupied by the polymeric impregnant. However, the production of PIC requires more elaborate facilities and at present it has little regular use. PCCs show reduced permeability and may have somewhat enhanced strength. Equally important are changes in the properties of the fresh wet material: changes in flow and slump and improved adhesion to old concrete. Industrial flooring, concrete repair (including marine and underwater work) and grouts are prominent uses. By way of example, a PCC developed for making rapid repairs to concrete road surfaces and bridge decks is based on a thermoset PF resin (25 per cent phenol + 5 per cent formaldehyde), with 50 per cent sand, 15 per cent Portland cement and 5 per cent water. Calcium ions released in the hydration of

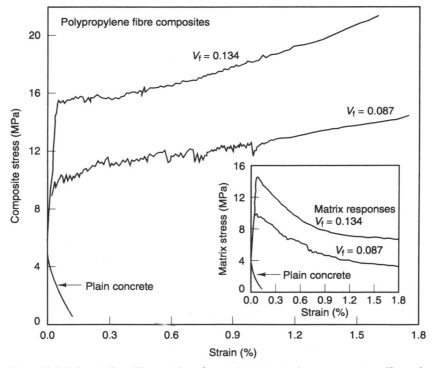

Figure 33.3 *Polypropylene fibre-reinforced cement: stress–strain response at two fibre volume fractions* V_f *(adapted from Shah and Ouyang, 1991)*

the Portland cement catalyse the polymerisation of the PF and the formulation is very tolerant of wet conditions.

Polymeric MF admixtures are used as superplasticisers for high-strength or high-slump concrete. Polymers are also added to concrete in the form of fibre. Chopped fibrillated PP improves impact resistance and the high modulus aramid fibres and carbon fibres are used to some degree as reinforcements, mainly to achieve improved load distribution and to increase toughness. The fibres used are typically about 25 mm in length (similar to the maximum aggregate size). Only about 2 per cent by volume can be incorporated easily with normal mixing equipment, but special mixing techniques allow volumes up to 15 per cent to be achieved. Even such high volumes of fibre have little effect on the initial elastic response but greatly alter the behaviour at higher strains and lead to a significant increase in tensile strength (figure 33.3). The fibres stabilise microcracks at high strains and such cement composites have much higher bending capacity than plain concrete. Grids originally developed for soil reinforcement are also beginning to be used as reinforcement for concrete for similar purposes, especially to distribute stress and control cracking in exposed concrete surfaces such as sewer linings or marine structures. Corrosion resistance especially to salt is a valuable property of such polymer grids.

33.7 Other uses

Expansion bearings and resilient mountings

A small but significant use of polymers, particularly in bridge and pipeline construction, is in forming expansion bearing pads, for which PTFE is almost invariably employed. Commonly, the PTFE pad slides against a polished stainless steel countersurface, on which it may exhibit a coefficient of friction as low as 0.05, and to which it has little tendency to adhere. Alternatively the countersurface may be another polymer, such as a polyamide (nylon) containing molybdenum disulphide solid lubricant. At compressive stresses up to about 3.5 N mm^{-2} unreinforced solid PTFE is a satisfactory material, but at higher stresses (up to about 10 N mm^{-2}) PTFE is compounded with a glass filler. In design calculations creep, wear and weathering as well as friction properties need to be considered. PTFE is itself highly resistant to environmental degradation but the penetration of ultra-violet radiation to the back of the pad can degrade the adhesive bond to the substrate. This may be controlled by compounding the PTFE with an ultra-violet absorber.

Elastomeric mountings may be used to reduce the transmission of sound and vibration through building structures. For this purpose polychloroprene CR and polyisoprene IR are used.

Chemical grouts for soils

Acrylamide is a water-soluble monomer which undergoes delayed polymerisation after injection into unstable or porous ground or permeable rocks. The acrylamide solution has low viscosity and penetrates well into fine-pored materials. The insoluble polyacrylamide gel formed is effective in reducing permeability and in stabilising soils.

References

ASCE (1984) *Structural Plastics Design Manual*, American Society of Civil Engineers, New York.

ASCE (1985) *Structural Plastics Selection Manual*, American Society of Civil Engineers, New York.

ASTM D2836 Standard for measuring the oxygen index of plastics.

ASTM E84 Standard test method for surface burning characteristics of building materials.

BS 476 Fire test methods on building materials and structures.

BS 2782 Methods of testing plastics.

BS 6076 Tubular polyethylene film for use as protective sleeving for buried iron pipes and fittings.

Chambers, R. E. and McGrath, T. J. (1981) Structural design of buried pipe, *Proceedings of the International Conference on Underground Plastic Pipe*, ed. B. J. Schrock, American Society of Civil Engineers, New York, pp. 10–25.

Clarke, J. L. (1993) *Alternative Materials for the Reinforcement and Prestressing of Concrete*, Blackie, Glasgow.

Clifton, J. R., Beeghly, H. F. and Mathey, R. G. (1975) *Nonmetallic Coatings for Concrete Reinforcing Bars*, National Bureau of Standards, Building Science Series No. 65, US Government Printing Office, Washington D. C.

CP 312 Plastics pipework (thermoplastics material).

De Rosa, P. J., Hoffman, J. M. and Olliff, J. L. (1988) *Pipe Materials Selection Manual: Water Mains*, UK Water Industry Research Limited, Swindon.

Hall, C. (1989) *Polymer Materials: An Introduction for Scientists and Technologists*, 2nd edition, Macmillan, London.

Hollaway, L. (Ed.) (1994) *Handbook of Polymer Composites for Engineers*, Woodhead, Cambridge.

Hull, D. (1981) *Introduction to Composite Materials*, Cambridge University Press.

Iremonger, M. J. and Lawler, J. P. (1980) Relationship between modulus and density for high density closed-cell thermoplastic foams, *J. Appl. Polymer Sci.*, Vol. 25, pp. 809–819.

Küster, E. (1979) Biological degradation of synthetic polymers, *J. Appl. Polymer Sci.: Appl. Polymer Symp.*, Vol. 35, pp. 395–404.

Matthews, F. L. and Rawlings, R. D. (1994) *Composite Materials: Engineering and Science*, Chapman and Hall, London.

Shah, S. P. and Ouyang, C. (1991) Mechanical behaviour of fiber-reinforced cement-based composites, *J. Amer. Ceramic Soc.*, Vol. 74, pp. 2947–2953.

Vincent, P. I. (1971) *Impact Tests and Service Performance of Thermoplastics*, Plastics Institute, London.

Waterman, N. A. and Ashby, M. F. (1991) *Elsevier Materials Selector, Vols 1 and 3*, Elsevier Applied Science, London.

Further reading

Adams, R. D. and Wake, W. C. (1984) *Structural Adhesive Joints in Engineering*, Elsevier Applied Science, London.

Bentur, A. and Mindess, S. (1990) *Fiber-Reinforced Cementitious Composites*, Elsevier, London.

Birley, A. W., Haworth, B. and Batchelor, J. (1992) *Physics of Plastics: Processing, Properties and Materials Engineering*, Hanser, Munich.

Brandrup, B. and Immergut, E. H. (1989) *Polymer Handbook*, 3rd edition, Wiley, New York.

Brockbank, J. W. (1989) Offshore applications of polymers, *Rapra Review Reports* (No. 23), Vol. 2, No. 4.

Brydson, J. (1989) *Plastics Materials*, 5th edition, Butterworth, London.

BS 4618 Recommendations for the presentation of plastics design data.

BS 4994: 1987 Specification for design and construction of vessels and tanks in reinforced plastics.

BS 5480: 1990 Specification for glass reinforced plastics (GRP) pipes, joints and fittings for use for water supply or sewerage.

BS 5481: 1977 (1989) Specification for unplasticized PVC pipe and fittings for gravity sewers.

BS 6177: 1982 Guide to selection and use of elastomeric bearings for vibration isolation of buildings.

BS 7291 Thermoplastic pipes and associated fittings for hot and cold water for domestic purposes and heating installations in buildings.

BS EN 302 Adhesives for load-bearing timber structures: test methods.

Ellis, B. (Ed.) (1992) *Chemistry and Technology of Epoxy Resins*, Chapman and Hall, London.

Fardell, P. J. (1993) Toxicity of plastics and rubbers in fire, *Rapra Review Reports* (No. 69), Vol. 6, No. 9.

Hancox, N. L. and Mayer, R. M. (1994) *Design Data for Reinforced Plastics: a Guide for Engineers and Designers*, Chapman and Hall, London.

Hollaway, L. (Ed.) (1990) *Polymers and Polymer Composites in Construction*, Thomas Telford, London.

Mays, G. C. and Hutchinson, A. R. (1992) *Adhesives in Civil Engineering*, Cambridge University Press.

Mills, N. (1993) *Plastics: Microstructure and Engineering Applications*, 2nd edition, Edward Arnold, London.

Nelson, G. L. (Ed.) (1990) *Fire and Polymers: Hazards Identification and Prevention*, American Chemical Society, Washington D. C.

Packham, D. E. (Ed.) (1992) *Handbook of Adhesion*, Longman, Harlow.

Powell, P. C. (1994) *Engineering with Fibre–Polymer Laminates*, Chapman and Hall, London.

Rankilor, P. R. (1981) *Membranes in Ground Engineering*, Wiley, Chichester.

RILEM (1994) *Technical Recommendations for the Testing and Use of Construction Materials*, Spon, London.

Shields, J. (1984) *Adhesives Handbook*, 3rd edition, Butterworth, London.

Turner, G. P. A. (1988) *Introduction to Paint Chemistry and Principles of Paint Technology*, 3rd edition, Chapman and Hall, London.

Ward, I. M. and Hadley, D. W. (1993) *An Introduction to the Mechanical Properties of Solid Polymers*, Wiley, Chichester.

Woods, G. (1990) *The ICI Polyurethanes Book*, 2nd edition, Wiley, Chichester.
Yang, H. H. (1993) *Kevlar Aramid Fiber*, Wiley, Chichester.
Numerical databases for polymer materials, properties and suppliers include: IPS (DATA
    Business Publishing); PDLCOM (Plastics Design Library, USA), chemical and
    environmental compatibility; Plaspec (D & S Data Resources); Polymat (Deutsches
    Kunststoffinstitut, Germany); Rapra (Rapra Technology, UK).

VII BRICKS AND BLOCKS

Introduction

One characteristic of all the materials discussed elsewhere in this text is that their shape is not one of their intrinsic properties. Bricks and blocks differ in this respect as they invariably have an approximately cuboidal shape and indeed the words *brick* and *block* are associated in the minds of many with a shape rather than a material. A study of their material properties is important however as they are widely used in the construction industry.

The value of small building bricks or blocks has been recognised, in one form or another, for many centuries. Largely mechanised processes are used in the brick and block-making industries today, although clay bricks are still made by hand in many parts of the world.

Clay, calcium silicate and concrete products all include both bricks and blocks, the distinction between the two being primarily one of size, blocks being larger than bricks. Clay and calcium silicate *bricks* and concrete *blocks*, whose manufacture and properties are considered here, are in this context the most widely used form for each material. In the UK some 5000 million clay bricks, 250 million calcium silicate bricks, 400 million concrete bricks and about 70 million square metres of various kinds of concrete blocks are manufactured each year.

All bricks and blocks have broadly similar uses, although their properties differ in some important respects depending on the raw materials used and the method of manufacture. An understanding of these properties is essential if bricks and blocks are to be used effectively, as these properties have an important bearing on the appropriate application of each product and on the precautions to be taken to ensure their satisfactory behaviour in service.

34

Clay Brick

34.1 Raw materials and manufacture

Clay bricks are made by shaping suitable clays and shales to units of standard size, which are then fired to a temperature in the range 900 to 1200°C. The fired product (see figure 34.1) is a ceramic composed predominantly of silica SiO_2 (generally between 55 and 65 per cent by weight) and alumina Al_2O_3 (10 to 25 per cent) combined with as much as 25 per cent of other constituents.

Sand facings and face textures may be applied before firing, and key slots formed for plaster finishes. Many bricks are perforated and pressed bricks commonly have *frogs* (see figure 34.1); both features reduce brick weight. The normal

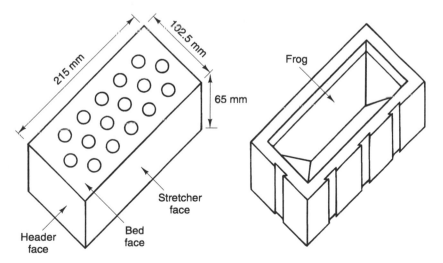

Figure 34.1 *Two typical bricks: one perforated, the other frogged with plastering keys on one stretcher and one header face. The dimensions ('work size') are those of a standard format brick as defined in BS 3921*

495

size of the building brick is 215 mm × 102.5 mm × 65 mm. Allowing for 10 mm mortar joints, this corresponds to the standard format or 'coordinating size' of 225 mm × 112.5 mm × 75 mm as defined in BS 3921. Some bricks are also available in a metric modular format 200 mm × 100 mm × 75 mm (BS 6649). Besides standard rectangular bricks, a large number of so-called *standard special* shapes are commonly available; these range from simple half bricks (*bats* or *closers*) to special shapes for the construction of arches, plinths and other elaborate details. They are fully defined in BS 4729.

Brick-making clays

The brick industry uses a great variety of clays, laid down at different geological periods and ranging from soft, easily moulded glacial deposits to much older, relatively harder shales (Keeling, 1963; Prentice, 1990). This geological diversity reflects itself in the varied composition and mineralogy of brick-making clays. Crystalline aluminosilicate clay minerals, most commonly kaolinite and illite, less commonly montmorillonite and the chlorites, are present as major constituents together with associated water. These minerals are in the form of fine particles (with typical dimensions of 2 μm and less) and are responsible for the cohesion and plasticity of the moist clay (see also Chapter 21). However the clays rarely exceed 30 to 45 per cent of the total composition. In addition there are coarser particles (silt and sand components, predominantly quartz and micas) together with minor accessory minerals such as gypsum $CaSO_4.2H_2O$, pyrites FeS_2, iron oxides and calcite $CaCO_3$. Organic matter may also be present; this is a particular feature of Lower Oxford Clay which contains 5 to 6 per cent by weight of dispersed hydrocarbons, the combustion of which provides about two-thirds of the energy required during firing. This contributes greatly to fuel economy in the UK fletton brick industry, which produces bricks (*flettons*) from clays of this kind in the Bedford–Peterborough area. Waste organic matter may also be added to the clay to assist firing.

Brick-forming processes

After the raw clay has been screened and crushed, machine-made bricks are formed either by *extrusion* and cutting, or by *pressing* (Clews, 1969). Plastic clays may be extruded as a continuous column of rectangular section (with or without perforations) which is cut into individual bricks by a wire, acting like a cheese cutter, as it emerges from the die. Such bricks are commonly called *wirecuts*. The harder raw materials may alternatively be shaped by pressing the clay into individual moulds. There are two variations: in the *stiff plastic* process, extensively used for Coal Measures clays, a clot of clay is roughly shaped in one mould and pressed in a second; in the *semi-dry* process used for flettons the shaly clay enters the mould as granules, and is moulded at higher pressures. A few bricks are still made from soft clays by hand-moulding.

Firing

Firing transforms the raw clay brick into a rigid, continuous (although usually porous) ceramic by way of a complicated succession of physical and chemical changes. The green clay even after preliminary drying contains as much as 10 per cent by weight of free water which is lost rapidly as the kiln temperature rises above 100°C. The clay minerals illite, montmorillonite and halloysite also contain weakly bound water within their lattice structures, which is readily lost (150 to 200°C). Further water is evolved as the clay minerals themselves decompose between about 400°C and 700°C, leaving a residue of preponderantly non-crystalline material, mainly silica and alumina. At about 900°C crystalline silica, alumina and spinel compounds appear and the mineral mullite $3Al_2O_3.2SiO_2$ forms above about 1000°C. The minor oxide constituents Na_2O, K_2O, MgO, CaO and FeO produce relatively low melting eutectic mixtures with principal components SiO_2 and Al_2O_3, so that some melting may occur below 1000°C. This marks the onset of vitrification which promotes sintering of individual clay particles. During this period dimensional shrinking of up to 15 per cent occurs. The temperature range for vitrification and the viscosity of the liquid phase depend on the clay composition. Firing ultimately produces a consolidated but porous mass which contains both microcrystalline mullite and vitreous material, together with unchanged quartz.

Other chemical changes occur during firing, some of which are complex and incompletely understood. Among the most important are those which involve sulfates (see section 34.2). Mineral gypsum $CaSO_4.2H_2O$ is a common impurity in clays and although its hydrate water is lost at an early stage in firing, anhydrous $CaSO_4$ may survive the highest kiln temperatures. It may however react with Mg, K or Na from the clay minerals to form $MgSO_4$, K_2SO_4 or Na_2SO_4. Calcium and magnesium carbonate mineral impurities are also found in certain brick-making clays, and these decompose at kiln temperatures to form the corresponding oxides. These basic oxides can absorb acidic sulfur gases produced in the combustion of the kiln fuel to yield once again the corresponding sulfates. The alkali metal sulfates Na_2SO_4 and K_2SO_4 however are themselves somewhat unstable above about 700 to 800°C and they may be reduced in quantity by hard firing.

Clay composition and chemical changes during firing also influence the final colour of bricks. The main strongly coloured constituent of brick ceramic is ionic iron; after kiln firing under oxidising conditions the iron is present in the ferric Fe(III) oxidation state and red brick colours are obtained. If reducing conditions exist during the last stages of firing because of limited air supply or deliberate injection of fuel into the kiln atmosphere to remove oxygen, iron is present in the ferrous Fe(II) oxidation state or more probably in the mixed Fe(II)/Fe(III) states, and blue or blue-black ceramic colours are produced. These colours may be masked, muted or otherwise modified by other constituents, notably lime. Large amounts of lime produce buffs and yellows. Considerable variation of colour may be found on individual bricks and within a single batch of bricks, which may be sorted for colour after drawing from the kiln. Surface or body colour is also commonly imparted by inorganic pigments added to the raw clay.

34.2 Properties of clay bricks

The engineer is concerned primarily with mechanical behaviour, water absorption and permeability, and durability. All these properties are controlled in some measure by the porous nature of brick ceramic.

Microstructure of brick

Figure 34.2 shows the pore structures of three clay bricks as revealed at fracture surfaces under high magnification. The individual particles of the unfired clay have been transformed into an apparently uniform ceramic matrix, throughout which there runs a continuous network of voids.

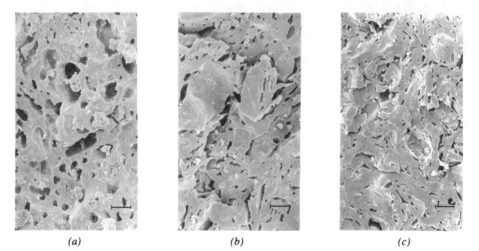

(a)   (b)   (c)

Figure 34.2 *Scanning electron micrographs of brick ceramic pore structures: (a) pressed fletton common water (absorption WA 17 per cent); (b) wirecut (WA 7.5 per cent); (c) pressed engineering class A (WA 4 per cent). The length of the scale bar is 5 μm (courtesy of C. Hall, W.D. Hoff and M. Skeldon, UMIST)*

The volume fraction *porosity* (defined in Chapter 22) of commercially available clay bricks varies greatly, from about 1 per cent to at least 50 per cent. The porosity depends both on clay composition and on the duration and temperature of firing. The pores have dimensions typically about 1 to 10 μm. There appear to be almost no micropores (< 0.002 μm) in clay bricks. As a consequence, the specific surface of brick ceramic is relatively low, 0.2 to 5.0 $m^2 g^{-1}$.

Porosity, water absorption and suction

The existence of minute pores confers marked capillary properties on brick ceramic. In particular almost all bricks absorb water by capillarity.

The volume fraction porosity n can be determined accurately and directly by measuring the weight gain on saturating with water an initially dry brick after evacuation to remove the air from the pore network. (An alternative method of filling the entire void space with water is to immerse the brick in boiling water for several hours.) The water absorption WA (by either method) is expressed in BS 3921 as a weight per cent. WA may be converted simply to a volume basis porosity: $n = (WA)\rho/100\rho_w$, where ρ is the dry brick bulk density and ρ_w the density of water.

Simple immersion without prior evacuation or boiling invariably leads to incomplete saturation because air is trapped within the pore network; this air slowly disappears by diffusion to the faces of the brick over months or years.

The *rate* at which a brick absorbs water (frequently called its suction rate) may be determined by immersing one face in a tray of water, and measuring the gain in weight with time. A test of this kind is specified in BS 3921 to help in the specification of the brick/mortar bond in highly stressed masonry. The one-minute water uptake (*initial rate of absorption*) is taken as the suction rate. In tests extending over longer periods it is found that the total weight of water absorbed per unit area, w, increases as the square root of the elapsed time t; thus $w = At^2$, where A is defined as the *water absorption coefficient* in CIB and RILEM recommendations (RILEM, 1972). Closely related to A is the sorptivity S which equals A/ρ_w (see Gummerson *et al.*, 1980). The sorptivity of brick ranges from about 0.1 mm min$^{-1/2}$ or less for engineering bricks of low porosity to 2–3 mm min$^{-1/2}$ for high porosity bricks ($n > 0.3$). However, the sorptivity is not a simple function of the porosity n alone but depends on the size and shape of the pores.

The existence of fine pores in brick ceramic means that the suction exerted by dry brick is generally relatively strong (although bricks of low porosity have little capacity). Suction curves similar to that given in Chapter 25 (figure 25.4) for a silty soil show this clearly, see figure 34.3. The rate of movement of water is described by the extended Darcy equation

$$u = -K\nabla\Psi$$

where u is the fluid velocity, Ψ is the suction (hydraulic potential) and K is the permeability (hydraulic conductivity). Both K and Ψ depend strongly on the water content. Figure 34.3 shows clearly the effect of wetting in reducing the suction exerted by a brick ceramic.

Density

The solid density ρ_s of brick ceramic depends on the clay composition and varies from about 2250 kg m^{-3} to about 2800 kg m^{-3}, most commonly lying close to 2600 kg m^{-3}. For an individual brick material, the bulk density $\rho = \rho_s(1 - n)$ exactly, were n is the porosity. Figure 34.4 shows the general inverse correlation between bulk density and porosity for a diverse sample of different clay bricks, among which ρ_s also varies. Since n varies widely among bricks of different kinds and since frogs and perforations may be present, standard format bricks vary greatly in weight, from about 1.8 kg to about 3.8 kg.

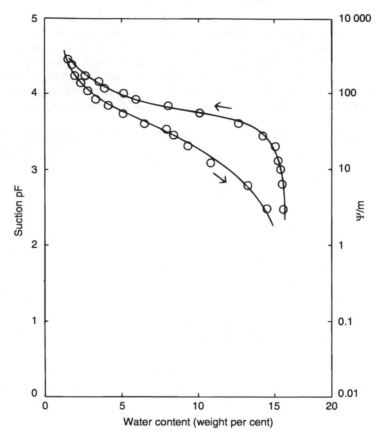

Figure 34.3 *Suction (hydraulic potential) of a clay brick ceramic (porosity n = 0.30) plotted against water content*

Compressive strength and other mechanical properties

The compressive strength is the only mechanical property used in brick specification; it is the failure stress measured normal to the bed face. Bricks are tested wet, normally with frogs filled with hardened mortar. A considerable variation is found between individual bricks and a batch of ten is tested to obtain a mean strength. Full details are laid down in BS 3921; the tests prescribed in other national standards differ somewhat. Generally, compressive strength decreases with increasing porosity (figure 34.5), but strength is also influenced by clay composition and firing. The compressive strength is limited by brittle fracture and is sensitive to individual flaws in the sample under test, including those associated with large particles, fissures formed during shaping, and shrinkage cracks.

The Young's modulus of elasticity of brick ceramic lies usually in the range 5 to 30 kN mm^{-2}. Brick ceramic itself is not subject to creep at normal temperatures although creep may occur in brickwork (see Chapter 37).

The flexural strengths of brick materials are sometimes required in calculations of the lateral strength of brickwork. A three-point bending test may be used, for

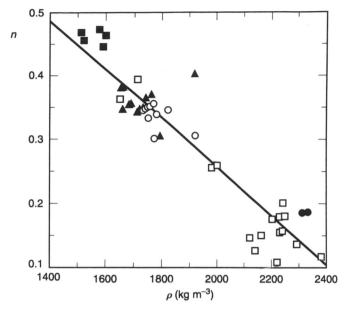

Figure 34.4 *Volume fraction porosity n plotted against bulk density ρ for a sample of 61 clay bricks of several different types (each type denoted by a different symbol). From Hall et al. (1992)*

example that described in the RILEM procedure for autoclaved aerated concrete (RILEM, 1994). Brick ceramic is relatively weak in *tension* and the flexural strength (or modulus of rupture) is typically only 5–10 per cent of the compressive strength.

Efflorescence and soluble salts content

Brickwork (especially new work) sometimes develops an efflorescence of white salts brought to the surface by water and deposited by evaporation. These salts may have an external origin (for example in soil water in contact with the brickwork) or may derive from the mortar; however the salts frequently originate in the bricks themselves. As already noted fired brick ceramic may contain soluble salts, in which the sulfates of sodium, potassium, magnesium and calcium usually predominate. The total soluble salts content of bricks may be as high as 5 per cent by weight, although it is more commonly 0.1 to 1 per cent. Visible efflorescence can be formed from very small amounts of salts and the total salt content is not a reliable guide to efflorescence liability. This is also affected by the mobility of the salts (which in turn depends on ceramic composition and the pore structure of the brick) and is best assessed directly by repeated wetting and drying of a test brick.

Efflorescence may be disfiguring but it is often harmless and disappears after a few seasons. However, efflorescent salts usually contain a high proportion of sulfates, and the engineer should be alert to the possibility of sulfate attack (see also Chapter 15) on cement mortar joints. This risk is the greater if the situation is

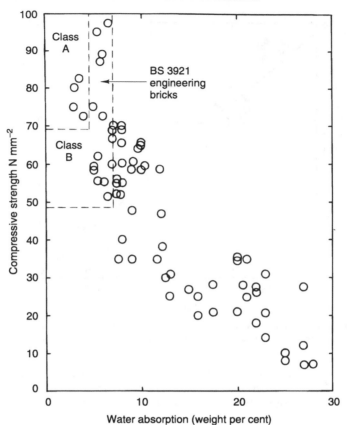

Figure 34.5 *The correlation of compressive strength and water absorption in seventy commercial bricks*

exposed and the brickwork is persistently wet (Harrison and Bowler, 1990; Bowler, 1993; see also Chapter 37).

The total sulfate content of the brick is a fair guide to the risk of sulfate attack in mortars with which they are in contact. Accordingly, the soluble salts content is used (together with frost resistance) in BS 3921 as a guide to durability (see section 34.3). Bricks of designation L (low soluble salts content) are not permitted to contain more than 0.50 per cent by weight of soluble sulfate ion, or more than 0.03 per cent of each magnesium, potassium or sodium ions. Bricks of designation N (normal) may not contain more than 1.6 per cent of soluble sulfate or more than 0.25 per cent total magnesium, potassium and sodium ions.

Moisture expansion

Fired brick ceramics exhibit a long-term expansion on exposure to moist air. *Moisture expansion* is progressive and continues indefinitely, although at a diminishing rate, such that the total expansion increases roughly as log(time). Thus flettons fired at 1050°C showed increases in length of 0.02 per cent at 10 days, 0.04 per cent at 100 days and 0.06 per cent at 1000 days (Smith, 1974). The exact cause of moisture expansion is still unclear, but it apparently involves some degree of *irreversible* recombination of water with amorphous or glassy constituents of the brick ceramic. The expansion rate depends on the mineralogy of the fired clay and to a marked extent on the porosity and the maximum firing temperature, but it is not greatly influenced by exposure conditions. Bricks made from clays with high lime contents (notably Gault clays) generally give low expansions as little glassy material is present in the fired ceramic.

In a study of ten brick clays Smith (1973, 1993) found moisture expansions after 28 years ranging from 0.20 per cent to as low as 0.02 per cent. Long-term (commonly 50-year) moisture expansion in service can now be predicted from short-term steam exposure tests (Lomax and Ford, 1988; Smith, 1993). Clay bricks may be assigned to one of three classes according to the estimated 50-year moisture expansion: low ($<$ 0.04 per cent), medium (0.04 – 0.08 per cent) and high ($>$ 0.08 per cent). It is now recognised that the long-term contribution of moisture expansion to movement in clay brick masonry can be considerable and it is essential to allow for this in design (BRE, 1979, Part 2; Foster and Johnson, 1982; Shrive, 1991). *Reversible* changes in dimensions on wetting and drying brick ceramic are less than 0.01 per cent and are negligible for most purposes.

Frost damage

Frost damage is the physical deterioration to which wet bricks may be liable when exposed to freezing conditions. It represents a major cause of failure in certain bricks under conditions of severe exposure and one which should receive full attention at the time of specification and selection.

Frost damage arises from the stresses created within bricks by the freezing of water within the pores, stresses which lead to cracking and splitting, often with spalling of the brick face, either by popping or delamination. The factors which lead to susceptibility to frost damage in particular bricks are not yet well understood. Strong bricks of low porosity are generally resistant to frost damage, but so also are many porous bricks of lower strength. At present no entirely satisfactory laboratory test is available, and liability to frost damage is assessed when required by observation of field performance. Resistance to frost damage is a major criterion of durability and is one of the factors used in the classification of bricks in BS 3921 (see section 34.3).

Thermal properties

The thermal conductivity of brick ceramic is controlled by the proportions of crystalline and glassy constituents, and the porosity. Dry brick of bulk density 2400 kg m^{-3} ($n \approx 0.07$) has a conductivity of about 1.2 W m^{-1} K^{-1}, but the conductivity falls to about 0.4 W m^{-1} K^{-1} at a brick density of 1600 kg m^{-3}. However, the thermal conductivity rises sharply with increasing moisture content, by about 60 per cent at 3 per cent volumetric water content and by about 135 per cent at 15 per cent water content. Data are given by Arnold (1969) and Clews (1969).

The coefficient of thermal expansion of most clay bricks lies in the range 5 to 7 $\times$ 10^{-6} K^{-1}.

Resistance to chemical attack

Brick ceramic is generally very resistant to alkalis, acids and most commonly encountered chemicals and is attacked only under extreme conditions. However bricks required to perform under severe acid conditions, for example in chemical plant, and clay pipes for acid effluents, are specially selected (see acid resistance test, BS 65).

Behaviour under fire conditions

Because it is itself a fired material, the performance of brick ceramic under fire conditions is generally excellent. Thermal stresses may produce some spalling in certain types of bricks, and in severe fires, temperatures may approach the vitrification range of brick, causing slight fusion of exposed faces. Neither of these effects seriously diminishes fire resistance. The fire resistance of perforated brick and cellular brick masonry is somewhat lower than that of the same thickness of solid brick, which generally has greater resistance to thermal shock and better resistance to the transmission of heat at high temperatures.

34.3 Classification of bricks

The building industry broadly recognises three kinds of brick, which are differentiated on the basis of function. These are: *common*, for general building purposes; *facing*, manufactured for acceptable appearance; and *engineering*, for use where high strength and/or low water absorption are required. Bricks manufactured to BS 3921 must meet certain requirements on size, compressive strength, durability and soluble salts content. For some purposes there is a requirement on water absorption. Table 34.1 shows the classification of bricks by compressive strength and water absorption. Table 34.2 shows the designations for durability. Frost resistant (F) bricks are durable in all building situations including extremely exposed conditions where brickwork may be persistently wet (for example in retaining walls, parapets and pavings). Moderately frost resistant (M) bricks are durable except when saturated and subject to repeated freezing and thawing.

TABLE 34.1
Classification of bricks by compressive strength and water absorption
(BS 3921)

Class	Compressive strength $(N\,mm^{-2})$	Water absorption (per cent by mass)
Engineering A	≥ 70	≤ 4.5
Engineering B	≥ 50	≤ 7.0
Damp-proof course 1	≥ 5	≤ 4.5
Damp-proof course 2	≥ 5	≤ 7.0
All others	≥ 5	No limits

TABLE 34.2
Durability designations for clay bricks (BS 3921)

Designation	Frost resistance	Soluble salts content
FL	Frost resistant (F)	Low (L)
FN	Frost resistant (F)	Normal (N)
ML	Moderately frost resistant (M)	Low (L)
MN	Moderately frost resistant (M)	Normal (N)
OL	Not frost resistant (O)	Low (L)
ON	Not frost resistant (O)	Normal (N)

Engineering bricks are designated class A and class B on the basis of strength and water absorption.

For the purpose of BS 5628 load-bearing bricks are classified as shown in table 34.1. Bricks for use as damp-proof courses are required to meet class B engineering brick standard for water absorption (see also BS 743).

Other fired clay products

Besides brick, the heavy clay industry produces clay drain and sewer pipes and tiles for roofing and flooring. Hollow clay blocks for load-bearing partitions and suspended floors are also manufactured; clay units of this kind are widely used in several countries, although not in the UK. In all cases, the material properties resemble those of brick.

35

Calcium Silicate Brick

The manufacture, properties and classification of calcium silicate brick are outlined below.

35.1 Constituent materials and manufacture

The raw materials used in the manufacture of calcium silicate bricks are a siliceous aggregate, a high calcium lime and water. A very fine aggregate is generally used, with the majority passing a 1.15 mm sieve, the ratio, by weight, of aggregate to lime being in the range 10 to 20. Inert and stable inorganic pigments are also added when different coloured bricks are required. The materials are first intimately mixed in the required proportions and are then conveyed to an automatic press where they are moulded to the required size and shape. At normal temperatures, hydrated lime when mixed with water does not *set*, unlike Portland cement (see Chapter 12), and for this reason dry mixes and high moulding pressures are required to ensure that the freshly moulded bricks possess sufficient strength to permit immediate handling without damage.

From the press, the bricks are transferred to an autoclave where high-pressure steam curing for some hours results in the combination of the lime with part of the siliceous aggregate to produce a hydrous calcium silicate, known as tobermorite, which forms the binding medium in the finished brick. Tobermorite is also the binding medium in a number of autoclaved products including aerated concrete blocks (see Chapter 36).

35.2 Properties of calcium silicate brick

The size, strength and drying shrinkage of calcium silicate brick are included among the properties whose limits are specified in BS 187. Some of these and other properties are considered here. For a more detailed discussion of these the reader should refer to the various publications listed at the end of this Part.

Size

Standard brick dimensions are 215 × 102.5 × 65 mm, these being respectively the length, thickness and height of the brick. Metric modular sizes, which are not commonly used, are 190 × 90 × 90 or 65 mm and 290 × 90 × 90 or 65 mm.

Absorption

The water absorption of calcium silicate brick varies between about 6 and 16 per cent, by weight.

Density

This varies between about 1700 kg m^{-3} and 2100 kg m^{-3} depending on the composition of the brick and the manufacturer. Where the density of brickwork using a particular brick is critical, for example for sound insulation, this information should be obtained from the manufacturer.

Strength

Compressive strength is the usual criterion for brick but as with any other engineering material, the strength of calcium silicate brick is not a property which can be specified without due recognition of the variability of strength between nominally identical samples of brick. This is recognised for classification purposes (see section 35.3). Typically the range of mean compressive strengths for bricks in general use is 14–27.5 N mm^{-2} depending on the quality of brick being produced, although strengths greater than 48.5 N mm^{-2} can be achieved by some manufacturers.

Drying shrinkage

Calcium silicate brick shrinks on drying out in a similar manner to concrete products although the magnitude of this drying shrinkage, 0.01–0.04 per cent, is about half that associated with the latter.

Frost resistance

The effect of repeated freezing and thawing on the integrity of calcium silicate brick is slight, under normal circumstances, and except for very severe conditions, for example when subjected to repeated freezing and thawing whilst continually wet, class 2 bricks (see section 35.3) are quite adequate. For the more severe conditions class 3 or 4 bricks should be used.

Expansion

A largely reversible expansion occurs after the initial drying shrinkage has taken place if calcium silicate brick becomes moist but this movement is, in normal usage, somewhat less than that associated with the initial drying shrinkage. Expansion, or contraction, also occurs with temperature change, the coefficient of thermal expansion being in the range 8 to 14×10^{-6} K^{-1}, with the actual value depending on the aggregate used and the mix proportions.

Modulus of elasticity

The modulus of elasticity of calcium silicate brickwork is related to its compressive strength with values generally in the range of 14–18 kN mm^{-2}.

Efflorescence

This manifests itself as a white deposit on the surface of bricks and blocks and is a result of the crystallisation of soluble salts in the product concerned. Calcium silicate brick is virtually free of soluble salts and, unless contaminated from elsewhere, is not subject to efflorescence.

Chemical attack

Sulfates

Calcium silicate brick is generally not susceptible to attack from sulphates in soil or ground water although as an added precaution class 3 or 4 bricks should be used where sulfates are present.

Salt solutions

If calcium silicate brick is saturated by salt solutions, for example sodium or calcium chloride, sodium sulfate or sodium carbonate, and is subjected to alternate wetting and drying or to frost action, it will deteriorate. Wherever these conditions may occur, calcium silicate brick should not be used. This also applies in circumstances where the brick may be exposed to acid fumes or acid solutions.

Thermal conductivity

This depends on both the density of the brick and its moisture content. Dry calcium silicate brick has a thermal conductivity ranging from about 0.5 to 0.8 W m^{-1} K^{-1}.

35.3 Classification of calcium silicate bricks

Compressive strength forms the basis for the classification of calcium silicate brick subject to the other requirements of BS 187 being satisfied. One of these is that the average drying shrinkage measured under standard testing conditions shall not exceed 0.04 per cent, except where the bricks are to be used under permanently damp conditions, when no limit is specified.

For each class of brick, BS 187 specifies the *minimum mean compressive strength* of samples of ten bricks and the *minimum predicted lower limit of compressive strength* of a sample. The predicted lower limit is based on the standard deviation (see Chapter 17) obtained from the test results during normal quality control testing and is that strength below which the strength of not more than 1 in 40 of the samples might be expected to fall.

The minimum mean compressive strength or *minimum mean class strength* as it is commonly known, and the *minimum predicted lower limit of strength*, for each class of brick are given in table 35.1.

The durability of calcium silicate brick is closely related to its compressive strength, that is to say the higher the class of brick the greater its durability, and the selection of calcium silicate brick for any given use is therefore generally based on the class of brick (see table 37.2).

TABLE 35.1
Classification of calcium silicate bricks

	Class of brick					
	2	3	4	5	6	7
Minimum mean class strength (N mm^{-2})	14.0	20.5	27.5	34.5	41.5	48.5
Minimum predicted lower limit (N mm^{-2})	10.0	15.5	21.5	28.0	34.5	40.5

36

Concrete Block

36.1 Constituent materials

The essential materials for making concrete blocks are a hydraulic binder, water and aggregate, and, in the case of aerated concrete blocks only, a reactive foaming agent to produce their characteristic cellular structure. Additives and/or admixtures are also used sometimes to extend the product range.

The binder is a Portland cement either alone or mixed with lime, pulverised-fuel ash or ground-granulated blastfurnace slag. The Portland cement is usually of class N or R; others are used only when their special properties are required in specific applications. A very wide range of aggregates is used; they may be of dense or lightweight type, normally with a maximum size not exceeding 10 mm. The most common additives are pigments used to impart different colours; admixtures, such as retarders, accelerators and water-reducing agents, are used to facilitate the block manufacturing processes or to impart certain special properties to the finished blocks. The constituent materials recommended by BS 6073: Part 1 for concrete block manufacturing are summarised in figure 36.1, together with the British Standards with which they should comply.

36.2 Manufacture

There are two main types of concrete blocks, aggregate concrete blocks and aerated concrete blocks, and their respective methods of manufacture are distinctly different.

Aggregate concrete block

The manufacturing process involves compaction of the newly mixed constituent materials (basically binder, water and aggregate) in a mould followed immediately by extrusion of the pressed block so that the mould can be used repeatedly.

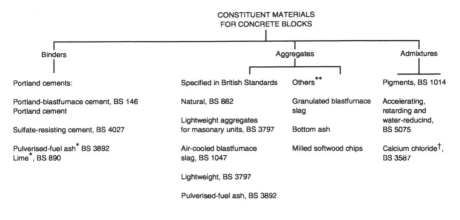

CONSTITUENT MATERIALS
FOR CONCRETE BLOCKS

Binders	Aggregates		Admixtures
Portland cements:	Specified in British Standards	Others**	Pigments, BS 1014
Portland-blastfurnace cement, BS 146 Portland cement	Natural, BS 882	Granulated blastfurnace slag	Accelerating, retarding and water-reducind,
Sulfate-resisting cement, BS 4027	Lightweight aggregates for masonary units, BS 3797	Bottom ash	BS 5075
Pulverised-fuel ash* BS 3892 Lime*, BS 890	Air-cooled blastfurnace slag, BS 1047	Milled softwood chips	Calcium chloride†, BS 3587
	Lightweight, BS 3797		
	Pulverised-fuel ash, BS 3892		

*   With Portland cement in mass ratios with respect to the cement not exceeding: ground-granulated blastfurnace
    slag, 65%, pulverised-fuel ash, 35%; lime, 10%.
** For compliance requirements, see BS 6073: Part 1.
†   BS 3587 withdrawn, BS 6073: Part 1 not yet amended

Figure 36.1 *Constituent materials for manufacturing concrete blocks recommended in BS 6073: Part 1*

Since the finished blocks are required to be self-supporting and able to withstand any movement and vibration from the moment they are extruded, very much drier, higher fine aggregate content and leaner mixes are used than in normal concrete work.

The two basic types of block-making machine commonly used are known as egg-laying and static machines. In the first case, the blocks (usually 44 at a time) are produced directly on to a horizontal surface as the machine moves along; the blocks are left in this position to harden sufficiently, before they are stacked for final air-curing for up to 28 days, depending on climatic conditions. In the second case the blocks (usually 12 or 26 at a time) are stamped and released either directly on to pallets or are hand-loaded on to pallets which are transferred immediately to curing chambers where the blocks are normally subjected to steam curing at atmospheric pressure for about 12 hours and then stacked outside for final curing, usually up to the age of 7 days. Autoclaving (high-pressure steam curing) is sometimes used, in which case the blocks are ready for use immediately after they are withdrawn from the curing chamber.

Aerated concrete block

In this case, a slurry of binder, pre-heated water and siliceous materials mixed together with aluminium powder is first cast as a 'cake' in large moulds (usually 1 m × 2 m × 1 m high). The aluminium powder reacts with lime in the cement producing a mass of minute hydrogen bubbles within the mix, which thus expands to fill the mould. As the mix sets, the hydrogen within the now cellular structure diffuses and is replaced by air. After the initial set, while the aerated 'cake' is still in its plastic stage, the mould shutters are stripped off and the 'cake' is cut into the required block sizes by thin wires on a cutting machine. This cut cake is then placed in an autoclave for high-pressure steam curing for about 24 hours, when the blocks are ready for use as soon as they have cooled to the ambient temperature.

The binder can be Portland cement or lime, or the two combined. The siliceous materials, such as silica sand, pulverised-fuel ash, ground-granulated blastfurnace slag or a combination of these, are used to take advantage of the fact that at autoclaving temperatures they combine with lime to form cementitious products.

36.3 Form and size

There are three basic forms of concrete block, solid, cellular and hollow as shown in figure 36.2, and within each type a variety of products are available, thus providing versatility to blockwork construction both in style and function. A solid block has no formed holes or cavities other than those inherent in the material, although it may contain transverse slots to facilitate cutting, while cellular and hollow blocks have one or more formed holes or cavities which in hollow blocks pass right through them (BS 6073: Part 1). It should be noted that all three forms of block can be produced using aggregate concrete but with aerated concrete only the solid form can be manufactured. Concrete blocks are commonly referred to as common, facing and special blocks and they are available in both normal and insulating types (figure 36.2). Common blocks, which have an open texture, are for general use for both load-bearing and non-load-bearing walling. Facing blocks have a closed texture and are suitable for direct or exposed work internally and are produced in a wide range of colours and finishes. Special blocks are produced essentially to simplify construction and include cavity, corner and lintel blocks, as well as half and three-quarters blocks (figure 36.2).

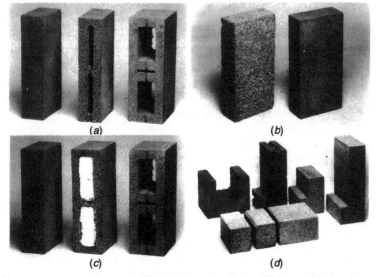

Figure 36.2 *Different types of concrete block: (a) solid, cellular and hollow blocks; (b) common and facing blocks; (c) normal and insulating blocks; (d) lintel, quoin, cavity closure and multi-purpose cutting blocks (Courtesy of Butterley Aglite Ltd, Denby, England)*

British Standard BS 6073: Part 1 defines a block as a masonry unit of a larger size in all dimensions than specified for bricks but no dimension should exceed 650 mm nor should the height (in its normal aspect) exceed either its length or six times its thickness. There are now no specified limits on formed voidage in relation to overall block volume except that the external shell wall thickness should not be less than 15 mm or 1.75 times the nominal maximum size of the aggregate used, whichever is the greater.

Concrete blocks are available in a wide range of sizes, a selection of which is given in table 36.1 which is taken from BS 6073: Part 2. However, the dimensions of blocks in the UK are related generally to those of bricks, with a typical block face area being equivalent to the face area of six bricks including their mortar joints. The most commonly used block has a work face of 440 mm × 215 mm and is usually 100 mm thick.

TABLE 36.1
Typical work sizes of concrete blocks

Face		Thickness (mm)
Length* (mm)	Height* (mm)	
390	190	
440	140	
440	190	
440	215	60, 75, 90, 100, 140,
440	290	150, 190, 200, 215
590	140	
590	190	
590	215	

*To obtain co-ordinating sizes, add 10 mm for mortar joint to the length and height dimensions but not to the thickness.

36.4 Properties

The properties of concrete blocks depend to a varying degree on the type and proportions of the constituent materials, the manufacturing process, and the mode and duration of curing employed, as well as on the form and size of the block itself. Since all of these can vary greatly, it is advisable to consult the manufacturer's literature for specific information on any particular product.

Density

The density of concrete blocks is largely a function of the aggregate density, size and grading, degree of compaction or aeration and the block form. The typical range for dry density is 500–2100 kg m^{-3} with aerated and solid dense aggregate concrete blocks being on the lighter and heavier end of the scale respectively, and lightweight and dense aggregate concrete blocks of cellular and hollow form falling in the middle of the range. The test method for measuring concrete block density is given in BS 6073: Part 2.

Strength

In addition to size, compressive strength is the basic requirement of concrete blocks, except for non-load-bearing blocks with a thickness less than 75 mm which are required to comply with the transverse breaking strength test for handling.

The compressive strength of concrete blocks is dependent mainly on their mix composition (in particular binder content), degree of compaction (or aeration) and to a lesser extent on the aggregate type and curing normally used. In general, for a given set of materials the strength of a concrete block will increase with its density. The range of strengths specified in BS 6073: Part 2 is 2.8–35 N mm^{-2} but from considerations of cost the more normal practical upper limit is about 20 N mm^{-2}, and the most commonly used blocks fall within a much smaller strength band of 3.5–10 N mm^{-2}.

The compressive strength of concrete blocks tested in accordance with BS 6073: Part 1 is specified as a minimum average value of ten blocks of which no single value should fall below 80 per cent of the permitted average value. It should also be noted that since all forms of block are tested for strength in the same manner, strength being calculated as crushing load divided by the gross area, it follows that for a given strength the form of block (solid, cellular or hollow) will not affect its load-carrying capacity.

Modulus of elasticity

As a general rule, the modulus of elasticity of concrete blocks can be assumed to be related to their strength, increasing as strength increases. However, in the case of aggregate concrete blocks it is greatly influenced by the type of aggregate used and, in the case of autoclaved aerated concrete, by the degree of aeration.

The modulus of elasticity is not normally quoted and when this information is required the manufacturer should be asked to furnish it.

Dimensional changes

Concrete blocks will undergo some dimensional changes owing to variations in the ambient moisture and temperature conditions. The magnitude of such movements, to a varying degree, is largely influenced by the constituent materials (mainly the aggregate), mix proportions and the process of block-manufacturing adopted. Drying shrinkage is considered to be the most important in normal applications and BS 6073: Part 1 specifies maximum permissible values of 0.06 and 0.09 per cent for aggregate concrete and autoclaved aerated concrete blocks respectively, tested in accordance with the specified procedure. It should be noted, however, that the drying shrinkage of concrete blocks can be reduced significantly by ensuring that the units are properly matured and by preventing them from becoming excessively wet on site prior to their use.

The thermal coefficient of expansion of concrete blocks ranges between 8 and 12×10^{-6} K^{-1}, with autoclaved aerated and artificial lightweight aggregate concrete blocks usually having a value around 8×10^{-6} K^{-1}.

Durability

In general, concrete blocks are adequately durable for most normal applications. As a general rule, in extreme conditions of pollution (chemical attack) and weather (frost attack), fair faced blocks with strength in excess of 7 N mm^{-2} should be used. It should be noted that open-texture blocks are no more susceptible to frost attack than other blocks owing to the freedom with which water can move within the block on freezing. The use of autoclaved blocks and the use of sulfate-resisting Portland cement and/or pulverised-fuel ash increase the resistance to sulfate attack.

Efflorescence

Efflorescence of the type found in clay bricks is rarely a problem with concrete blocks. Such efflorescence as occurs in concrete blocks normally consists of sodium, potassium and calcium carbonates formed as a result of a reaction between the corresponding free hydroxides brought to the surface and atmospheric carbon dioxide. Its occurrence can be further reduced by autoclaving, using a pozzolana in the mix, avoiding premature drying shrinkage and unsuitable curing of blocks, and proper design, construction and maintenance of blockwork. While not normally detrimental to the blockwork, in some cases salts may form sulfates which can cause deterioration of the blocks.

Fire resistance

In general, concrete blocks have good fire-resistance properties. However, their actual fire-endurance is controlled by numerous factors, for example the type and grading of aggregate and the cement content in the mix, the form, weight and thickness of the block and its moisture content. As a general rule, most concrete blocks of 100 mm thickness can provide an adequate resistance to fire for up to 2 hours if load-bearing or up to 4 hours if non-load-bearing but specific information should be obtained from the manufacturer.

Thermal conductivity

The thermal conductivity of a concrete block is largely dependent on its block density as can be seen from the results reported by Arnold (1969) shown in figure 36.3, which shows a similar relationship to that originally proposed by Jacob (1949). Thus, in general, autoclaved aerated concrete and lightweight concrete

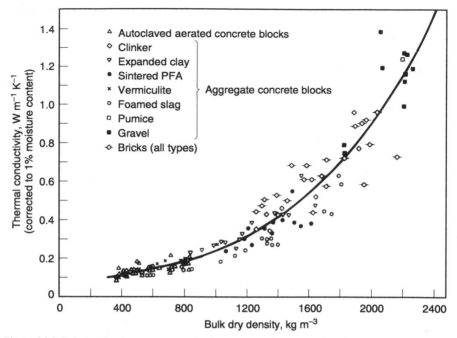

Figure 36.3 *Relationship between thermal conductivity and density (BRE: crown copyright)*

blocks have relatively low thermal conductivities. Similarly cellular and hollow blocks, because of their lower net density, have lower thermal conductivities than their solid counterparts. The thermal conductivity of a concrete block is further affected by its moisture content, increasing as the moisture content increases. Appropriate corrections, such as shown in table 36.2 (Jacob, 1949), should be made for moisture content.

TABLE 36.2
Moisture correction factors for thermal conductivity

Moisture content % (by volume)	0	1	3	5	10	15	20	25
Correction factor	1.0	1.3	1.6	1.75	2.10	2.35	2.55	2.75

37

Brickwork and Blockwork

Individual bricks and blocks when bonded together with mortar are referred to collectively as brickwork and blockwork, or alternatively brick and block masonry. The mortars used and the properties and applications of masonry are described in this chapter.

For a more detailed discussion of the application of brickwork and blockwork in civil engineering and building works the reader is referred to Sutherland (1981), and for further information on their properties to the references listed at the end of this chapter.

37.1 Mortars

Typical mortars used in both brickwork and blockwork are given in table 37.1. Related standards are BS 12, BS 146, BS 890, BS1199, BS1200, BS 4027, BS 4248 and BS 4887.

TABLE 37.1
Recommended mortar mixes

Type of mortar	Mix proportions by volume			
Cement : lime : sand	$1 : 0$ to $\frac{1}{4} : 3$	$1 : \frac{1}{2} : 4$ to $4\frac{1}{2}$	$1 : 1 : 5$ to 6	$1 : 2 : 8$ to 9
Masonry cement : sand	–	$1 : 2\frac{1}{2}$ to $3\frac{1}{2}$	$1 : 4$ to 5	$1 : 5\frac{1}{2}$ to $6\frac{1}{2}$
Cement : sand with plasticiser	–	$1 : 3$ to 4	$1 : 5$ to 6	$1 : 7$ to 8
Mortar designation	(i)	(ii)	(iii)	(iv)

increasing strength
decreasing ability to accommodate movement

Cement–lime–sand mortars have improved working qualities compared with ordinary cement–sand mortars as a result of the effective replacement of part of the cement with lime.

Masonry cement contains a mixture of Portland cement, a very fine mineral filler and an air-entraining agent, and it is the plasticising effects of the filler and the entrained air which impart to *masonry cement mortars* their extremely good working qualities.

Sand–cement mortars with an air-entraining agent (mortar plasticiser) have both improved working qualities and greater resistance to damage by frost action owing to the presence of entrained air (see Chapter 12).

The mix proportions shown in table 37.1 for different types of mortars, within a single mortar designation, will give mortars having generally similar compressive strengths although, for the higher-strength mortars particularly, those containing entrained air will tend to have lower strengths than the equivalent cement–lime–sand mortars. Whilst mortar designations are related primarily to mortar strength it is only for heavily loaded brickwork in load-bearing construction that mortar strength will be the overriding consideration in the choice of mortar. Owing to the tendency of air-entrained mortars to have somewhat lower strengths than the cement–lime–sand mortars, the use of these mortars in calculated load-bearing masonry is normally only permissible when a high degree of quality control is exercised and full information on their strength characteristics is available.

More generally, the choice of mortar designation and type of mortar will be influenced primarily by *durability* requirements. Cement–sand mortars with plasticisers, masonry cement mortars and cement–lime–sand mortars, in that order, have a decreasing resistance to damage by frost action, and increasingly improved bond characteristics and hence resistance to rain penetration through the mortar joints in masonry. However, these differences between the different types of mortar can be reduced; for example air-entrainment of cement–lime–sand mortars will increase their resistance to frost action although some reduction in strength might be expected. Desirable minimum qualities of mortar for durability in different situations are given in table 37.2 these being based on the recommendations in the British Standards for brickwork and blockwork. In general, the recommended mortars are weaker than the bricks or blocks with which they are associated. This is desirable so that the small movements associated with moisture and temperature changes and also the normally anticipated differential settlements of the foundations can be accommodated within the mortar joints themselves. If this is not the case then the bricks or blocks may be subjected to tensile stresses which will result in cracking of these units. Such cracks are unsightly and can seriously affect their durability, including resistance to rain penetration, whereas the smaller cracks associated with movement within the weaker mortar joints are much less detrimental and, should this become necessary, more easily repaired than cracks in the bricks or blocks themselves.

37.2 Applications and properties of brickwork and blockwork

Masonry construction today is used principally for non-load-bearing and load-bearing walls, the distinction between these being that the latter carry vertical roof and/or floor loads, as in cross-wall construction, whilst the former carry only their self-weight. Both may be subjected to wind loading. Masonry may be prestressed,

TABLE 37.2
Desirable minimum quality of brick, block and mortar for durability

Location	Clay bricks	Calcium silicate bricks (class)	Concrete blocks (type)^c	Mortar designation NF^a	Mortar designation PF^a
Internal walls, inner leaf of external cavity walls, backing (inner face) of external solid walls	NF^a (O) PF^a (M)	2	T	(iv)	(iii)
External walls, outer leaf of external cavity walls	(1)^b(M) (2) (M) (3) (F)	2 3 3	T R R^d	(iv)^e (iii)^g (ii)^g	(iii)^e (iii)^{g,i} (ii)^{g,i}
Facing (outer face) of external solid construction	(F) —	— (see External walls)	—	(ii)^h (see External walls)	(ii)^h
Damp-proof course (d.p.c.)	d.p.c.	(not applicable)		(i)	(i)
External freestanding walls (with coping)	(F)	3	S	(iii)	(iii)
Earth retaining walls (backfilled with free-draining material)	(F)	4	R^d	(ii)^{f,g}	(ii)^{f,g}
Sewerage work (foul water)	A	7^j	—	(ii)^{f,h}	(ii)^{f,h}

^{a} NF, no risk of frost during construction; PF, possibility of frost during construction.

^{b} (1) Above d.p.c. (2) Below d.p.c. but not closer than 150 mm to finished ground level. (3) Below 150 mm above finished ground level.

^{c} R: minimum density 1500 kg m^{-3} and average compressive strength $\geqslant 7$ N mm^{-2}. S: minimum density 1500 kg m^{-3} *and* average compressive strength $\geqslant 3.5$ N mm^{-2} *or* average compressive strength $\geqslant 7$ N mm^{-2}. T: all types except that blocks less than 75 mm thick generally only suitable for internal partitions.

^{d} Concrete blocks are generally not suitable for use in contact with ground from which there is any danger of sulfate attack.

^{e} For clay brickwork, not less than mortar designation (iii) should be used; if this brickwork is to be rendered, see footnote h.

^{f} For clay bricks, mortar designation (i) should be used.

^{g} Where sulfates are present in the soil or ground water, see footnote h.

^{h} Cement mortars with sulfate-resisting cement in place of ordinary Portland cement should be used.

^{i} Where clay bricks inadvertently become wet, the use of plasticised mortars is desirable.

^{j} Some calcium silicate bricks are not suitable for this work.

reinforced or unreinforced and its many applications include retaining walls, parapets and sewerage works.

The choice of a particular type of brick or block for any given application and the method of construction will depend very largely on the properties of the associated brickwork, or blockwork. These properties are related to the properties of its components, namely the individual bricks or blocks (units) and the mortar used to bind the units together.

Strength

The *compressive* strength of bricks or blocks is determined from standard tests on individual units and, for the reasons discussed in Chapter 14, the measured strengths will be dependent on the ratio of the height to least lateral dimension of the units. The most common form of load-bearing masonry construction, a wall, invariably has a height many times its least lateral dimension. A direct consequence of this is that the actual wall strength will generally be less than that of the individual units. Typical ratios of wall strength to unit strength, excluding any slenderness effects, for masonry constructed using a class (i) mortar are given in table 37.3.

TABLE 37.3
Typical ratios of wall compressive strength to unit strength for class (i) mortar

Ratio of height to least lateral dimension for unit	Wall strength/unit strength with unit compressive strength (N mm^{-2})				
	5	10	20	35	70
0.6	0.5	0.44	0.37	0.32	0.27
2.0 (hollow block)	1.0	0.61	0.37	0.32	—
2.0 (solid block)	1.0	0.88	0.74	0.65	—

Typically a class (i) mortar may have a cube compressive strength of about 12 N mm^{-2} with corresponding strengths for class (iii) and (iv) mortars of about 4 and 2 N mm^{-2} respectively. Using class (iii) and (iv) mortars, the ratios of wall to unit strength for unit strengths of 5, 10, 20 and 35 mm^{-2} will be approximately 0, 10, 20 and 25 per cent and 10, 20, 30 and 35 per cent respectively less than the values given in table 37.3 for class (i) mortar. The reason for these relatively small reductions in wall strength, compared with the significant reductions in mortar strength, is related to the different stress conditions in the mortar in the standard cube test and in the mortar bed between adjacent units. The relatively thin mortar bed is subjected to considerable biaxial restraint from the bricks or blocks above and below and this increases its effective *compressive* strength to many times its cube strength. Failure of an axially loaded wall, in fact, generally occurs following splitting of the units caused by the tensile stresses induced in them by the restraint they provide to the squeezing out of the mortar.

Flexural strength is another important property of masonry and for unreinforced masonry this normally means the effective flexural *tensile* strength which is used when assessing the ability of walls to withstand lateral wind forces.

Flexural strengths about a *horizontal axis*, with tensile stresses acting normal to the bed joints, are independent of unit strength, irrespective of the type of unit, with failure occurring along the bed joint. Stronger mortars tend to give higher flexural strengths. In the case of clay brickwork the increased mortar bond associated with clay bricks having low absorption characteristics produces a marked increase in flexural strength.

Flexural strengths about a *vertical axis* are about three times those about a horizontal axis. For brickwork, with failure about a vertical axis generally occurring as a result of shearing across the bed joints, the factors affecting the flexural strength are similar to those for bending about a horizontal axis. For concrete blockwork however, where the unit strength is generally significantly less than for brickwork, flexural failure about a vertical axis frequently occurs with tension cracks passing through the blocks themselves. In these circumstances, the flexural strength of blockwork is largely dependent on the strength of the blocks, with the influence of mortar strength only becoming significant for the higher block strengths.

Chemical attack

The susceptibility of brickwork and blockwork to chemical attack depends on the resistance to attack of both units and mortar. Cement-based mortars are susceptible to sulfate attack and sulfate-resisting cement should be used wherever sulfate attack is possible. In this regard it should be noted that, with clay brickwork, cement-based mortars may be attacked by sulfates derived from the clay bricks themselves. This also applies to cement renderings on clay brickwork. Sulfate attack, which is gradual, only occurs when brickwork remains damp for long periods.

Durability

In many situations, particularly for non-load-bearing walls, durability not strength is the overriding criterion. Desirable minimum qualities of units and mortars, based on recommendations in the relevant British Standards, for different applications are given in table 37.2. These reflect the properties of the different types of unit and classes of mortar.

Movement

Brickwork and blockwork, if unrestrained, will expand and/or shrink as a result of the inherent properties of the masonry units and the mortar, and the normal variations in moisture content and temperature. Load-bearing masonry also undergoes elastic and creep deformations under permanent load. Movements associated with the irreversible shrinkage and expansion of masonry units and with creep are time-dependent and these may continue, at a diminishing rate, for several years.

Some restraint to movement will always be present whenever there is any difference between the potential unrestrained movements of immediately adjacent materials. The associated induced stresses can cause serious cracking and/or buckling in masonry construction unless measures are taken to minimise such restraints or in the case of tension cracking to minimise their effects by, for example, the provision of bed-joint reinforcement in block walls.

Two examples of potential problem areas are cavity walls with an outer skin of clay brickwork and an inner skin of concrete blockwork (wall ties must provide minimal restraint to differential (in-plane) movement of the two skins) and framed structures supporting non-load-bearing wall panels. In the latter case horizontal and vertical movement joints must be provided to prevent transfer of load to the wall panels and to minimise restraint to the inherent shrinkage or expansion of the wall panels.

Adequate attention to detail in all the above respects will ensure that the effects of differential movements will not interfere with the satisfactory performance of masonry construction in service.

Fire resistance

Brick and block masonry has good fire-resistant properties irrespective of the type of unit. For different masonry units there are differences between the thickness of wall required for the same fire rating but these are relatively small and fire resistance is not normally an important consideration in the choice of unit.

Fire-resistance test procedures are described in BS 476: Part 8 and the required minimum thickness of different types of wall construction for specified periods of fire resistance (fire ratings) are given in various Building Regulations.

Thermal insulation

An increasingly important consideration in building design is the thermal insulation provided by the external envelope. This is regulated by specifying maximum permissive U-values for the roof and walls of a building. The U-value is a measure of the thermal transmittance through, for example, a wall and is expressed as the rate of transmission of energy through unit area of the wall surface for one degree difference in temperature between its internal and external surface (W m^{-2} K^{-1}).

The U-value is directly related to the combined thermal resistance (R, expressed in W^{-1} m^2 K) of the materials used in the wall construction, of any air spaces and of the internal and external surfaces of the wall. Typical values of thermal resistance (R) for internal and external surfaces and for a 50 mm air space, in a cavity wall, are 0.12, 0.06 and 0.18 respectively. The thermal resistance (R) of the individual thickness (t, in metres) of each different material with thermal conductivity (k, expressed in W m^{-1} K^{-1}) is obtained from the simple relationship $R = tk^{-1}$. The

TABLE 37.4
Example of U-value calculation for wall construction

Wall section	Thickness (mm)	Thermal conductivity (k) (W m^{-1} K^{-1})	Thermal resistance (R) (W^{-1} m^2 K)	Thermal transmittance (U) (W m^{-2} K^{-1})
External surface	—	—	0.060	
Masonry (A)	103	0.84	0.123	$U = \dfrac{1}{\Sigma R} = \dfrac{1}{1.119}$
Cavity	50	—	0.180	
Masonry (B)	100	0.18	0.555	$= 0.89$
Lightweight plaster	13	0.16	0.081	
Internal surface	—	—	0.120	
	Total air to air resistance		1.119	

combined thermal resistance (ΣR) of the wall as a whole is obtained by summing the individual thermal resistances thus obtained. The U-value = $(\Sigma R)^{-1}$.

A typical calculation for a cavity wall, with an inner leaf of lightweight masonry, is shown in table 37.4.

The introduction of a suitable insulation material, that is a material with a very low thermal conductivity, into wall construction will result in an appreciable reduction in the corresponding U-value, although the location of the insulation and its effects on possible moisture penetration and condensation must be given careful consideration.

Sound insulation

Another important design consideration for buildings is that of sound insulation and reference should be made to Building Regulations for sound-insulation requirements.

Sound insulation is a function of both sound absorption and sound transmission. Sound absorption is a measure of the ability of a surface to reduce the reflection of incident sound; smooth surfaces have very little sound absorption, whilst porous or otherwise irregular surfaces are more effective in this respect.

The resistance of masonry to airborne sound transmission generally increases with increasing density and wall thickness or more specifically with increasing mass per unit area (known as a mass law). Tests for determining sound-transmission properties are described in BS 2750.

It may be noted that whilst an increase in density of the masonry units will improve the resistance to sound transmission it will at the same time reduce its effectiveness for thermal insulation. This type of conflict frequently occurs in design and only when this is undertaken with a thorough knowledge of the properties of the available construction materials can a successful compromise be assured.

References

Arnold, P.J. (1969) Thermal conductivity of masonry materials, *JIHVE*, Vol. 37, pp. 101–108, 117.

ASTM C62-89a Standard specification for building brick (solid masonry units made from clay or shale).

Bowler, G.K. (1993) Deterioration of mortar by chemical and physical action, *Masonry International*, Vol. 6, pp. 78–82.

Building Research Establishment (1979) *Estimation of Thermal and Moisture Movements and Stress, Part 1*, Digest 227, and *Part 2*, Digest 228, HMSO, London.

BS 12: 1991 Specification for Portland cement.

BS 65: 1991 Specification for vitrified clay pipes, fittings and ducts.

BS 146: Part 2: 1991 Specification for Portland-blastfurnace cement.

BS 187: 1978 Specification for calcium silicate (sandlime and flintlime) bricks.

BS 476: Part 20: 1987 Method for determination of the fire resistance of elements of construction (general principles).

BS 743: 1970 Specification for materials for damp-proof courses.

BS 877: Part 2: 1973 (1977) Specification for foamed or expanded blastfurnace slag lightweight aggregate for concrete.

BS 882: 1992 Specification for aggregates from natural sources for concrete.

BS 890: 1972 Specification for building limes.

BS 1014: 1992 Specification for pigments for Portland cement and Portland cement products.

BS 1047: 1983 Specification for air-cooled blastfurnace slag aggregate for use in construction.

BS 1165: 1985 Specification for clinker and furnace bottom ash aggregate for concrete.

BS 1199 and 1200: 1976 Specifications for building sands from natural sources.

BS 2750: 1980 Measurement of sound insulation in buildings and of building elements.

BS 3587: 1963 Specifications for calcium chloride (technical).

BS 3797: 1990 Specification for lightweight aggregates for masonry units and structural concrete.

BS 3892: Part 1: 1982 Specification for pulverized-fuel ash for use as a cementitious component in structural concrete; Part 2: 1984 Specification for pulverized-fuel ash and for miscellaneous use in concrete.

BS 3921: 1985 Specification for clay bricks.

BS 4027: 1991 Specification for sulphate-resisting Portland cement.

BS 4248: 1974 Specification of supersulphated cement.

BS 4729: 1990 Specification for dimensions of bricks of special shapes and sizes.

BS 4887: Mortar admixtures, Part 1: 1986 Specification for air-entraining (plasticizing) admixtures.

BS 5075: Part 1: 1982 Specification for accelerating admixtures, retarding admixtures and water-reducing admixtures.

BS 5628: Code of practice for use of masonry; Part 1: 1992 Structural use of unreinforced masonry; Part 2: 1985 Structural use of reinforced and prestressed masonry; Part 3: 1985 Materials and components, design and workmanship.

BS 6073: 1981 Precast concrete masonry units; Part 1: Specification for precast concrete masonry units; Part 2: Method for specifying precast concrete masonry units.

BS 6649: 1985 Specification for clay and calcium silicate modular bricks.

BS 6677: 1986 Clay and calcium silicate pavers for flexible pavements.

BS 6699: 1992 Specification for ground granulated blastfurnace slag for use with Portland cement.

Clews, F.H. (1969) *Heavy Clay Technology*, 2nd edition, Academic Press, London.

Foster, D. and Johnson, G.D. (1982) Design for movement in clay brickwork in the U.K., *Proc. Brit. Ceram. Soc.*, No. 30, pp. 1–12.

Gummerson, R.J., Hall, C. and Hoff, W.D. (1980) Water movement in porous building materials – II. Hydraulic suction and sorptivity of brick and other masonry materials, *Build. Environ.*, Vol. 15, pp. 101–108.

Hall, C., Hoff, W.D. and Prout, W. (1992) Sorptivity–porosity relations in clay brick ceramic, *Amer. Ceram. Soc. Bull.*, Vol. 71, pp. 1112–1116.

Harrison, W.H. and Bowler, G.K. (1990) Aspects of mortar durability, *Brit. Ceram. Trans. J.*, Vol. 89, pp. 93–101.

Jacob, M. (1949) *Heat Transfer, Part 1*, Chapman and Hall, London.

Keeling, P.S. (1963) *The Geology and Mineralogy of Brick Clays*, Brick Development Association monograph.

Lomax, J. and Ford, R.W. (1988) A method for assessing the long term moisture expansion characteristics of clay bricks, in *Proceedings of the Eighth International Brick and Block Masonry Conference,* Dublin (ed. J.W. de Gourcy), Elsevier, London, p. 102.

Prentice, J.E. (1990) *Geology of Construction Materials*, Chapman and Hall, London.

RILEM (1972) Natural and Artificial Stones Technical Commission, Testing methods. II Baked clay masonry units, *Mater. Struct.*, Vol. 5, pp. 231–259.

RILEM (1994) *Technical Recommendations for the Testing and Use of Construction Materials*, Spon, London.

Shrive, N.G. (1991) Materials and material properties, in *Reinforced and Prestressed Masonry* (ed. A.W. Hendry), Longman, London, pp. 25–57.

Smith, R.G. (1973) Moisture expansion of structural ceramics. III Long term unrestrained expansion of test bricks, *Trans. Brit. Ceram. Soc.*, Vol. 72, pp. 1–5.

Smith, R.G. (1974) Moisture expansion of structural ceramics. IV Expansion of unrestrained fletton brickwork, *Trans. Brit. Ceram. Soc.*, Vol. 73, pp. 191–198.

Smith, R.G. (1993) Moisture expansion of structural ceramics. V. 28 years of expansion. *Brit. Ceram. Trans. J.*, Vol. 92, pp. 233–238.

Sutherland, R.J.M. (1981) Brick and block masonry in engineering, *Proc. Inst. Civ. Engrs., Part 1*, Vol. 70, pp. 31–63; also see Discussion, Part 1, Vol. 70, pp. 811–828.

Further reading

Brownell, W.E. (1976) *Structural Clay Products*. Springer-Verlag, Vienna.

Building Research Establishment (1966) *Cracking in Buildings*, Digest 75, HMSO, London.

Building Research Establishment (1968) *Sulphate Attack on Brickwork*, Digest 89, HMSO, London.

Building Research Establishment (1970) *Lightweight Aggregate Concrete*, Digest 123, HMSO, London.

Building Research Establishment (1974) *Clay Brickwork, Part 1*, Digest 164, and *Part 2*, Digest 165, HMSO, London.

Building Research Establishment (1975) *Autoclaved Aerated Concrete*, Digest 178, HMSO, London.

Building Research Establishment (1981) *Strength of Brickwork and Blockwork Walls: Design for Vertical Load*, Digest 246, HMSO, London.

Building Research Establishment (1983) *Perforated Clay Bricks*, Digest 273, HMSO, London.

Building Research Establishment (1991a) *Building Mortar*, Digest 362, HMSO, London.

Building Research Establishment (1991b) *Repairing Brick and Block Masonry*, Digest 359, HMSO, London.

Building Research Establishment (1992) *Calcium Silicate (Sandlime, Flintlime) Brickwork*, Digest 157, HMSO, London.

Butterworth, B. (1964) Frost resistance of bricks and tiles: a review, *Brit. Ceram. Soc. J.*, Vol. 1, p. 203.

Curtin, W.G. (1980) Brick diaphragm walls – development, application, design and future development, *Struct. Eng.*, Vol 58A, pp. 41–48.

Curtin, W.G. (1990) Brickwork, in *Specification, Vol. 1 (Technical)*, MBC Architectural Press, London, pp. 65–88.

Curtin, W.G., Shaw, G. and Beck, K.J.K. (1988) *Design of Reinforced and Prestressed Masonry*, Thomas Telford, London.

de Vekey, R.C. and West, H.W.H. (1980) The flexural strength of concrete blockwork, *Mag, Conc. Res.*, Vol. 32, pp. 206–218.

Fisher, K. (1982) Fire resistance of brickwork: regulatory requirements and test performances, *Proc. Brit. Ceram. Soc.*, No. 30, pp. 65–80.

Haseltine, B.A. and West, H.W.H. (1982) The fire behaviour of masonry structures, *Proc. Brit. Ceram. Soc.,* No. 30, pp. 81–89.

Hendry, A.W. (1981) *Structural Brickwork*, Macmillan, London.

Litvan, G.G. (1975) Testing the frost susceptibility of bricks, *Masonry Past and Present*, ASTM Special Technical Publication No. 589, pp. 123–132.

Litvan, G.G. (1980) Freeze–thaw durability of porous building materials, *Durability of Building Materials and Components*, ASTM Special Technical Publication No. 691, pp. 455–463.

Malhotra, H.L. (1966) Fire resistance of brick and block walls, *Fire Note No. 6, Ministry of Technology and Fire Offices' Committee Joint Fire Research Organisation*, HMSO, London.

Orton, A. (1986) *Structural Design of Masonry*, Longman, London.

Prout, W. and Hoff, W.D. (1991) Fundamental studies of frost damage in clay brick, in *Durability of Building Materials and Components* (eds J.M. Baker, P.J. Nixon, A.J. Majumdar and H. Davies), Spon, London.

RILEM (1994) *Technical Recommendations for the Testing and Use of Construction Materials,* Spon, London.

Stoaling, D. (1981) Blockwork in diaphragm wall construction, *Concrete*, Vol. 15, pp. 16–18.

West, H.W.H. (1970) Clay products, *Weathering and Performance of Building Materials* (eds. J.W. Simpson and P.J. Horrobin), MTP, Aylesbury, pp. 105–133.

West, H.W.H. *et al.* (1977) The resistance of brickwork to lateral loading – Part 1: Experimental methods and results of tests on small specimens and full sized walls, *Struct. Engr.*, Vol. 55, pp. 411–421.

West, H.W.H *et al.* (1979) *The Resistance to Lateral Loads of Walls Built of Calcium Silicate Bricks*, Technical Note 288, British Ceramic Research Association, Stoke-on-Trent.

(Also various publications by British Ceramic Research, Brick Development Association, Building Research Establishment, Calcium Silicate Brick Association and British Cement Association.)

Index